Methods in Molecular Biology

Series Editor
John M. Walker
School of Life and Medical Sciences
University of Hertfordshire
Hatfield, Hertfordshire, AL10 9AB, UK

For further volumes:
http://www.springer.com/series/7651

Proteomics for Drug Discovery

Methods and Protocols

Edited by

Iulia M. Lazar

Department of Biological Sciences
Virginia Tech
Blacksburg, VA, USA

Maria Kontoyianni

Department of Pharmaceutical Sciences,
School of Pharmacy
Southern Illinois University Edwardsville
Edwardsville, IL, USA

Alexandru C. Lazar

ImmunoGen
Waltham, Massachusetts, USA

Humana Press

Editors
Iulia M. Lazar
Department of Biological Sciences
Virginia Tech
Blacksburg, VA, USA

Maria Kontoyianni
Department of Pharmaceutical Sciences, School of Pharmacy
Southern Illinois University Edwardsville
Edwardsville, IL, USA

Alexandru C. Lazar
ImmunoGen
Waltham, Massachusetts, USA

ISSN 1064-3745 ISSN 1940-6029 (electronic)
Methods in Molecular Biology
ISBN 978-1-4939-7200-5 ISBN 978-1-4939-7201-2 (eBook)
DOI 10.1007/978-1-4939-7201-2

Library of Congress Control Number: 2017944435

Printed on acid-free paper

This Humana Press imprint is published by Springer Nature
The registered company is Springer Science+Business Media LLC
The registered company address is: 233 Spring Street, New York, NY 10013, U.S.A.

Preface

Drug and drug target discovery are lengthy and costly enterprises that are motivated by a never-ending desire to contribute to the improvement of human health. From academic to government, clinical, or corporate laboratories, the field draws thousands of scientists who aspire to contribute with scientific and streamlined technological advances toward a deeper understanding of disease and better patient care. As proteins continue to represent the vast majority of drug targets, it is not surprising that proteomics methods have attracted much attention from the drug development community with research endeavors pertaining to target identification and validation.

Proteomics for Drug Discovery is a book aimed at providing the scientist with the necessary knowledge for the implementation of some basic experimental and bioinformatics protocols in the drug discovery research laboratory. It is not an encyclopedia that covers all available proteomic protocols; neither does it attempt to be a comprehensive presentation, but rather a selection of methods, techniques, and platforms that have been proven to accelerate the advancement of new drug discovery paradigms. The book is targeting an audience ranging from novice to seasoned students and researchers—biochemists, chemists, molecular biologists, bioinformaticians—who either wish to learn about, or expand their knowledge of, proteomic technologies.

The book is comprised of chapters with a focus on experimental and bioinformatics approaches. The experimental chapters describe wet lab and instrument operation procedures, and follow the format characteristic to the *Methods in Molecular Biology* book series (i.e., introduction, materials, methods, and notes). The focus of these chapters is on select protocols and methodologies that address the analysis of posttranslational modifications, targeted protein quantification, protein–protein, protein–lipid, or protein–ATP interactions, protein arrays, tissue and cell extract preparation, labeling, chemoproteomics, and drug efficacy assessment. Emphasis was placed not just on methods that describe large-scale protein analysis but also on ones that are amenable to future implementation in a high-throughput format. In addition, most experimental chapters introduce the reader to the use of advanced technologies such as mass spectrometry, microarrays, microfluidics, and electron microscopy. The bioinformatics chapters address methods that support in-silico drug discovery or that describe the basic principles of virtual screening and processing of mass spectrometry data.

The editors would like to extend a special "thank you" to all contributors who dedicated their time to the development of this book. It is a testimony to their commitment to training the new generation of scientists who will advance drug discovery to new heights. As new modalities are entering the drug discovery sector, uncharted territories will always benefit from a deep knowledge of not just the fundamentals of disease biology but also of the means that were created for probing the complex biological milieu.

Blacksburg, VA	*Iulia M. Lazar*
Edwardsville, IL	*Maria Kontoyianni*
Waltham, MA	*Alexandru C. Lazar*

Contents

Contributors

ELIJAH BALL • *Department of Pharmaceutical Sciences, School of Pharmacy, Southern Illinois University Edwardsville, Edwardsville, IL, USA*
MARCUS BANTSCHEFF • *Cellzome, A GSK Company, Heidelberg, Germany*
JOSIP BLONDER • *Antibody Characterization Laboratory, Cancer Research Technology Program, Leidos Biomedical Research Inc., Frederick National Laboratory for Cancer Research, Frederick, MD, USA*
SCOTT M. BRITTAIN • *Novartis Institutes for Biomedical Research, Cambridge, MA, USA*
DANIEL G.S. CAPELLUTO • *Protein Signaling Domains Laboratory, Department of Biological Sciences, Biocomplexity Institute, Virginia Tech, Blacksburg, VA, USA; Center for Soft Matter and Biological Physics, Virginia Tech, Blacksburg, VA, USA*
NEIL O. CARRAGHER • *Cancer Research UK Edinburgh Centre, Institute of Genetics and Molecular Medicine, University of Edinburgh, Edinburgh, UK*
KING C. CHAN • *Protein Characterization Laboratory, Cancer Research Technology Program, Leidos Biomedical Research Inc., Frederick National Laboratory for Cancer Research, Frederick, MD, USA*
YI CHEN • *H. Lee Moffitt Cancer Center and Research Institute, University of South Florida, Tampa, FL, USA*
YIAN A. CHEN • *H. Lee Moffitt Cancer Center and Research Institute, University of South Florida, Tampa, FL, USA*
DOMENICO COPPOLA • *H. Lee Moffitt Cancer Center and Research Institute, University of South Florida, Tampa, FL, USA*
DANETTE L. DANIELS • *Promega Corporation, Madison, WI, USA*
WILLIAM DEARNALEY • *Virginia Tech Carilion Research Institute, Roanoke, VA, USA*
HANS CHRISTIAN EBERL • *Cellzome, A GSK Company, Heidelberg, Germany*
AVITAL EISENBERG-LERNER • *Department of Immunology, The Weizmann Institute of Science, Rehovot, Israel*
CARLA V. FINKIELSTEIN • *Integrated Cellular Responses Laboratory, Department of Biological Sciences, Biocomplexity Institute, Virginia Tech, Blacksburg, VA, USA*
KATE J. FISHER • *H. Lee Moffitt Cancer Center and Research Institute, University of South Florida, Tampa, FL, USA*
ALEXANDER GONCEARENCO • *National Center for Biotechnology Information, National Institutes of Health, Bethesda, MD, USA*
LISA GUAY-WOODFORD • *Center for Translational Science, Children's National Health System, Washington, DC, USA; Department of Pediatrics, George Washington University School of Medicine & Health Sciences, Washington, DC, USA*
NAOE HARAFUJI • *Center for Translational Science, Children's National Health System, Washington, DC, USA; Department of Pediatrics, George Washington University School of Medicine & Health Sciences, Washington, DC, USA*
SONJA HESS • *Proteome Exploration Laboratory, California Institute of Technology, Pasadena, CA, USA; MedImmune, Gaithersburg, MD, USA*
RISHI K. JAIN • *Novartis Institutes for Biomedical Research, Cambridge, MA, USA*
YUEMENG JIA • *Department of Pathology, School of Medicine, University of Virginia, Charlottesville, VA, USA*

DONALD J. JOHANN JR. • *Wintrop P. Rockefeller Cancer Institute, University of Arkansas for Medical Sciences, Little Rock, AR, USA*
JATINDER S. JOSAN • *Department of Chemistry, Virginia Tech, Blacksburg, VA, USA*
MATHIAS KALXDORF • *Cellzome, A GSK Company, Heidelberg, Germany*
DEBORAH F. KELLY • *Virginia Tech Carilion Research Institute, Roanoke, VA, USA; Department of Biological Sciences, Virginia Tech, Blacksburg, VA, USA*
MARIA KONTOYIANNI • *Department of Pharmaceutical Sciences, School of Pharmacy, Southern Illinois University Edwardsville, Edwardsville, IL, USA*
JOHN M. KOOMEN • *Molecular Oncology and Proteomics, H. Lee Moffitt Cancer Center and Research Institute, University of South Florida, Tampa, FL, USA*
SHAILESH KUMAR • *Department of Pathology, School of Medicine, University of Virginia, Charlottesville, VA, USA*
LESLIE E.W. LACONTE • *Virginia Tech Carilion Research Institute, Roanoke, VA, USA*
IULIA M. LAZAR • *Department of Biological Sciences, Virginia Tech, Blacksburg, VA, USA*
HUI LI • *Department of Pathology, School of Medicine, University of Virginia, Charlottesville, VA, USA; Department of Biochemistry and Molecular Genetics, School of Medicine, University of Virginia, Charlottesville, VA, USA*
MINGHUI LI • *National Center for Biotechnology Information, National Institutes of Health, Bethesda, MD, USA; School of Biology and Basic Medical Sciences, Soochow University, Suzhou, China*
JENNIFER LIPPS • *Novartis Institutes for Biomedical Research, Cambridge, MA, USA*
LUIS LLAMAS • *Novartis Institutes for Biomedical Research, Cambridge, MA, USA*
MARK LLOYD • *H. Lee Moffitt Cancer Center and Research Institute, University of South Florida, Tampa, FL, USA*
KENNETH G. MACLEOD • *Cancer Research UK Edinburgh Centre, Institute of Genetics and Molecular Medicine, University of Edinburgh, Edinburgh, UK*
JACQUI L. MENDEZ-JOHNSON • *Promega Corporation, Madison, WI, USA*
YIFAT MERBL • *Department of Immunology, The Weizmann Institute of Science, Rehovot, Israel*
ANNIE MORADIAN • *Proteome Exploration Laboratory, California Institute of Technology, Pasadena, CA, USA*
KONARK MUKHERJEE • *Virginia Tech Carilion Research Institute, Roanoke, VA, USA; Department of Biological Sciences, Virginia Tech, Blacksburg, VA, USA*
RACHEL FRIEDMAN OHANA • *Promega Corporation, Madison, WI, USA*
ANNA R. PANCHENKO • *National Center for Biotechnology Information, National Institutes of Health, Bethesda, MD, USA*
DARUE A. PRIETO • *Protein Characterization Laboratory, Cancer Research Technology Program, Leidos Biomedical Research Inc., Frederick National Laboratory for Cancer Research, Frederick, MD, USA*
IFAT REGEV • *Department of Immunology, The Weizmann Institute of Science, Rehovot, Israel*
JOSÉ A. RODRÍGUEZ-CORRALES • *Department of Chemistry, Virginia Tech, Blacksburg, VA, USA*
MARKUS SCHIRLE • *Novartis Institutes for Biomedical Research, Cambridge, MA, USA*
BRYAN SERRELS • *Cancer Research UK Edinburgh Centre, Institute of Genetics and Molecular Medicine, University of Edinburgh, Edinburgh, UK*

DAVID SHIBATA • *Department of Surgery, Health Science Center, The University of Tennessee, Memphis, TN, USA; H. Lee Moffitt Cancer Center and Research Institute, University of South Florida, Tampa, FL, USA*

BENJAMIN A. SHOEMAKER • *National Center for Biotechnology Information, National Institutes of Health, Bethesda, MD, USA*

ERIN SIEGEL • *H. Lee Moffitt Cancer Center and Research Institute, University of South Florida, Tampa, FL, USA*

FRANCO L. SIMONETTI • *Leloir Institute Foundation, Buenos Aires, Argentina*

SARIKA SRIVASTAVA • *Virginia Tech Carilion Research Institute, Roanoke, VA, USA*

MICHAEL J. SWEREDOSKI • *Proteome Exploration Laboratory, California Institute of Technology, Pasadena, CA, USA*

GREGORY TAKACS • *Department of Pharmaceutical Sciences, School of Pharmacy, Southern Illinois University Edwardsville, Edwardsville, IL, USA*

TUO-XIAN TANG • *Protein Signaling Domains Laboratory, Department of Biological Sciences, Biocomplexity Institute, Virginia Tech, Blacksburg, VA, USA; Center for Soft Matter and Biological Physics, Virginia Tech, Blacksburg, VA, USA*

JASON R. THOMAS • *Novartis Institutes for Biomedical Research, Cambridge, MA, USA*

MARJETA URH • *Promega Corporation, Madison, WI, USA*

A. CAMERON VARANO • *Virginia Tech Carilion Research Institute, Roanoke, VA, USA; Translational Biology, Medicine and Health Graduate Program, Virginia Tech, Blacksburg, VA, USA*

GORDON WHITELY • *Antibody Characterization Laboratory, Cancer Research Technology Program, Leidos Biomedical Research Inc., Frederick National Laboratory for Cancer Research, Frederick, MD, USA*

BRENTSEN WOLF • *Department of Pharmaceutical Sciences, School of Pharmacy, Southern Illinois University Edwardsville, Edwardsville, IL, USA*

ELIZABETH R. WOOD • *H. Lee Moffitt Cancer Center and Research Institute, University of South Florida, Tampa, FL, USA*

RONALD WORTHINGTON • *Department of Pharmaceutical Sciences, School of Pharmacy, Southern Illinois University Edwardsville, Edwardsville, IL, USA*

ZHONGQIU XIE • *Department of Pathology, School of Medicine, University of Virginia, Charlottesville, VA, USA*

WEN XIONG • *Protein Signaling Domains Laboratory, Department of Biological Sciences, Biocomplexity Institute, Virginia Tech, Blacksburg, VA, USA; Center for Soft Matter and Biological Physics, Virginia Tech, Blacksburg, VA, USA*

XIAOYING YE • *Antibody Characterization Laboratory, Cancer Research Technology Program, Leidos Biomedical Research Inc., Frederick National Laboratory for Cancer Research, Frederick, MD, USA*

ALINA D. ZAMFIR • *Mass Spectrometry Laboratory, National Institute for Research and Development in Electrochemistry and Condensed Matter, Timisoara, Romania*

Chapter 1

A Photoaffinity Labeling-Based Chemoproteomics Strategy for Unbiased Target Deconvolution of Small Molecule Drug Candidates

Jason R. Thomas, Scott M. Brittain, Jennifer Lipps, Luis Llamas, Rishi K. Jain, and Markus Schirle

Abstract

The combination of photoaffinity labeling (PAL) and quantitative chemoproteomics enables the comprehensive, unbiased determination of protein interaction profiles to support target identification of bioactive small molecules. This approach is amenable to cells in culture and compatible with pharmacologically relevant transmembrane target classes like G-protein coupled receptors and ions channels which have been notoriously hard to access by conventional chemoproteomics approaches. Here, we describe a strategy that combines PAL probe titration and competition with excess parental compounds with the goal of enabling the identification of specific interactors as well as assessing the functional relevance of a binding event for the phenotype under investigation.

Key words Photoaffinity labeling, Chemoproteomics, Quantitative proteomics, Target identification, Isobaric mass tags, G-protein coupled receptor

1 Introduction

Phenotypic screening is an approach to both drug and biological discovery that utilizes screenable cellular models or model organisms that reflect a given aspect of a biological pathway or disease state. These models are then used to screen collections of small molecules to identify compounds that modulate the pathway or disease state in the desired manner in a target-agnostic manner. In part reflecting the recent resurgence of phenotype-based drug discovery [1], a large number of complementary approaches have been introduced to support follow-up studies to identify the efficacy target, i.e., the protein through which the compound elicits its phenotypic effect via direct binding and modulation (*see* ref. 2 for a

Jason R. Thomas and Scott M. Brittain contributed equally to this work.

Iulia M. Lazar et al. (eds.), *Proteomics for Drug Discovery: Methods and Protocols*, Methods in Molecular Biology, vol. 1647, DOI 10.1007/978-1-4939-7201-2_1, © Springer Science+Business Media LLC 2017

recent review). Among these approaches, chemoproteomic strategies that generate efficacy target hypotheses by defining the cellular protein interactome of a compound by coupling an affinity enrichment step to quantitative mass spectrometry are among the most popular and powerful [3]. Most chemoproteomic strategies are based on affinity tool compounds, biologically active derivatives of the compound of interest that have been designed to allow for immobilization and subsequent enrichment of protein interactors. In photoaffinity labeling (PAL), the compound is derivatized to contain (a) an ultraviolet (UV) light-activated photocrosslinking moiety and (b) an enrichment handle such as biotin. Upon UV activation, a highly reactive intermediate is produced allowing the probe to irreversibly react with its target; the enrichment handle facilitates enrichment of probe-labeled proteins. These probes can be used in whole cell lysates or applied directly to cells in culture. For the latter scenario, the enrichment handle is often installed only after cell lysis using bioorthogonal chemistry to avoid potential issues with cell permeability. Through its ability to interrogate binding events in live cells, PAL has emerged as an attractive strategy for target deconvolution under in situ conditions particularly because it allows access to pharmacologically relevant target classes that are frequently missed by the noncovalent, lysate-based chemoproteomics approaches. On the other hand, there are known challenges with PAL which may reduce its overall effectiveness, in particular when applied to unbiased de novo target deconvolution as opposed to direct target hypothesis testing. First, the crosslinking efficiency is notoriously low, even when using purified proteins. Second, there is a significant degree of nonspecific labeling which can be at least in part explained by the proximity-based nature of the crosslinking step.

Here, we describe a quantitative mass spectrometry-based PAL strategy (Fig. 1) that addresses a key challenge in PAL, namely low overall signal-to-noise, to facilitate the identification of specific interactors of a small molecule of interest, as well as provides a first step toward assessing if an interactor is the potential efficacy target. In order to discriminate specific from nonspecific binding, a combination of titrating probe concentration and competition with excess parental compound is performed. Proteins that bind specifically to the PAL probe are defined by the ability of a protein to bind to the PAL probe in a dose-responsive and saturable manner (Fig. 1a). Proteins that bind to the parent compound are determined by the ability of the excess parental compound to compete PAL probe labeling. After the identification of proteins that bind specifically, the relevance of the binding event to the phenotype under investigation can then be determined through competition with a series of structurally related compound with varying degrees of phenotypic activity (structure activity relationship, SAR). The goal here is to determine if a correlation can be established between binding and phenotypic activity (Fig. 1b). Competition experiments in general also provide a means to deprioritize proteins that preferentially interact with the pharmacologically irrelevant photocrosslinker portion of the PAL probe: These proteins will show specific enrichment in the probe titration but should not show significant competition by a competition compound that lacks the photocrosslinker moiety. With regard to proteomic readouts, several variants of quantitative mass spectrometry are available [4, 5],

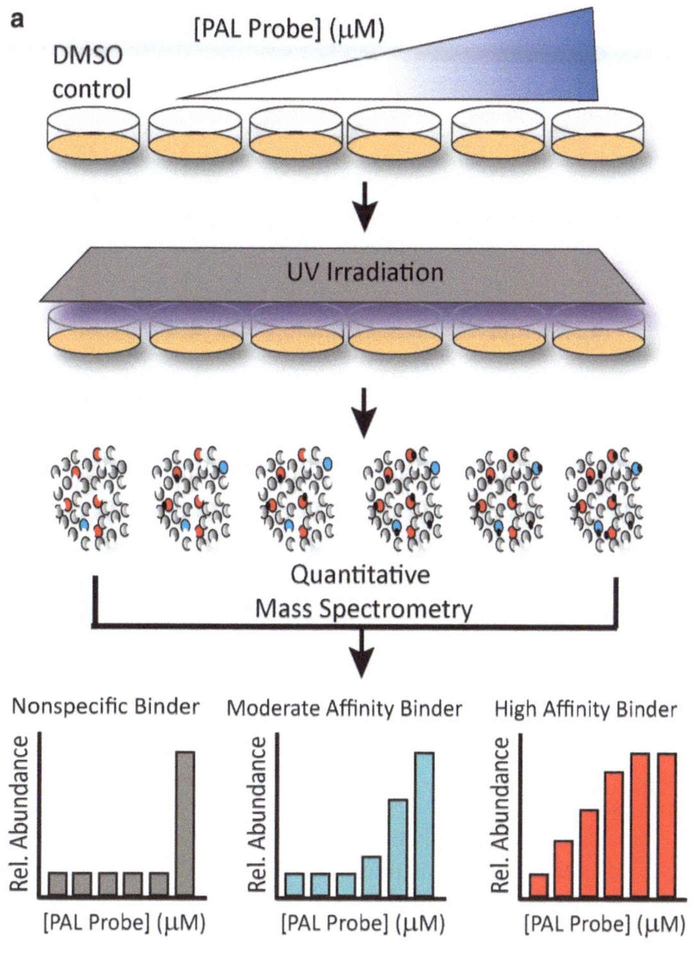

Fig. 1 (**a**) Overview of the PAL probe titration workflow: key goal for this workflow is the identification of specific binders showing dose-responsive, saturable binding to differentiate from nonspecific background. (**b**) Overview of the compound competition workflow with constant probe concentration and single dose competition with a series of structurally related compounds that differ in their activity in the phenotypic assay (SAR series): key goal for this workflow is the identification of specific binders to the pharmacophore that are relevant for the phenotype under investigation. Workflows are depicted separately for clarity and with six conditions (TMT Sixplex™ reagent), but it should be noted that the multiplexing capability of TMT reagents for up to ten conditions (TMT 10plex™) allows for mixed workflows, e.g., inclusion of a competition channel in a probe titration experiment as in the presented study

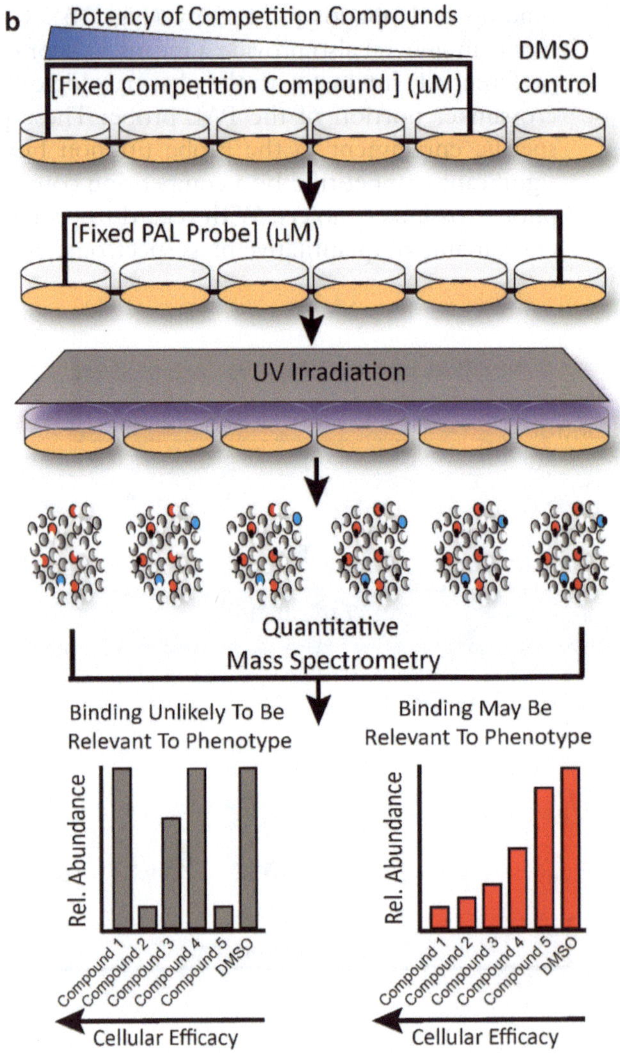

Fig. 1 (continued)

but the flexibility and multiplexing capabilities of isobaric mass tags such as tandem mass tags (TMT™) make them attractive options for complex workflows such as the extended PAL probe titration and SAR correlation experiments described here.

In the presented example, we apply this workflow in the context of a previous study that had determined that expression of the G-protein coupled receptor (GPCR) GPR39 is required for the inhibitory activity of the cyclohexyl-methyl aminopyrimidine (CMAP) chemotype on Hedgehog (Hh) signaling [6]. Noncovalent, lysate-based chemoproteomics had failed to provide proof of direct interaction which suggested a target that may require an intact cellular environment for binding capability including integral transmembrane target families such as GPCR and ion channels. We were able

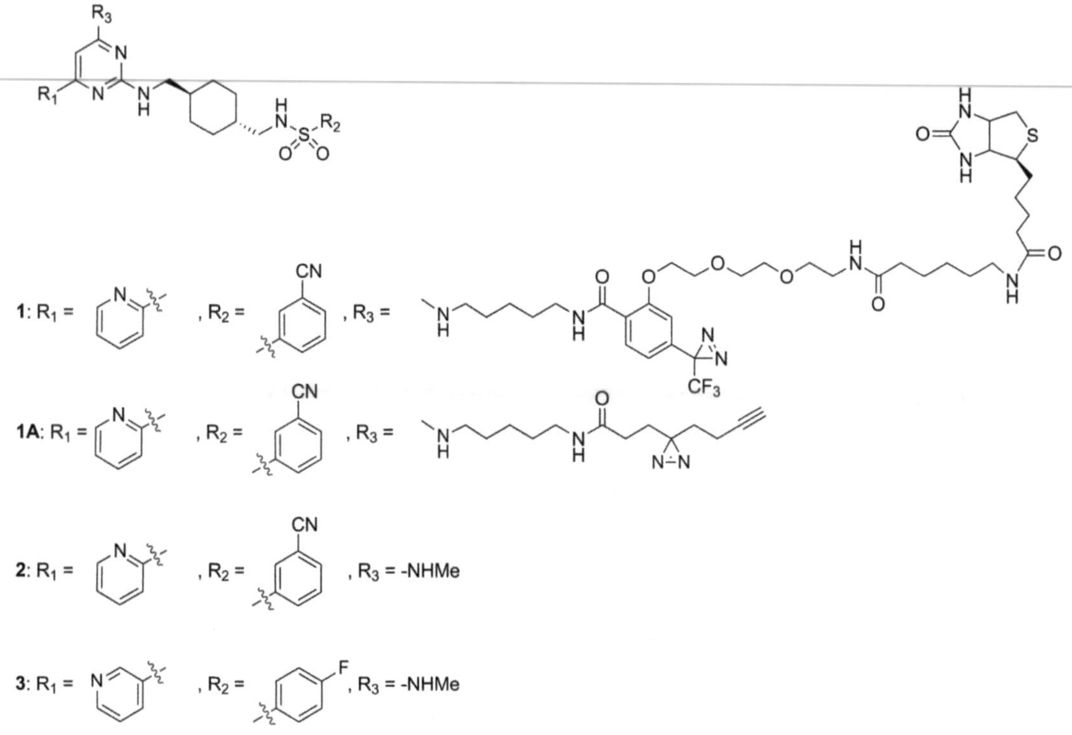

Fig. 2 PAL probe **1** and the competition compounds **2, 3** used in this study as well as an alternative PAL probe **1A**. Hedgehog pathway activities (Gliluc) are 0.04 μM (compound **1**), 0.015 μM (**2**) and 4.5 μM (**3**)

to generate a PAL probe **1** that retained comparable activity (40 nM) to the active parent compound **2** (15 nM) in a luciferase-based Hedgehog reporter gene assay (Gliluc) in murine TM3 cells. Using the comprehensive PAL-based strategy described above and HEK293T cells expressing murine GPR39 (mGpr39) as a model system (the CMAP scaffold does not show appreciable activity in human cells), we are able to provide evidence for direct interaction of the CMAP-based PAL probe with mGpr39 by probe titration as well as competition by compound **2** (*see* Fig. 2 for compounds, Fig. 3a, b for data). In addition, we illustrate how the dual criteria of saturable labeling and SAR-dependent competition narrows down the list of protein interactors to those that are likely functionally relevant to Hedgehog signaling. This also further corroborates mGpr39 as an efficacy target of this chemotype (Fig. 3c). While the presented data are based on probe **1** which contains a preinstalled biotin enrichment handle, we have also included the experimental steps for the introduction of the biotin moiety after cell lysis to an exemplary alkyne-bearing probe **1A** using copper-based click chemistry [7]. We routinely use the general experimental strategy described here for unbiased target deconvolution for small molecule hits from phenotypic screens.

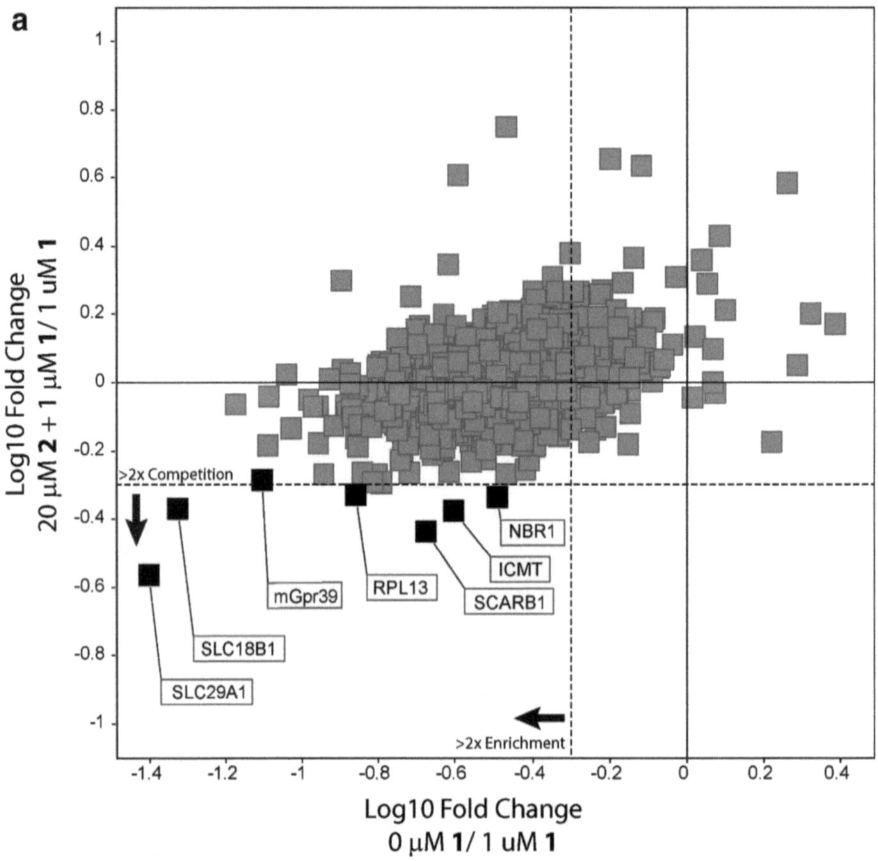

Fig. 3 (**a**) Scatter plot showing protein enrichment by probe **1** on the X-axis (depicted as log10 fold change for 0 μM probe **1** over 1 μM probe **1**) and competition by 20 μM **2** at 1 μM probe **1** on the Y-axis (depicted as log10 fold change over DMSO at 1 μM probe **1**). Several proteins, including murine GPR39 (mGpr39), show significant enrichment as well as competition with compound **2** indicative of specific binding to the pharmacophore. (**b**) Representative full quantitative profiles for selected proteins demonstrating the combination of probe titration curves (0, 0.01, 0.05, 0.1, 0.5, 1, 5, 10 μM PAL probe, *gray circles*) with competition by 20 μM compound **2** at 1 μM probe **1** (*white circle*). mGpr39, SLC18B1, and SLC29A1 show saturable labeling already at low PAL probe concentrations as well as competition by excess compound, indicative of specific binding to the pharmacophore. This is in contrast to ICMT, NBR1, RPL13, and SCARB1 which do not exhibit saturable labeling by probe **1**. Given the presence of proteins that exhibit saturable labeling, these latter proteins could be deprioritized as putative efficacy targets. (**c**) Table showing residual labeling for selected protein hits from Fig. 1a upon competition with 20 μM of compounds **2** and **3** as representative competitor compounds with high and significantly lower activity in the Gliluc assay (1 μM PAL probe in both cases). mGpr39 competition correlates with activity, in line with this being a compound efficacy target. Unlike mGpr39, SLC18B1 exhibits similar degree of competition upon treatment with compound **2** or **3**. The fact that SLC18B1 binds with comparable affinity to each compound despite the large difference in cellular activity would suggest that SLC18B1 is not responsible for the observed activity on Hedgehog signaling. Similar to mGpr39, SLC29A1 shows a competition profile that correlates with activity on Hedgehog signaling. For further hit triaging in such cases, an extended SAR series would typically be used

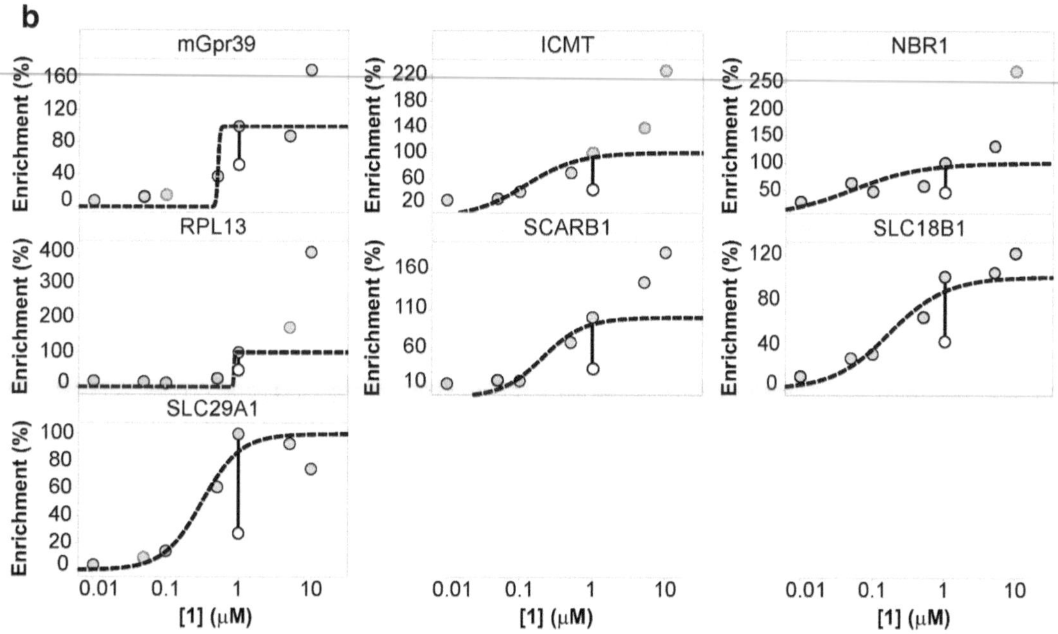

c

	2 (highly active)	3 (less active)
mGRP39	52%	78%
ICMT	42%	77%
NBR1	46%	107%
RPL13	47%	58%
SCARB1	37%	62%
SLC18B1	43%	45%
SLC29A1	27%	55%

Fig. 3 (continued)

2 Materials

Use only high purity reagents and prepare all solutions using distilled, deionized water. All organic solvents should be HPLC grade or higher.

2.1 Cell Culture, Compound Treatment, and Irradiation

1. HEK293T (ATCC) cells engineered to stably express V5-tagged mGpr39.
2. Gibco Dulbecco's phosphate buffered saline (DPBS) (Thermo Fisher Scientific).
3. Gibco Opti-MEM™ reduced serum media, no phenol red (Thermo Fisher Scientific).
4. Compound 1: 10 mM PAL probe in DMSO (*see* **Note 1**).

5. Compound 2 (competition compound with potent cellular activity): 10 mM in DMSO.
6. Compound 3 (competition compound with weak cellular activity): 10 mM in DMSO.
7. Blak-Ray XX-40BLB UV bench lamp, 40 Watt (UVP).
8. Blak-Ray UV intensity meter Model J-221 (UVP).
9. Cell culture dish 140 × 20 mm.
10. Conical centrifuge tubes (5 mL).

2.2 Cell Lysis

1. HEPES buffer stock solution: 50 mM, pH 7.5.
2. Cell lysis buffer: 50 mM HEPES (pH 7.5), 150 mM NaCl, 1.5 mM $MgCl_2$, 5% glycerol. Immediately before use, add 0.8% NP40 and Halt protease inhibitor cocktail (Thermo Fisher Scientific).
3. Barocycler NEP2320, FT500 PULSE tubes and tube tool (Pressure Biosciences, Inc.).

2.3 Click Reaction

This step is only required if the PAL probe does not already contain a biotin affinity handle.

1. Sodium dodecyl sulfate (SDS) solution: 10% in water (w/v).
2. Biotin-PEG3-Azide (Click Chemistry Tools, Scottsdale, AZ): 5 mM in DMSO.
3. Tris[(1-benzyl-1H-1,2,3-triazol-4-yl)methyl]amine (TBTA) solution: 9 mg/mL in butanol:DMSO (4:1, v/v).
4. Cu(II)SO_4 solution: 50 mM in water.
5. Tris(2-carboxyethyl)phosphine hydrochloride (TCEP) solution: 50 mM in water.
6. Base buffer: 50 mM HEPES (pH 7.5), 150 mM NaCl, 1.5 mM $MgCl_2$, 5% glycerol (v/v).
7. Heating block shaker for microcentrifuge tubes (e.g., Eppendorf thermomixer).
8. Sonicator (e.g., Q700, Qsonica Llc).

2.4 Affinity Enrichment

1. Pierce NeutrAvidin Agarose beads, 50% aqueous slurry (Thermo Fisher Scientific).
2. Wash buffer I: 1 mM DTT, 0.4% NP40 in base buffer (*see* Subheading 2.3, **item 6**). Prepare fresh before use.
3. Wash buffer II: 1 mM DTT in base buffer (*see* Subheading 2.3, **item 6**). Prepare fresh before use.
4. Pierce 660 nm protein assay, ionic detergent compatibility reagent (Thermo Fisher Scientific).
5. Microcentrifuge tubes (5 mL).

6. Chromatography spin columns, 1 mL, with 90 μm pore size frit and Luer-lock connector (e.g., Mobicol, MoBiTec, Germany).
7. NuPAGE LDS (Lithium dodecyl sulfate) sample buffer (4×) (Thermo Fisher Scientific).

2.5 Sample Preparation, Visualization, and Proteolysis

1. Iodoacetamide solution: 200 mg/mL in water. Prepare fresh before use and protect from light.
2. Criterion™ XT Bis-Tris Gel Pierce Silver Stain Kit, 4–12% (Bio-Rad).
3. Trypsin protease, MS-grade (Thermo Fisher Scientific).
4. Detergent removal spin columns (Thermo Fisher Scientific).

2.6 Isobaric Labeling

1. TMT Sixplex™ or 10plex™ reagent kit (Thermo Fisher Scientific).
2. Triethylammonium bicarbonate (TEAB) solution: 1 M in water.
3. SpeedVac (e.g., HT-4X Evaporator, Genevac Inc.).

2.7 High pH Reversed Phase Separation

1. Dionex UltiMate™ 3000 HPLC system with fraction collection operating Chromeleon v. 6.8 (Thermo Fisher Scientific) or similar performance HPLC system.
2. Waters Xbridge™ C18 3.5 μm 2.1 × 150 mm HPLC column (Waters Corporation) or equivalent C18 HPLC column.
3. Low volume autosampler vials and septa.
4. U-bottom 96-well plate.
5. V-bottom 96-well plates.
6. High pH mobile phase A: water (HPLC grade).
7. High pH mobile phase B: acetonitrile (HPLC grade).
8. High pH mobile phase C: 200 mM ammonium acetate, pH 10.
9. Trifluoro-acetic acid (TFA) solution: 0.5% in water (v/v).

2.8 Protein Identification and Quantitation by nanoLC-MS/MS

1. Q Exactive mass spectrometer (Thermo Fisher Scientific; Xcalibur 3.0.63, Tune 2.5 SP1 software) equipped with Picoview 550 Nanospray source (New Objective).
2. EasyLC 1000 HPLC system (Thermo Fisher Scientific), equipped with a 20 μL loop, set up for 96-well plates, or similar performance HPLC system.
3. Spraying capillary: fused silica capillary (75 μm ID × 360 μm OD) pulled to a tip diameter of approximately 8–10 μm using a P-2000 capillary puller (Sutter Instruments) and packed, e.g., with ReproSil-Pur 120 C18-AQ, 3 μm material, 160 mm bed length (Dr. Maisch GmbH, Germany).

4. Kasil-fritted trapping column (75 μm ID) packed with ReproSil-Pur 120 C18-AQ, 5 μm material (15 mm bed length), or equivalent trapping column (*see* **Note 2**).
5. Mobile phase A: 0.1% formic acid/acetonitrile (98:2, v/v).
6. Mobile phase B: 0.1% formic acid/98% acetonitrile (2:98, v/v).

2.9 Data Processing and Analysis

1. Mascot v 2.5.1 (Matrix Science).
2. Transproteomic pipeline (TPP) v3.3sqall, available on the Internet from the Seattle Proteome Center at the Institute of System Biology (Seattle, USA).
3. Uniprot protein database (canonical human sequences, EBI, Cambridge, UK) supplemented with sequences of mGpr39, common contaminants as well as reversed entries for all forward entries.
4. In-house developed software for parsing and uploading of information from TPP-derived xml files and Mascot result files to a relational database (Oracle 10 g; Oracle).
5. Spotfire 6.5.3 software for data visualization and analysis (TIBCO Software Inc.).

3 Methods

3.1 Cell Culture, Compound Treatment, and Irradiation

1. Seed 20 × 10^6 HEK293T cells stably expressing V5-tagged mGpr39 into a 140 × 20 mm cell culture dish and allow cells to recover for 12–16 h at 37 °C with 5% CO_2. Ten plates in total will be needed (*see* **Note 3**).
2. Prepare DMSO-treated media by adding 10 mL Opti-MEM™ (37 °C) with 0% FBS and 20 μL DMSO to 8–15 mL tubes. Incubate tubes at 37 °C until ready for use (*see* **Note 4**).
3. Prepare highly active competition compound-treated media by adding 10 mL Opti-MEM™ (37 °C) with 0% FBS and 20 μL competition compound **2** (10 mM) to a 15 mL tube. Incubate tubes at 37 °C until ready for use.
4. Prepare less active competition compound-treated media by adding 10 mL Opti-MEM™ (37 °C) with 0% FBS and 20 μL competition compound **3** (10 mM) to a 15 mL tube. Incubate tubes at 37 °C until ready for use.
5. Prepare a serial dilution of the PAL probe (compound **1**) from 10 mM to 0.01 mM and from 5 mM to 0.05 mM in DMSO. Each dilution should have a final volume of 25 μL.
6. Wash each plate twice with 10 mL DPBS (37 °C).
7. To one plate of cells add compound **2**-treated media, to another plate of cells add compound **3**-treated media, and to

the remaining plates add DMSO-treated media. Incubate the cells for 60 min at 37 °C.

8. To one plate containing DMSO-treated media add 10 μL DMSO.
9. To one plate containing DMSO-treated media add 10 μL compound **1** (0.01 mM).
10. To one plate containing DMSO-treated media add 10 μL compound **1** (0.05 mM).
11. To one plate containing DMSO-treated media add 10 μL compound **1** (0.1 mM).
12. To one plate containing DMSO-treated media add 10 μL compound **1** (0.5 mM).
13. To one plate containing DMSO-treated media add 10 μL compound **1** (1 mM).
14. To one plate containing DMSO-treated media add 10 μL compound **1** (5 mM).
15. To one plate containing DMSO-treated media add 10 μL compound **1** (10 mM).
16. To the plate containing compound **2**-treated media add 10 μL compound **1** (1 mM).
17. To the plate containing compound **3**-treated media add 10 μL compound **1** (1 mM).
18. Incubate cells for 60 min at 37 °C (*see* **Note 5**).
19. During 60 min incubation turn on the UV bench lamp (*see* **Notes 6** and **7**).
20. Remove media via aspiration (leave ~5 mL behind).
21. Irradiate (368 nm) cells at 4 °C for 20 min.
22. Harvest cells by scraping, transfer to individual 15 mL tubes, and spin in a prechilled centrifuge at 200 × *g* for 5 min at 4 °C.
23. Remove the supernatant via aspiration and store treated cell pellets at −80 °C.

3.2 Cell Lysis

1. Suspend cell pellet in 1 mL of chilled lysis buffer and transfer to a PULSE tube.
2. Add 400 μL of chilled lysis buffer to each PULSE tube (final volume for each cell pellet should be 1.4 mL).
3. Lyse cells in a chilled Barocycler using 5 cycles of 35 kpsi for 20 s followed by 20 s at atmospheric pressure. Transfer lysed cells to individual 1.5 mL tubes.
4. Spin samples at 800 × *g* in a prechilled centrifuge for 20 min and collect supernatant ("lysate").

3.3 Click Reaction

This step is only required if the PAL probe does not already contain a biotin affinity handle.

1. Aliquot 975 μL lysate from each sample in a 1.5 mL tube.
2. Add 125 μL 10% SDS solution.
3. Add 25 μL of 5 mM Biotin-PEG3-Azide in DMSO.
4. Prepare Catalyst Mix by combining 1.35 mL of butanol: DMSO (4:1, v/v), 150 μL of TBTA stock, 0.5 mL 50 mM Cu(II)SO_4 solution, and 0.5 mL 50 mM TCEP.
5. Vortex Catalyst Mix and add 125 μL to each sample.
6. Vortex each reaction vigorously.
7. Using an Eppendorf thermomixer, incubate for 2 h at 37 °C with gentle agitation (300 rpm).
8. During incubation, add 6 mL cold acetone to individual 15 mL tubes. Store at −20 °C until ready for use.
9. After the 2 h incubation, transfer the entire contents of each click reaction to individual 15 mL tubes containing 6 mL cold acetone.
10. Incubate for 4–16 h at −20 °C.
11. Spin precipitated protein in a prechilled centrifuge at 1250 × *g* for 10 min.
12. Remove excess acetone by aspiration and reconstitute protein pellet in 1 mL base buffer containing 1% SDS by probe sonication using 20 consecutive 1 s pulses (12 W).

3.4 Affinity Enrichment (See Note 8)

1. Aliquot 100 μL of the 50% NeutrAvidin agarose bead slurry, wash agarose three times with 1 mL of wash buffer I and three times with 1 mL of wash buffer II, and transfer the aliquots to 5 mL tubes. After each addition of buffer the beads are spun at 90 × *g* for 2 min.
2. Determine total protein concentration for each sample using Pierce 660 nm protein assay and adjust to a protein concentration of 5 mg/mL.
3. For each sample, add 4 mL of base buffer to the 50 μL of NeutrAvidin agarose bead bed in the 5 mL tubes, and then add 1 mL of sample per tube.
4. Incubate samples on agarose beads overnight on a rocking platform at room temperature.
5. Spin samples at 90 × *g* for 2 min at room temperature.
6. Remove the supernatant, leave ~1 mL supernatant above the resin bed.
7. Resuspend agarose beads in residual supernatant and transfer samples to Mobicol spin columns.

8. Wash beads with 3 mL wash buffer I.
9. Wash beads with 3 mL wash buffer II.
10. Spin samples at 100 × *g* for 30 s, place plug in the bottom of the Mobicol spin column, and replace the Luer-lock lid with the screw cap lid.
11. Add 72 μL 2× LDS sample buffer to each sample and heat samples to 95 °C for 5 min (*see* **Note 9**).
12. Remove bottom plug and spin samples at 14,000 × *g* for 2 min to collect eluent.

3.5 Sample Preparation, Visualization, and Proteolysis

1. To alkylate the cysteine sulfhydryls, add 5.5 μL of 200 mg/mL iodoacetamide solution per 50 μL of collected eluent and vortex.
2. Incubate samples for 30 min at room temperature in the dark.
3. Load 10 μL of each sample onto a 4–12% Bis-Tris NuPAGE gel, run gel at 100 V until the ladder begins to separate and run gel at 160 V for ~45 min or until desired resolution is achieved. Prepare gel for silver staining to assess total protein levels using the Pierce Silver Stain Kit (*see* **Note 10**).
4. Remove bottom stopper from detergent removal spin columns.
5. Use 2 mL collection tubes to wash the columns; spin the storage tubes for 1 min at 1500 × *g* to remove the storage buffer.
6. Wash each column with 400 μL 10 mM HEPES, pH 7.8 and spin for 1 min at 1500 × *g*.
7. Discard eluate and repeat **step 6** three times.
8. Carefully add alkylated eluents without touching the resin of the column and incubate for 2 min.
9. Centrifuge for 2 min at 1500 and then add 4 mL of base buffer to each tube and spin for 2 min at 1500 × *g* to collect detergent-free sample.
10. Add 135 μL of 10 mM HEPES buffer, pH 7.8, and wash the resin by centrifuging for 2 min at 1500 × *g*.
11. Add 1 mL 10 mM HEPES buffer, pH 7.8, to 20 μg trypsin (*see* **Note 11**).
12. Add 5 μL trypsin solution to each detergent-free sample and allow digestion to proceed overnight at 37 °C.

3.6 Isobaric Labeling (See Note 12)

1. Selecting a different TMT reagent for each condition, allow one vial of each TMT reagent to come to room temperature, add 200 μL of acetonitrile to each vial, and vortex several times for 30–60 s over a 5–10 min period.

2. Add 30 μL 1 M TEAB solution directly to each digested sample and vortex.
3. Add 450 μL acetonitrile directly to each digested sample and vortex.
4. Add 100 μL TMT reagent to each sample, and incubate at room temperature for 1 h.
5. Dry samples in SpeedVac, then reconstitute each sample in 50 μL of a 0.1% acetic acid/acetonitrile (30:70, v/v) solution.
6. Combine all the reconstituted samples into one vial and dry samples in SpeedVac.
7. Reconstitute consolidated sample in 100 μL of a 0.1% trifluoroacetic acid/acetonitrile (98:2, v/v) solution for off-line high-pH reversed phase separation.

3.7 High-pH Reversed Phase Separation (See Note 13)

1. Add consolidated TMT-labeled sample to low-volume autosampler vial and place in HLPC autosampler.
2. Peptide separation is achieved using a flow rate of 250 μL/min and the following ternary gradient: high-pH mobile phase C is held at 10% throughout gradient; starting conditions are 89% high-pH A and 1% high-pH mobile phase B, ramping high-pH mobile phase B to 50% over 65 min, and returning to 1% B from 65 to 75 min.
3. Add 20 μL of 0.5% TFA to each well of the collection plate prior to collection
4. Inject 50 μL of sample and collect 1 min fractions in a U-bottom 96-well plate from 15 through 73 min.
5. Dry U-bottom well plate with collected fractions in SpeedVac.
6. Pool every ~3 fractions with 40 μL mobile phase A to achieve 16 fractions and transfer into a V-bottom 96-well plate for nanoLC-MS/MS analysis.

3.8 Peptide Identification and Quantitation by nanoLC-MS/MS

1. Load 18 μL of sample on the trapping column using mobile phase A at a flow rate of 2.5 μL/min. Peptides are eluted using an 80 min gradient (2% mobile phase B for 5 min, 2–40% B from 5 to 65 min, followed by 70% B from 65 to 70 min, then returning to 2% B from 70 to 80 min) at a flowrate of 250 nL/min on the capillary separation column with direct spraying into the mass spectrometer.
2. Acquire LC-MS/MS data by operating the Q-Exactive mass spectrometer in a data-dependent mode, typical settings are: MS/MS are collected for the 12 most intense ions within a mass-to-charge ratio *(m/z)* range of 300–1250 detected in the full MS survey scan event. MS scans are acquired at 35,000 mass resolution (R) at m/z 200, using a target value of 1×10^6 ions and a maximum fill time of 100 ms. MS/MS scans are

acquired using a target value of 5×10^4 ions, maximum fill time of 100 ms, and isolation window of 2 Da. MS intensity threshold for ions to be selected for MS/MS is 1×10^5 ions and normalized collision energy is set to 28%. Further criteria for ions to be selected for MS/MS are charge state z of 2–4, and a typical exclusion time of 30 s before ions can be selected again for MS/MS. Spray voltage is set to 2.0 kV, heated capillary temperature is 250 °C, and S-lens RF level is 50. No sheath or auxiliary gas flow is applied.

3.9 Data Processing and Analysis

1. Conversion of Xcalibur Rawfiles to Mascot-compatible mgf-files as well as subsequent generation of mzXML, pepXML, protXML and Interact files are done using the corresponding TPP-modules.
2. Mgf-files are searched against the protein sequence database using Mascot with peptide mass tolerance set to 10 ppm, fragment tolerance set to 0.1 Da, and trypsin cleavage specificity (cleavage at K, R except if followed by P) allowing for two missed cleavages. Carbamidomethylation of cysteine is set as fixed modification, methionine oxidation, and TMT-modification of N-termini and lysine residues are set as variable modifications.
3. Data validation of peptide and protein identifications is done at the level of the complete dataset consisting of combined Mascot search results for all individual samples per experiment via the PeptideProphet and ProteinProphet modules in TPP using the decoy database function, nonparametric model, and the standard peptide probability threshold of 0.05.
4. TPP output (Interact) files containing peptide and protein information and corresponding TMT reporter ion peak intensity information for each peptide-to-spectrum match (as parsed from the Mascot result files) are uploaded to Oracle using in-house software.
5. For each peptide sequence and modification state, reporter ion signal intensities from all spectral matches are summed for each reporter ion type to account for lower precision of reporter ion ratios at low absolute signal intensities. Summed intensities are corrected according to the isotope correction factors given by the manufacturer to account for isotope impurities. Only peptide-to-spectrum matches that are unique assignments to a given identified protein within the total dataset are considered for protein quantitation.
6. High confidence protein identifications are reported based on protein probability cut-off values based on a false positive prediction of <1% by ProteinProphet.

7. For each protein in the high confidence dataset, individual peptide fold changes are calculated by normalizing each channel to the control channel (here: 1 μM probe, no competition compound, *see* **Notes 14** and **15**).
8. Protein fold changes over control are derived as median peptide fold change per protein.
9. Data for proteins with ≥2 quantified peptides are visualized in Spotfire (*see* the scatter plot in Fig. 3a) and the curve fitting function in Spotfire is used for the generation of probe titration curves and determination of concentrations where 50% enrichment is observed (Fig. 3b).

4 Notes

1. We have found that the different PAL moieties are quite stable under general laboratory conditions and handling. In an abundance of caution, we do recommend storing DMSO solutions protected from light; that is, in amber vials/tubes or wrapped with protective foil. Extreme precautions, such as working in as little as possible light, have not made a difference in our hands either with regard to compound stability or assay performance.
2. To generate Kasil fritted capillaries, mix 450 μL Kasil®1624 potassium silicate and 88 μL formamide in a glass vial and vortex. Dip fused silica capillary into the mixture for a few seconds to draw the mixture into capillary. Polymerize at 100 °C for 1 h and trim the frit to 1 mm.
3. The exact number of plates needed for assessing phenotypic relevance of specific binders will depend on the number of compounds in the SAR series. One plate of cells will be needed for each compound to be assessed; additionally, one plate for a DMSO control and one plate for a no PAL probe control will also be needed.
4. The HEK293T cells stably expressing V5-tagged mGpr39 used here can be incubated in media with 0% FBS without any apparent deleterious effects; e.g., cells lifting from the plate or change in cell morphology. However, other cell lines may require the presence of FBS to remain healthy. The concentration of FBS should be kept as low as possible to reduce nonspecific labeling of serum proteins by the PAL probe.
5. For de novo target deconvolution, we have found 60 min preincubation with competition compound followed by 60 min photoaffinity labeling probe treatment to be a good general starting point.
6. The UV bench lamp is generally allowed to warm up for at least 20 min prior to use. We have found it useful to monitor the UV

bench lamp performance by checking the UV intensity prior to irradiating cells using a Blak-Ray UV intensity Meter (Model J-221). Under optimal conditions the UV intensity will be out of range for the J-221 even at the highest setting. If the UV intensity falls within the range of the J-221 the UV bulbs are replaced immediately.

7. While there is no Federal or State regulatory limit for ultraviolet (UV) radiation exposure in the United States, it is worth emphasizing the importance of wearing full personal protection gear (lab coat, gloves, and protective eyewear) while working with the UV bench lamp. Operating the UV bench lamp in a cold room helps to keep the cells cold during the irradiation process and keeps UV exposure limited to samples.
8. Keratins from hair and skin are frequent contaminants in proteomics samples. Use gloves, tie back hair, and wear full-length lab coat during all the following procedures and sample handling. Use fresh buffers whenever possible.
9. Incubating the samples at 95 °C for 5 min creates a significant amount of pressure in the Mobicol columns. It is important to loosen the screw caps (~1/4 turn) before removing the bottom plug.
10. For probe titration experiments, due to contributions from nonspecific background, the total protein levels should generally increase with increasing probe concentrations. For competition experiments, the total protein levels should generally be consistent, i.e., independent of the concentration of competitor compound used in the experiment.
11. For improved coverage of transmembrane proteins that often yield few tryptic peptides in the optimal *m/z* range for MS-based peptide sequencing, we have found that a double digestion strategy using a chymotrypsin predigestion step prior to standard trypsin digestion can be beneficial. However, it should be noted that due to lower reporter ion yield, chymotryptic peptides in a predominantly tryptic peptide background can show exacerbated fold change compression.
12. It is recommended to check TMT labeling efficiencies prior to further sample separation by analyzing aliquots of individual TMT reactions by LC-MS/MS, and comparing signal intensities (or peptide-to-spectrum matches) for the fully labeled peptide species to the corresponding non/under-labeled species. If low labeling efficiency is observed (<95%), check that lyophilized TMT reagents are fully solubilized and that the pH is ~8 throughout the reaction.
13. The benefits of this step are two-fold. Beyond increasing the analytical depth by reducing complexity of individual fractions, it reduces fold change compression observed in all isobaric

labeling strategies including TMT [8]. Alternatively, fold change compression can be addressed by MS3-based approaches to peptide quantitation as available, e.g., on Thermo Orbitrap Fusion Lumos mass spectrometers or to some extent computationally (e.g., [8]).

14. For competition experiments it is preferable to subsequently renormalize data using the median fold change of all quantified peptides or proteins to compensate for differences in total protein yield for each individual affinity purification.
15. We have found that a final concentration of 1 μM PAL probe is a good general probe concentration to use in competition experiments; however, in select cases, we have used the data from the PAL probe titration experiment to identify a more optimal PAL probe concentration. For example, if the target(s) of interest show (near-)saturable binding for a full log unit at 1 μM PAL probe, then a lower concentration of PAL probe could be more ideal.

Acknowledgments

We thank John Damask and Stephen Marshall for their crucial role in building the data processing and analysis environment.

References

1. Swinney DC, Anthony J (2011) How were new medicines discovered? Nat Rev Drug Discov 10 (7):507–519. doi:10.1038/nrd3480
2. Schirle M, Jenkins JL (2016) Identifying compound efficacy targets in phenotypic drug discovery. Drug Discov Today 21(1):82–89. doi:10.1016/j.drudis.2015.08.001
3. Rix U, Superti-Furga G (2009) Target profiling of small molecules by chemical proteomics. Nat Chem Biol 5(9):616–624. doi:10.1038/nchembio.216
4. Bantscheff M, Schirle M, Sweetman G et al (2007) Quantitative mass spectrometry in proteomics: a critical review. Anal Bioanal Chem 389(4): 1017–1031. doi:10.1007/s00216-007-1486-6
5. Bantscheff M, Lemeer S, Savitski MM, Kuster B (2012) Quantitative mass spectrometry in proteomics: critical review update from 2007 to the present. Anal Bioanal Chem 404(4):939–965. doi:10.1007/s00216-012-6203-4
6. Bassilana F, Carlson A, DaSilva JA et al (2014) Target identification for a Hedgehog pathway inhibitor reveals the receptor GPR39. Nat Chem Biol 10(5):343–349. doi:10.1038/nchembio.1481
7. Mackinnon AL, Taunton J (2009) Target identification by Diazirine photo-cross-linking and click chemistry. Curr Protoc Chem Biol 1:55–73. doi:10.1002/9780470559277.ch090167
8. Savitski MM, Mathieson T, Zinn N et al (2013) Measuring and managing ratio compression for accurate iTRAQ/TMT quantification. J Proteome Res 12(8):3586–3598. doi:10.1021/pr400098r

Chapter 2

Multiplexed Liquid Chromatography-Multiple Reaction Monitoring Mass Spectrometry Quantification of Cancer Signaling Proteins

Yi Chen, Kate J. Fisher, Mark Lloyd, Elizabeth R. Wood, Domenico Coppola, Erin Siegel, David Shibata, Yian A. Chen, and John M. Koomen

Abstract

Quantitative evaluation of protein expression across multiple cancer-related signaling pathways (e.g., Wnt/β-catenin, TGF-β, receptor tyrosine kinases (RTK), MAP kinases, NF-κB, and apoptosis) in tumor tissues may enable the development of a molecular profile for each individual tumor that can aid in the selection of appropriate targeted cancer therapies. Here, we describe the development of a broadly applicable protocol to develop and implement quantitative mass spectrometry assays using cell line models and frozen tissue specimens from colon cancer patients. Cell lines are used to develop peptide-based assays for protein quantification, which are incorporated into a method based on SDS-PAGE protein fractionation, in-gel digestion, and liquid chromatography-multiple reaction monitoring mass spectrometry (LC-MRM/MS). This analytical platform is then applied to frozen tumor tissues. This protocol can be broadly applied to the study of human disease using multiplexed LC-MRM assays.

Key words Liquid chromatography-multiple reaction monitoring mass spectrometry, Cancer, Signaling, Colorectal carcinoma, Targeted quantification, Protein expression

1 Introduction

Cancer is one of the most prevalent public health concerns in the United States [1]. Novel approaches to further our understanding of basic molecular mechanisms of cancer signaling [2–5], particularly in human tumor tissues, are necessary to improve molecular classification schemes for personalized or precision medicine. With landscapes of tumor types created by genomics [6–9] and discovery proteomics [10] as well as known cancer biology, numerous important biomarkers can be selected to help guide patient treatment. Liquid chromatography-multiple reaction monitoring mass spectrometry (LC-MRM) has a long track record of clinical use and

Iulia M. Lazar et al. (eds.), *Proteomics for Drug Discovery: Methods and Protocols*, Methods in Molecular Biology, vol. 1647, DOI 10.1007/978-1-4939-7201-2_2, © Springer Science+Business Media LLC 2017

provides a flexible approach to multiplexing numerous protein biomarker assays. This technique is now routinely used for biomarker verification from discovery proteomics datasets [11, 12]. Furthermore, the method excels at detection of target proteins in complex matrices [13] as well as high sensitivity and high precision quantification of low abundance proteins that contribute to intricate cancer signaling pathways [14–16]. Finally, this method has the capability to be translated to clinical samples [17–20].

In clinical sample analysis, techniques such as immunohistochemistry [21, 22] and more recently matrix-assisted laser desorption ionization (MALDI) mass spectrometry [23] or MALDI-MRM imaging [24] have been used for visual elucidation of "molecular images" on the tissue specimens. However, low abundance proteins are not detected with these strategies and require other approaches to quantify their expression in clinical samples. Gel fractionation prior to LC-MRM enables this approach to be compared directly to immunoblotting (Western) [25] and serves as an important pre-fractionation step to enrich the target proteins [26]. A variety of software applications are available to facilitate assay development (MRMer [27], SRM Builder (Thermo), MRMaid [28], Pinpoint (Thermo), and Skyline [29], *inter al.*) as well as to validate and publish SRM data (MRMAtlas [30], Panorama [31], MRMaid-DB [32], QuAD [33], *inter al.*), and perform computational processing and statistical validation of the data (e.g., mProphet [34]). These efforts have in turn supported sensitive high-throughput protein detection and quantification in a wide variety of biological samples (e.g., yeast [35–38], worms [39], cell lines [40], plasma [41], and human tissue, *inter al.*) as well as inter-lab assessment of the precision, portability, and reproducibility of those measurements [42, 43].

1.1 Experimental Design

The experimental procedure described in this protocol focuses on selected cancer-related signaling pathways for mechanism elucidation and targeted therapy selection in colon cancer, but could be adapted to investigate other human diseases. The workflow includes: (i) hypothesis-driven target selection; (ii) SDS-PAGE-LC-MRM assay development with stable isotope-labeled standards; (iii) multiplexed LC-MRM assay assembly; (iv) frozen tissue evaluation and LC-MRM quantification; and (v) data analysis. The rationale for target selection is based on known colon tumor biology and therefore includes low abundance proteins with roles in cancer-related signaling pathways (as opposed to targets generated from discovery proteomics). Clinical relevance includes proteins with matched targeted therapeutic agents (e.g., receptor tyrosine kinase or RTK inhibitors). Pathways include Wnt/β-catenin, TGF-β, RTKs, MAP kinases, NF-κB, and apoptosis. The relationships between the proteins in this multiplexed LC-MRM panel have been mapped to illustrate their interactions and begin to generate a

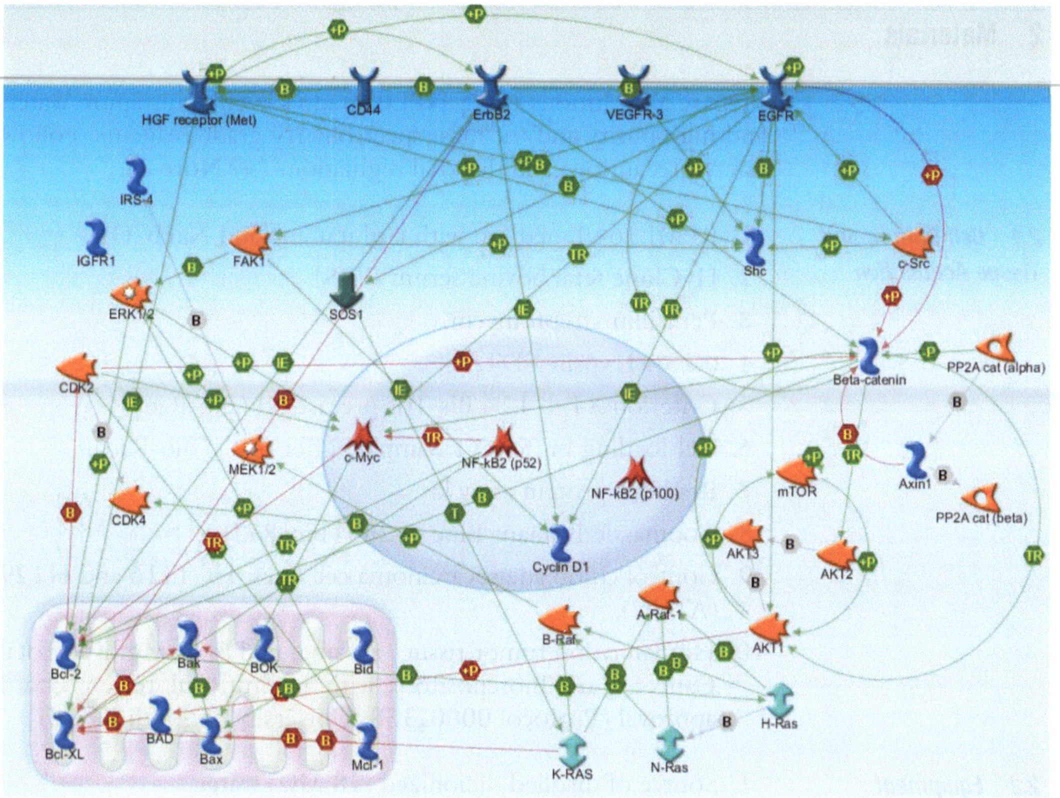

Fig. 1 Interaction map of selected targets from cancer signaling pathways. Selected functional relationships between proteins are displayed using the MapEditor function of GeneGO

system-level view for the LC-MRM results (Fig. 1). For assay development, SDS-PAGE protein fractionation is used for enrichment prior to LC-MRM screening. The final platform includes LC-MRM analysis of five distinct gel bands and is tested in cell line models prior to analysis of human tumor tissues.

In summary, the goal of this protocol is to enable investigation of cancer-related pathways in human tumors using quantitative mass spectrometry. The potential to achieve this goal is directly linked to recruitment of the scientific team that can contribute the necessary expertise and proper experimental design (*see* **Note 1**). Ultimately, we believe that this approach will improve our understanding of disease processes and elucidate mechanisms of response to therapy. The methods described frame the different steps (brief descriptions and estimated timelines are included in **Note 2**) required to include quantitative proteomics with medical history and histology for comprehensive analysis of tumor tissues.

2 Materials

Prepare all solutions using HPLC grade solvents, ultrapure water, and high purity and/or mass spectrometry grade reagents. Follow safety precautions and disposal regulations (*see* **Note 3**).

2.1 Cell Culture and Tissue Acquisition

1. RPMI 1640 medium with L-glutamine and $NaHCO_3$.
2. HyClone fetal bovine serum (FBS).
3. Penicillin-streptomycin.
4. 0.25% Trypsin-EDTA.
5. Criterion XT 4–12% Bis-Tris precast gel (Bio-Rad).
6. Gel-loading buffer: XT Sample buffer (4×) (Bio-Rad).
7. Bradford protein assay kit.
8. Coomassie Brilliant Blue G-250 (Bio-Rad).
9. Control colon adenocarcinoma cell lines: HCT116 and HT29 (ATCC).
10. Human colon tumor tissues are obtained from Moffitt's Total Cancer Care Biorepository with institutional review board approval (Protocol 00001138, University of South Florida).

2.2 Equipment

1. Source of distilled, deionized (18 MΩ) water.
2. pH Meter (Orion).
3. Spectrophotometer (UV-Visible, Beckman Coulter).
4. Microcentrifuge, 14,000 × *g*.
5. Sonic dismembrator.
6. Vacuum concentrator (SpeedVac).
7. Dri-Bath.
8. Microanalytical balance.
9. Midi-gel electrophoresis system.
10. Gel imager.
11. NanoUPLC system with refrigerated autosampler.
12. NanoUPLC trap column (PepMap, 100 μm ID × 2 cm, C18, 5 μm particle size, 100 Å pore size).
13. NanoUPLC separation column (PepMap, 75 μm ID × 25 cm, C18, 2 μm particle size, 100 Å pore size).
14. Triple quadrupole mass spectrometer with a nanoelectrospray ion source (e.g., TSQ Vantage or Quantiva, Thermo or equivalent).
15. Electrospray emitters pulled to 10 μm ID tips. For safety precautions, *see* **Note 3**.

16. MALDI MS and MS/MS (e.g., 4700, ABSciex) and semi-preparative HPLC may be required for peptide characterization and purification, unless purchasing the peptides from a vendor that will characterize them prior to delivery.
17. High-throughput whole slide scanning instrument with whole slide imaging software.
18. Skyline software or equivalent (https://skyline.gs.washington.edu/labkey/project/home/software/Skyline/begin.view) [29].
19. GeneGO software (Metacore, Thomson Reuters available at http://portal.genego.com or similar pathway mapping tool).
20. Matlab R2009b (http://www.mathworks.com).

2.3 Solutions and Buffers

1. Ammonium bicarbonate: 100 mM NH_4HCO_3 solution in water. Dissolve 0.79 g ammonium bicarbonate in 80 mL 18 MΩ cm water, verify the pH is 8, and bring volume up to 100 mL total. Dilute as needed with 18 MΩ cm water to make 50 mM and 30 mM solutions.
2. Dithiothreitol (DTT): 1.25 M DTT solution in water. Dissolve 192.5 mg DTT in 1 mL water.
3. Lysis buffer: 8 M urea in 100 mM ammonium bicarbonate buffer, pH 8. Store at room temperature for up to 1 week.
4. Protein denaturation for SDS-PAGE: Pipette out an aliquot (e.g., 50 μg) of cell lysate or tissue homogenate based on the total protein concentration determined by the Bradford assay. Add appropriate amount of 4× Criterion XT sample buffer, 1 μL of 1.25 M DTT and water to a total volume of 30–45 μL (dependent on the maximum volume per lane). Denature the proteins at 95 °C for 10 min and cool on ice prior to loading the gel.
5. Gel destaining buffer: 10% methanol, 5% acetic acid (v/v). Add 100 mL methanol and 50 mL glacial acetic acid to distilled, deionized water to make a total amount of 1000 mL.
6. Gel staining buffer: Dissolve 5 mg Coomassie Brilliant Blue G-250 in 100 mL gel destaining buffer.
7. Gel slice destaining buffer: 50 mM ammonium bicarbonate, 50% methanol. Mix 25 mL of 100 mM ammonium bicarbonate and 25 mL of methanol to make 50 mL gel slice destaining buffer.
8. Disulfide reduction stock solution (10×): 20 mM tris(2-carboxyethyl)phosphine hydrochloride (TCEP) in water. Dissolve 5.7 mg TCEP in 1 mL water. Prepare fresh. Cover with aluminum foil and store at 4 °C.

9. Cysteine alkylation stock solution (10×): 200 mM iodoacetamide (IAA) in water. Dissolve 37 mg IAA in 1 mL water. Prepare fresh. Cover with aluminum foil and store at 4 °C.
10. Trypsin proteolytic digestion solution: 20 ng/μL trypsin. Dissolve 20 μg sequencing grade modified trypsin in 1 mL of aqueous 50 mM acetic acid to prevent autolysis. Aliquot and store at −80 °C until use.
11. Peptide extraction buffer: 50% acetonitrile, 0.01% trifluoroacetic acid (TFA) (v/v). Add 10 μL TFA and 50 mL acetonitrile to water to make a total volume of 100 mL.
12. LC-MRM solvent A: 0.1% formic acid in HPLC grade or LC-MS grade water, 2% acetonitrile (v/v).
13. LC-MRM solvent B: 0.1% formic acid in HPLC grade or LC-MS grade water, 90% acetonitrile (v/v).
14. LC loading solvent: 0.1% formic acid or 0.04% TFA in HPLC grade or LC-MS grade water, 2% acetonitrile (v/v).
15. Autosampler syringe wash solvent: 0.1% formic acid in water, 2% acetonitrile (v/v). All the solvents must be degassed prior to use.

2.4 Instrument Configuration and Data Acquisition Parameters

1. LC-MRM instrument parameters: The UPLC contains a refrigerated autosampler, column compartment with switching valve, two pumps, and degasser/solvent rack. Samples (5 μL) are loaded on the trapping column at 6 μL/min and washed for 5 min using a capillary scale loading pump. Then, the two position six port valve is used to switch the trapping column in-line with the analytical column described above. The LC gradient program is delivered at 300 nL/min by the nanoflow pump according to the program described below (*see* Subheading 3.2, **step 19**).
2. The UPLC is connected to a triple quadrupole mass spectrometer, equipped with a nanoelectrospray ion source operated in the positive ion mode; specific settings include 2400 V spray voltage from 10 μm ID spray tips with 250 °C transfer tube temperature. The Q1 resolution is set to 0.4 *m/z* (LC-MRM screening) or 0.7 *m/z* (for the final scheduled method), and Q3 resolution is set to 0.7 *m/z*. Fragmentation is obtained with 1.5 mTorr argon. Each transition is monitored for 20 ms.
3. Software: The instrument vendor data acquisition software (e.g., XCalibur 2.1 and TSQ Vantage) and vendor or academic data analysis software (e.g., Skyline [29]) are required.

3 Methods

3.1 Target Selection and Refinement Using Literature Review, Sequence Analysis Tools, and In-House or Publically Available Data

1. Review literature to determine comprehensive list of proteins associated with pathway or process of interest (e.g., known protein-protein interactions, enzyme-substrate relationships, etc.) (*see* **Note 4**).
2. Use visualization software to build the network map of the targets (e.g., Map Editor, GeneGO).
3. Review literature of relevant biological information about each protein (e.g., highly homologous proteins, isoforms, splice variants, mutations, posttranslational modifications, etc.) (*see* **Note 4**).
4. Review publically available data about protein separation and available antibody reagents for comparative analysis/verification. Immunoprecipitation and western blot data from literature or vendors are needed to predict the region of protein migration in SDS-PAGE. If needed, adjacent regions are excised and evaluated with LC-MRM.
5. Obtain the protein sequence from UniProt (http://www.uniprot.org/) and publically available mass spectrometry data from PeptideAtlas (http://www.peptideatlas.org/) [44] or PRIDE/ProteomeXchange (https://www.ebi.ac.uk/pride/archive/) [45, 46] *inter al.* If sufficient MS data are available, **steps 6–30** can be omitted.
6. Use vendor or academic software (e.g., Skyline [29]) to predict tryptic peptides and transitions (precursor peptide and fragment ion pairs) from canonical sequences obtained from UniProt entries. Extensive tutorials are available online (https://skyline.gs.washington.edu/labkey/project/home/software/Skyline/begin.view).
7. Refine peptide selection to account for isoforms, modifications, and mutations of the proteins, as needed. Select unique peptides by filtering against the relevant background proteome (e.g., most current list of human entries in the UniProt database).
8. Select doubly charged peptides between 7 and 25 amino acids in length and exclude sequences containing the redox reactive residues, cysteine, and methionine, unless there are a few other choices. Additional selection criteria may eliminate sequences with adjacent or nearby tryptic cleavages sites or consensus sequences for glycosylation sites. Long peptides may have multiple charge states, which will result in signal splitting and poorer sensitivity; in addition, these peptides have more fragmentation channels and may not have significant amounts of signal in any one transition.

9. To select transitions for LC-MRM screening, we typically use *y* ions starting from either y_3 or y_x > peptide *m/z* and ending with $y_{(n-1)}$ for initial LC-MRM screens [32, 47].
10. Export the .csv file containing the transition list (precursor m/z, fragment m/z, collision energy, and retention time for scheduling) and import it into the instrument method using the vendor-provided software.

3.2 LC-MRM Assay Development

1. Select appropriate cell lines with high levels of protein expression for LC-MRM assay development. Acquire cell lines from accredited vendors with specific QA/QC metrics that routinely evaluate each cell line and culture them according to the manufacturer's instructions. As an example, these colon cancer cell lines are grown to 70% confluence in RPMI-1640 medium supplemented with 10% FBS and 1% penicillin-streptomycin (10,000 U/mL) and incubated in a 5% CO_2 atmosphere at 37 °C.
2. To provide material for multiple experiments, harvest ten million cells (~1 mg total protein) for each colon cancer cell line from two-dimensional culture after washing with cold PBS twice.
3. Lyse the cells in 8 M urea/100 mM ammonium bicarbonate buffer, pH 8, on ice.
4. Sonicate twice using 10 s pulses on ice and then centrifuge at 14,000 × *g* for 10 min at 4 °C.
5. Pipette the supernatant of the clarified cell lysate into a new microcentrifuge tube and measure the protein concentration by the Bradford assay (or micro-bichinchoninic acid/BCA assay).
6. Obtain an aliquot of the cell lysate (50 μg) and mix with an appropriate amount of concentrated Criterion XT sample buffer. Boil for 5 min for protein denaturation prior to gel loading.
7. Choose an appropriate gel to resolve the MW range of interest (*see* **Note 5**). In this experiment, separate proteins in 4–12% Bis-Tris gels (Criterion XT) for ~80 min until dye front reaches bottom of gel at 150 V. The best quality separation is needed to maximize protein enrichment and reduce the potential interference with the peptides selected for the initial LC-MRM screening. For GeLC-MRM analysis of tumor tissues, the gel separation is shortened and five regions are excised (*see* Subheading 3.3, **step 1**).
8. Stain the gel with Coomassie Brilliant Blue G-250 in an aqueous solution of 10% methanol/5% acetic acid (v/v) for 1 h, and

Table 1
LC gradient program

Time (min)	Duration (min)	B (%)
0	0	5
35	35	50
37	2	90
42	5	90
43	1	5
53	10	5

destain using the aqueous solution of 10% methanol/5% acetic acid for at least 3 h.

9. Excise the gel bands based on the predicted migration of the target proteins, and chop the bands into 1 mm^3 pieces.
10. Add 20 μL of disulfide reduction solution (20 mM TCEP) and 180 μL of 50 mM ammonium bicarbonate to the slices and incubate at 37 °C for 15 min, add another aliquot of TCEP and repeat incubation.
11. Add 20 μL of alkylation solution 200 mM IAA and 180 μL of 50 mM ammonium bicarbonate to the slices and incubate in the dark at room temperature for 20 min, add 2nd aliquot of IAA and repeat incubation.
12. Wash and dehydrate the gel slices with 200 μL 50 mM ammonium bicarbonate/50% methanol (v/v).
13. Add 20 μL of 20 ng/μL trypsin and 200 μL of 30 mM ammonium bicarbonate to each sample and incubate at 37 °C overnight. Check that gel cubes are submerged.
14. Recover the supernatant and extract gel slices twice with 100 μL 50% acetonitrile/0.01% TFA (v/v).
15. Concentrate the digested peptide solutions by vacuum centrifugation.
16. Resuspend in 30 μL of LC loading solvent for LC-MRM analysis.
17. Degas solvents A and B for nanoUPLC separation.
18. Load 5 μL of samples on the trap column at 6 μL/min and wash with loading solvent for 5 min.
19. The LC gradient program is provided in Table 1.
20. The triple quadrupole instrument settings are provided in Table 2.

Table 2
Triple quadrupole instrument settings

Nanoelectrospray voltage	2400 V
Emitter tip	10 μm (ID) 360 μm (OD)
Transfer tube temperature	250 °C
Q1 resolution	0.4 (screening) 0.7 (scheduled)
Q3 resolution	0.7
Collision gas pressure	1.5 mTorr
Scan time/transition	20 ms

21. Evaluate data from LC-MRM screening along with publically available data. As in discovery proteomics, the number of peptides identified in these screens is a direct metric for the confidence that the user should have in the results (*see* **Note 6** for additional details). The highest intensity peptides with multiple strong transitions will produce the most sensitive assays. In addition, the elution position in the LC gradient and observation of potential interferences are also useful in selecting the best peptide candidates for assay development.
22. Synthesize or purchase at least one stable isotope-labeled standard (SIS) peptide for each target protein based on the signal intensity and interference of the detected peptides. Synthesis of the unlabeled tryptic peptide sequences is useful for assay characterization, but not required (*see* **Note 7**).
23. Estimate the purity and confirm the sequence for each synthetic or SIS peptide with HPLC, MS, and MS/MS. Purify the peptides with a semi-preparative HPLC system, if necessary.
24. Perform amino acid analysis to quantify concentrations of stocks prepared for the synthetic and SIS peptides.
25. Perform manual infusion for the synthetic peptides on the triple quadrupole mass spectrometer to determine the charge states observed in MS^1, examine the MS/MS fragmentation pattern, choose transitions, and optimize collision energy. Select three to six fragment ions with high signal intensity for each peptide. Compare the selected fragment ions against those observed in the LC-MRM screens described above.
26. Characterize the peptides. If both the synthetic peptide and the SIS peptide were synthesized, calibration curves in buffer and matrix (preferably not containing the analyte of interest) with blanks and synthetic peptides from 10 amol to 100 fmol can

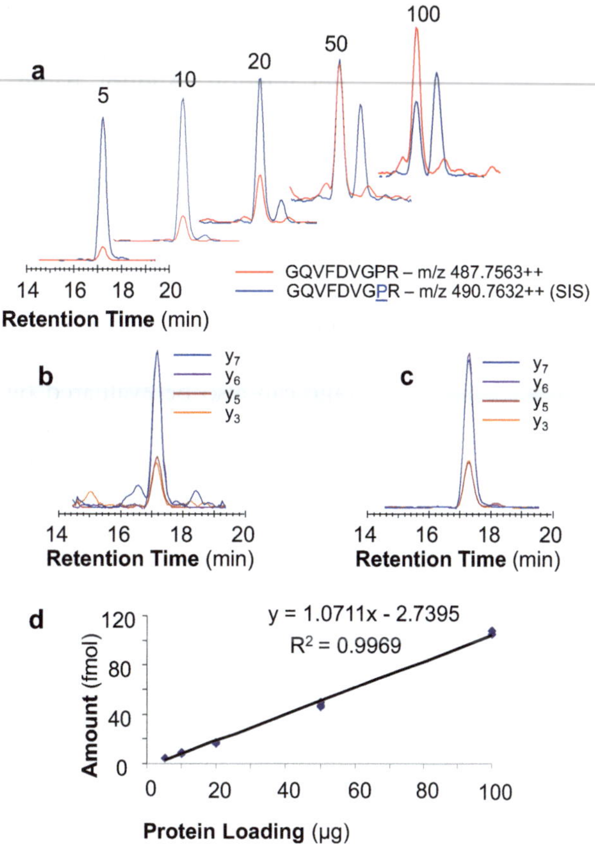

Fig. 2 MK01 quantification in HCT116 cell lysate serial dilutions. LC-MRM analysis of the GQVFDVGPR peptide (*red*) and the corresponding SIS peptide (*blue*) in serial dilutions of HCT116 cell lysate created by loading different amounts of total protein (from 5 to 100 μg) for SDS-PAGE (**a**). Relative intensities for each transition (y_3, y_5-y_7) observed for GQVFDVGPR (**b**) and the SIS peptide (**c**). Response curve and the equation for quantification of peptide GQVFDVGPR from MK01 (**d**)

help evaluate the limit of detection (LOD), upper and lower limits of quantification (ULOQ, LLOQ) and to decide on the amount of SIS peptide needed to mix with the biological samples as internal standard for quantification. Therefore, the amounts or concentrations used to make the calibration curve should include amounts expected from the biological samples. For example, ERK2 (MK01_HUMAN) is one of the 32 target proteins in gel region 4. We synthesized the SIS peptide with labeled proline, GQVFDVGPR (P8: $^{13}C_5{}^{15}N$), for the corresponding proteolytic peptide from the endogenous protein and monitored four transitions, y_3, y_5, y_6, and y_7, for both sequences to evaluate the assay performance (Fig. 2).

27. Mix internal standards into higher concentration stock solutions based on the proteins expected to be monitored in each gel fraction. Aliquot based on the expected scale of your batches for the quantitative experiment and store at −80 °C. As an example, if you expect to have 10 clinical samples with 2 controls in each batch, aliquot enough peptide mix for 15 experiments.
28. Check the stability of the SIS peptides. For example, the peak area of both the total and each transition of the peptides can be measured in five freeze-thaw cycle increments to a total of 15 cycles and compared to the fresh aliquot. Long-term storage can also be evaluated by comparing the results from the standard mix over time against a freshly prepared standard. The Pierce Retention Time Calibration peptide mix is useful for this task. The intensity of the individual peaks should meet specific QC metrics, as should the ratios between the SIS peptide peak intensities. These values will have to be determined for each set of analytes.

3.3 Assay Assembly and Verification in Cell Lines

1. The gel separation parameters for the final assay (particularly the number of excised regions or protein fractions) should balance throughput with effective separation. LC-MRM transitions are organized and concatenated into a single method for each SDS-PAGE region. Some proteins are monitored in the adjacent gel sections if they are expected to migrate at the edge of the excised region or have multiple isoforms with different MWs. Rather than excising individual gel bands, the gel run is shortened and five regions are excised using the MW markers and overall staining pattern of the cell lysates and tissue homogenates (Fig. 3).
2. In order to sample this number of peptides, the LC-MRM must be scheduled, meaning that peptides are not monitored for the entire experiment only during the expected elution time. If a peptide peak is expected to elute at 30 min, the instrument should be programmed to measure those transitions for a small window around that time. For cell lines, variability is less of an issue, so the time window could be less than 2 min. For complex clinical samples, the time window may need to be as wide as 5 min, depending on the quality of the samples and the reproducibility of the LC. We typically err on the side of expanding the time window, because signals not sampled lead to missingness in the data. In this project, the window is set to 5 min, because that enables sufficient sampling (>10 points in each peak) even for the largest number of measurements: 1150 transitions of 236 peptides from 32 target proteins in gel region 3 (MW 70-120, Fig. 3).

3. Prepare appropriate concentrations of cell lysate from each selected cell line and load 5, 10, 20, 50 and 100 μg of total protein into different gel lanes for fractionation and excision according to Subheading 3.3, **step 1**, following the steps in Subheading 3.2 to prepare protein fractions for LC-MRM. After concentrating the digested samples, resuspend the dried samples in 30 μL of the LC loading buffer containing the appropriate mix of SIS peptides for each gel fraction. For this project, highly characterized HCT116 and HT29 cell lines were selected for this experiment. Additional cell lines can be selected for larger scale verification of the multiplexed LC-MRM assay performance. We expanded our analysis to include SW620, SW480, HCT15, HCT116, DLD1, HT29, KM12SM, KM12C, Colo205, and KM12 (Fig. 4).

3.4 Tumor Tissue Selection, Sectioning, and Evaluation

1. Obtain the frozen tumor tissues from a tissue bank or via a prospective research protocol. A discussion of sample selection criteria is provided in **Note 8**. In this particular case for colon cancer research, colon tumor tissues are acquired from Moffitt's institutional Total Cancer Care Biorepository based on the following criteria: (i) adenocarcinoma of the colon; (ii) stage III or stage IV; (iii) age <90; (iv) collected at Moffitt; (v) >90% tumor; (vi) no prior treatment; (vii) >10 mg frozen tumor tissue banked; (viii) gene expression profile data complete ($n = 372$ available); (ix) cellularity >70%; (x) tumor/ malignancy >70%; (xi) processing time/ischemia <20 min.
2. Cut five serial sections from each frozen tissue block. Full embedding in optimal cutting temperature (OCT) medium is acceptable for these tissue sections, because it is removed prior to processing (*see* Subheading 3.5, **steps 2–4**). Tissue sections 1 and 5 will be prepared at 5 μm thickness and stained with hematoxylin and eosin (H&E) for pathology review to evaluate the consistency of histology though the sectioned depth of the tissue. Tissue sections 2–4 are cut with 25 μm thickness and saved in individual 0.5 mL Eppendorf tubes and stored at −80 °C for GeLC-MRM analysis. Additional sections can be excised based on the total material of the tumor tissue and experimental design for additional confirmatory tests such as immunohistochemistry of specific biomarkers or laser capture microdissection of tumor cells. Remaining material from each tumor could also be used for genomic characterization.
3. Hematoxylin and Eosin stained (H&E) slides for each tissue specimen must be reviewed by a board-certified pathologist (D. C.) of appropriate specialization to confirm the tumor type, grade, and other pertinent parameters. Proportions (%) of the following tissue components will be estimated: malignancy (tumor cells, stroma, necrosis, and pre-neoplastic lesions or

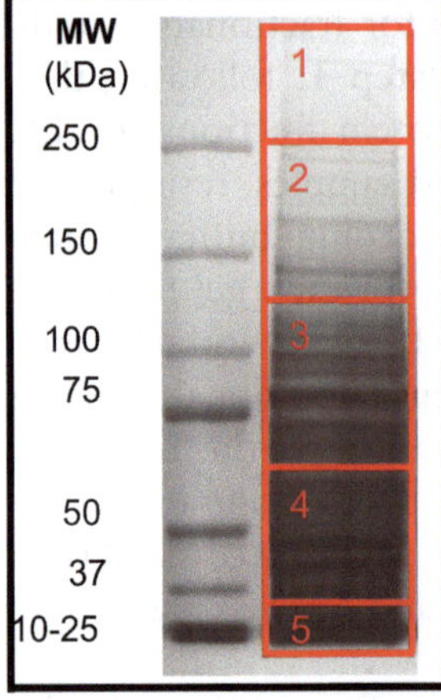

Gel	Proteins	Peptides	Transitions
1	3	33	174
2	17	144	695
3	32	236	1150
4	32	194	933
5	28	88	438

Regions 1-3: APC (YSDEQLNSGR)
Regions 1-2: ERBB3 (LTFQLEPNPHTK), TOR (LFDAPEAPLPSR)
Region 2: CD44 (FAGVFHVEK), ERBB2 (VLQGLPR, ELVSEFSR)
Regions 2-3: PGFRA (ELDIFGLNPADESTR), PGFRB (DQLVLGR), VGFR1 (AVSSFPDPALYPLGSR), VGFR2 (YLGYPPPEIK), VGFR3 (SGVDLADSNQK), SOS1 (EINSPNLLK), SOS2 (HAFELVSK), MET (VADFGLAR), EGFR (YLVIQGDER), IRS1 (HTQRPGEPEEGAR), IRS4 (EVSYNWDPK), IGF1R (SEILYIR)
Region 3: AXN1 (LLLETAAPR), AXN2 (STETVDSGYR), CADH1 (VTEPLDR), CTNA1 (SDALNSAIDK), CTNB1 (EGLLAIFK, AIPELTK), FAK1 (SLLDSVK), FAK2 (SNFELLEK), BRAF (IGDFGLATVKSR, IGDFGLATVK, NEVGVLR), RAF1 (APVSGTQEK), ARAF (IGTGSFGTVFR), NFκB1/p105 (AGADLSLLDR), NFκB2/p100 (AGAGAPELLR),
Regions 3-4: RELA/p65 (GSFSQADVHR), RELB (ADFSQADVHR), SRC8 (ANFENLAK), 2AAA (AVGPEITK), 2AAB (VLELDSVK), SHC1 (ALDFNTR)
Region 4: 2ABA (ILHTAWHPK), SRC (LLLNAENPR), CSK21 (QLYQTLTDYDIR), GSK3B (QTLPVIYVK), MYC (DQIPELENNEK), TCF7 (LPEPLEDGLK), KC1A (LFLIDFGLAK), CSK22 (VLGTEELYGYLK), MK01 (GQVFDVGPR), MK03 (ALDLLDR), MP2K1 (IPEQILGK), MP2K2 (IPEEILGK), PTEN (GVTIPSQR), MITF (ELGTLIPK), STK11 (IDSTEVIYQPR), AKT1 (SLLSGLLK), AKT2 (HPFLTALK), AKT3 (TDGSFIGYK), MCL1 (QSLEIISR), NFκB1/p50 (NIHLHAHSLVGK), NFκB2/p52 (QYAIVFR)
Regions 4-5: PP2A (YSFLQFDPAPR), MMP7 (DLPHITVDR), CDK2 (VVPPLDEDGR), CDK4 (VPNGGGGGGGLPISTVR), CCND1 (FLSLEPVK)
Region 5: CSK2B (FNLTGLNEQVPHYR), RASK (SFEDIHHYR), RASH (SYGIPYIETSAK), RASN (SFADINLYR), panRAS (LVVVGAGGVGK), CD2A1 (ALLEAGALPNAPNSYGR), Bcl-2 (FATVVEELFR), Bcl-X (EAGDEFELR), Bad (GLGPSPAGDGPSGSGK), Bim (IGDEFNAYYAR), Bok (AAFFVLLPER), Bak (QLAIIGDDINR), BAX (TGALLLQGFIQDR), Bid (IEADSESQEDIIR), Bfl-1 (VLQNVAFSVQK), BCL-w (AAGDEFETR)

Fig. 3 Gel band excision pattern using Bio-Rad prestained MW markers with corresponding numbers of monitored proteins, peptides, and transitions. Five gel regions are cut with the following MW ranges: *1*:

adenoma), normal tissue (smooth muscle, colon epithelium), abnormal tissue (acute or chronic inflammation, fibrosis, ulceration, etc.), and benign neoplastic tissue. Evaluate consistency of slides 1 and 5 to make sure that the tissue content and distribution do not change significantly throughout the tissue samples.

4. Scan the H&E stained tissues with a whole slide scanner (e.g., ScanScope XT, Aperio) for archiving and annotation.
5. Automated image analysis is optional, but it can be useful to calculate the areas of the tissue components from Subheading 3.4, **step 3** to supplement the pathologist's review. In addition, training a pattern recognition algorithm (Genie, Aperio) to differentiate cellular components, stromal components, and blank background enables rapid analysis of tissue content to further select the highest quality specimens. The algorithm has been applied to the entire slide's digital image to determine the percentage of each tissue type by area (μm^2) and to then calculate the amount of protein extracted per unit of tissue area for each specimen (Fig. 5).

3.5 Tumor Tissue Sample Preparation for LC-MRM

1. In this experiment, individual tumors are assessed, so the clinical samples are randomized. In paired experiment designs (e.g., tumor/normal or pre–/posttreatment samples from the same patient), linked samples should be blocked together and then randomized. Cell lines or tumor tissue pools, when available (frozen aliquots of the same biological sample), are analyzed with each batch of ten clinical samples as quality control samples for instrument performance. In general, these experiments should precede clinical sample analysis to enable interruption in the case of errors in sample preparation or poor instrument performance.
2. Keep the samples on ice to melt optimal cutting temperature (OCT) medium (<5 min). Note that this step should be modified for any measurement other than protein expression, because posttranslational modifications may be significantly impacted by this processing.
3. Centrifuge the samples at 14,000 × *g* for 10 min.
4. Discard the OCT supernatant and keep the tissues in original tubes. Additional washes with cold ethanol can further remove OCT without significant loss of protein.

Fig. 3 (Continued) >250 kDa, *2*: 120–250 kDa, *3*: 70–120 kDa, *4*: 30–70 kDa, and *5*: <30 kDa. Guided by the markers and the gel staining, horizontal cuts are made at the top of the 250 kDa MW marker, halfway between the 100 and 150 kDa markers, halfway between the 50 and 75 kDa markers, below the 37 kDa marker, and below the dye front, as shown by the *red* boxes overlaid on the gel image

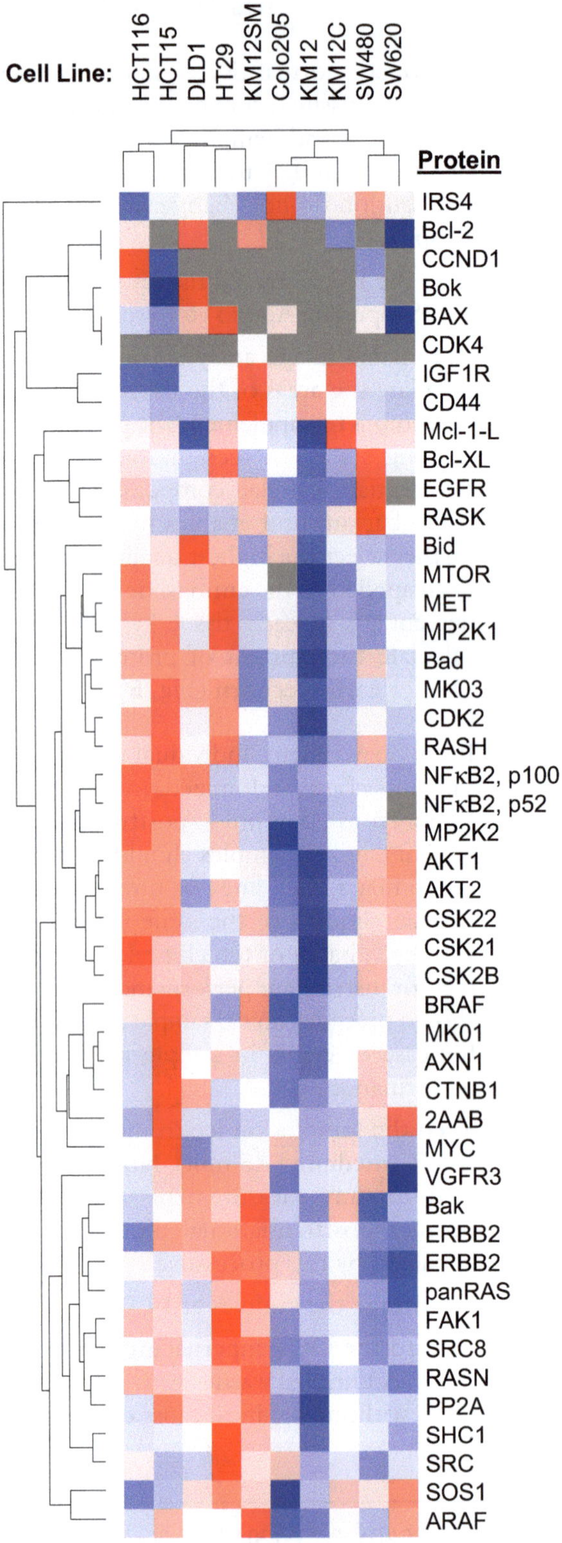

Fig. 4 Heat map and dendrograms for the comparison of protein expression in colon cancer cell lines. Lysates from SW620, SW480, HCT15, HCT116, DLD1, HT29, KM12SM, KM12C, Colo205, and KM12 cells were analyzed. *Blue* indicates lowest expression levels, while *red* indicates highest protein expression; *gray* indicates that the peptide was not observed

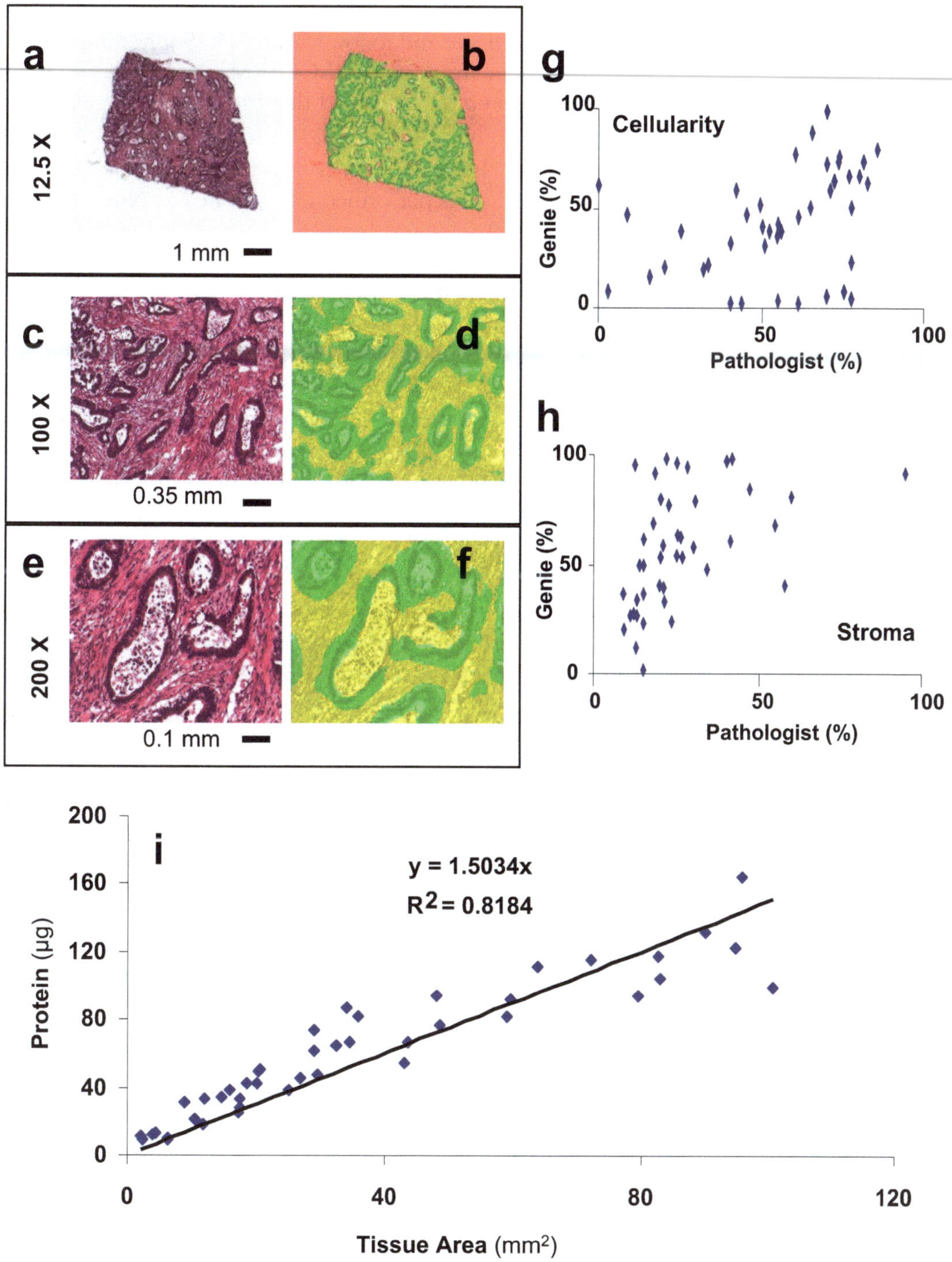

Fig. 5 Example tissue image and evaluation. An example H&E slide for a colon adenocarcinoma tissue (**a**) is shown. Automated whole slide tissue evaluation (**b**) by an in-house algorithm created using Genie (Aperio) indicates areas of cellularity (*green*), stroma (*yellow*), and background (*pink*). Additional views at higher magnification show the relative effectiveness of the assignment (**c**-**f**). Correlation plots are shown for the comparison of the pathologist's evaluations of tumor epithelium and stroma and the Genie recognition of cellularity (**g**) and stromal components (**h**). Tissue area is predictive of total protein obtained from tumor homogenate in 25 μm sections (**i**)

5. Add 50 μL 8 M urea/100 mM ammonium bicarbonate buffer, pH 8, to each tissue and follow the steps in Subheading 3.2 to prepare protein fractions for LC-MRM. After concentrating the digested samples, resuspend the dried samples in 30 μL of LC loading buffer with the appropriate SIS peptide mix for each gel fraction.
6. LC-MRM analysis is performed in triplicate (*see* **Note 9**).

3.6 Quality Control and Data Analysis

1. Import the raw data to analysis software (e.g., Skyline) and manually evaluate all peak selections.
2. Compare the retention time and fragmentation pattern (either by transition rank or relative intensity) of proteolytic peptides from the endogenous protein and the corresponding SIS peptides. Retention times should match exactly. Fragment ion ratios should be consistent with specific QC metrics (e.g., 5% variability), but transitions with significantly lower signal intensity may deviate more.
3. Export the peak area values for individual transitions and the summed totals to a .csv file for analysis.
4. Plot the peak areas for SIS peptides across the entire sample set. Examine any variations within a batch and between batches. Start by comparing the cell line or pooled standard samples included in each batch and continue to the individual clinical samples. Outliers should be marked and may need to be discarded based on poor instrument performance. If signal for the SIS peptides is less than 50% of the average in a given sample, it should likely be evaluated thoroughly and potentially removed from the dataset. If trends are observed, investigate possible causes (e.g., loss of instrument sensitivity over long analysis period or selection of samples with poor cellularity concentrated into a batch).
5. Measure ratios between the peak areas of SIS peptides or rank them by intensity to examine whether any of the SIS peptides is decreasing in ion signal.
6. Compare the data from cell line QC between batches to examine any changes in the performance of the assays.
7. Calculate the levels of protein expression using the ratio of proteolytic peptide and SIS peptide ion signals, the amount of the SIS peptide spiked into the sample, the fraction of the sample analyzed by LC-MRM, and the amount of protein loaded into the gel (output value in fmol/μg total protein). Sum the concentrations over multiple gel bands, when the peptide is detected in two GeLC-MRM fractions or treat these two measurements separately, when appropriate. Place values for proteolytic peptide signal, SIS peptide signal, and

calculated estimate of protein expression into three matrices for statistical analysis.

8. Using Matlab R2009b or an equivalent software program, examine data sampling distribution using scatter plots and histograms both in original scale of the raw data and after $\log_2$ transformation prior to determining whether parametric or nonparametric analyses should be applied.
9. Batch-to-batch normalization is optional, depending primarily on the data quality. While each SIS peptide is used to calculate the level of protein expression in every individual sample, batch-to-batch variation in the cell line data could be used as another normalizing factor, particularly if SIS peptides are not included or not observed for each analyte. Examine and normalize the clinical samples across multiple batches using internal standards and the peptide concentrations across the same cell line QC samples. These QC cell lines are evaluated in each batch and can be used to make sure the data are comparable between batches.
10. Lower limits of quantification (LLOQ) determined for each peptide in Subheading 3.2, **step 26** can be used to filter out protein expression levels below the LLOQ, which are detectable but not quantifiable.
11. Identify outliers from each set of three technical replicates for each peptide. A replicate is identified as an outlier using the following criteria: (i) the distance from the outlier to the median is over twice the distance of the median from the remaining data point or (ii) if the CV is >20%. The outliers were removed and treated as missing data.
12. Flag peptides with reliable detection in <50% of the (tissue or cell line) samples, as the conclusions drawn from these peptides should be interpreted with caution.
13. Perform quality control for samples (*see* **Note 10**). Central tendency, range, variability of all peptide expression levels and also the number of missing peptides are summarized for each sample. Overall expression levels across samples were visualized using boxplots heat maps. As an example, tissue samples with either lower protein expression levels or a high proportion of missing peptides (>30%) should be flagged for repeat analysis and review of the histology.
14. Calculate the average of technical triplicates which passed the QC for each biological sample. Log_2 average concentration from each biological sample is used in further analyses. Log_2 transformed peptide data are often normally distributed (and therefore parametric analyses can be applied). In addition, when comparing the expression levels between the two groups, the interpretation is straightforward.

15. Imputation of missing values is optional; alternatively, analytes with high missingness may need to be discarded or treated as a simple positive/negative readout. For analyses that are not capable of handling missing values, such as cluster analyses, when all three technical replicates of a sample are missing, they are replaced with the overall lowest detected concentration from all samples and peptides. However, other common imputation methods such as K-nearest neighbor method could be performed to impute missing values.
16. Perform hierarchical clustering analyses of the peptide expression levels and visualize the results in a heat map along with a dendrogram to explore relationships between proteins and tumor samples (Fig. 6). This helps to visualize peptides at all levels of expression across tissues/cell lines and examine reproducibility across technical replicates. Cluster 3.0 (http://bonsai.hgc.jp/~mdehoon/software/cluster/software.htm) and Java TreeView (http://jtreeview.sourceforge.net/) [48] are available freeware that can be used for this step. TreeView heat maps can be exported as PostScript files (.ps), which can be converted to PDF using GSview (http://pages.cs.wisc.edu/~ghost/gsview/) and then into images with Inkscape (https://inkscape.org/en/).
17. Perform statistical analyses (e.g., using *t*-tests or ANOVA) to test specific hypotheses, including, but not limited to mapping protein expression levels against transcript expression levels or known mutational status or relating protein expression levels with patient survival outcomes or responses to therapy. **Note 11** includes additional discussion of the expected outcomes.

4 Notes

1. Assembling the research team with representatives of each type of expertise required to address the research question is critical for success. In this case, the disciplines include colon cancer patient treatment, surgery, and relevant biology (DS), pathology (DC), epidemiology (ES), microscopy (ML), proteomics (YC, EW, and JK), and biostatistics (KF and YAC). Study design is also a critical step. The existing samples and available budget often determine the cohort that will be examined; these practical limitations need to be examined prior to experiment design and evaluated using power calculations as well as knowledge of the patient populations, histological tumor types, and molecular tumor subtypes. Placement of the study in the context of the patients' medical history information and genomics data is also critical. All the previous testing results that are relevant should be accrued at the beginning of the study.

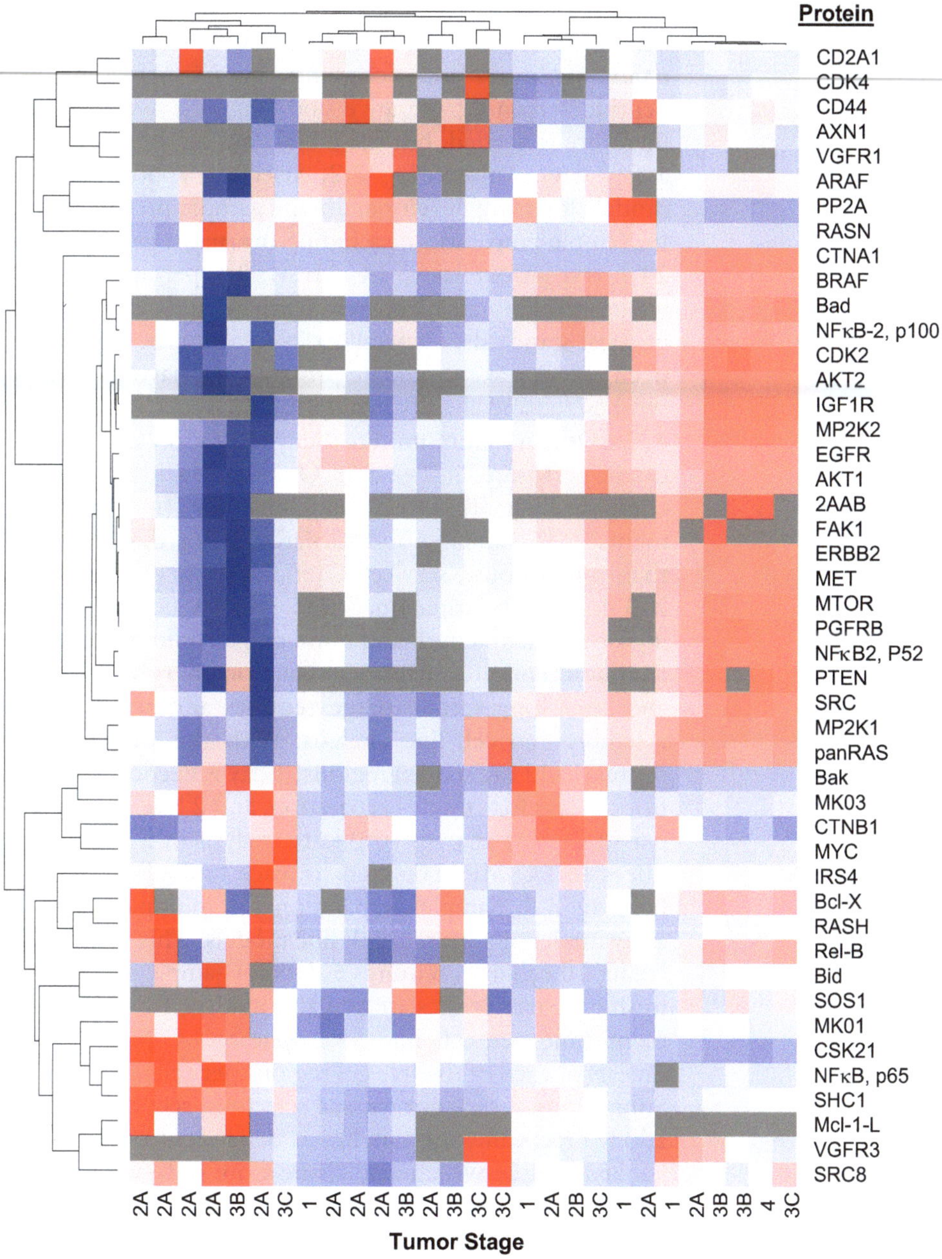

Fig. 6 Heat map and dendrograms for comparison of protein expression in colon tumor tissues. *Blue* indicates lowest expression levels, while *red* indicates highest protein expression; *gray* indicates that the peptide was not observed

Because these tumors are already banked, two to 5 years of follow-up information could be obtained. This data will enable combining the results from quantitative proteomics with medical history, histology, and gene expression profiles for comprehensive analysis of tumor tissues with the ultimate goal of establishing a molecular basis for personalized medicine.

2. Typical times for each step used in our laboratory are: Subheading 3.1, Selecting target proteins from cancer related signaling pathways and designing the LC-MRM method for peptide screening: iterative and ongoing assessment of the literature is needed, but the initial data can be gathered within a few days. Subheading 3.2, Selecting appropriate cell line models for assay development and sample preparation: 3 days for sample preparation, including overnight gel destaining and 24 h trypsin digestion. LC-MRM: 2 h per analysis including washes to eliminate carryover. Synthesis and characterization of internal standards: Generally a 2–4 week process, but we have been able to take candidate proteins and have the SIS peptide synthesized and characterized within 1 week. Synthesis requires 1–2 days at a minimum. Further purification and mass analysis can be performed in 1–2 days. Our amino acid analysis has been completed at Texas A&M University (*see* "Acknowledgments"), so shipping and turnaround for that data is longer than an in-house service. Calibration curves can be run in 2–3 days on the triple quadrupole mass spectrometer depending on gradient length and number of required data points and technical replicates. Peptide evaluation after freeze-thaw cycles and after long term storage requires additional time. Assay assembly and cell line analysis: With this number of target proteins and excised gel regions, 1 day is needed for gel separation, 2 days for sample preparation, and 2 h of instrument time per LC-MRM analysis (triplicate analysis of 5 gel regions produces 15 LC-MRM datasets for each biological sample). Subheading 3.3, Sample preparation for multiplexed LC-MRM assay implementation in cell line models: This experiment analyzing 10 cell lines requires 5 days for sample preparation after fresh or frozen lysates are available and then 1.5 days of instrument time per cell line for triplicate LC-MRM analysis of digests of each of the five gel regions per cell line (50 samples). Subheading 3.4, Tumor tissue selection, sectioning and evaluation: Practically, the timeline is dependent on IRB approval, banked sample availability, Tissue Core staff, whole slide scanning, and the pathologist's schedule. After several discussions, we used ~1 h per sample for pathology review and ~1 h per sample for pattern recognition software evaluation as well as 2 h for data reconciliation and interpretation. Subheading 3.5, Tumor tissue analysis: For each batch of ten tumor tissues with two cell line controls, 5 days was needed for sample preparation (60 samples total) with 1.5 days of instrument time per tissue/cell line for triplicate LC-MRM analysis. Subheading 3.6, Quality control and data analysis: For the first project, this process will be iterative and ongoing for weeks to months due to discussion of multiple relevant parameters; timeline is also limited by the other commitments of the

biostatisticians. After the research team has accumulated more experience, the time required will significantly decrease.

3. Safety concerns: When using toxic, highly flammable, or corrosive solvents and reagents, use protective clothing, gloves, safety goggles, or eye/face protection as appropriate, and handle the reagent in the fume hood. In addition, special care is needed when handling electrospray emitter tips; safety glasses are recommended. Solvents and waste must be properly disposed or recycled as per relevant local and federal regulations.
4. For gathering the preliminary information related to proteins, pathways, and biological processes, it is important to understand that our knowledge of biology is continuously growing and evolving. Additional protein interactions or novel post-translational modifications can be discovered that would influence the selection of peptides and the interpretation of the data from these LC-MRM assays. Literature review should be iterative throughout the assay development and implementation process.
5. The precast gel selection for protein fractionation prior to LC-MRM should be based on the overall list of proteins and expected MW ranges for their separation. This panel included large proteins >100 kDa (e.g., receptor tyrosine kinases) and small proteins <25 kDa (e.g., Bcl-2 family regulators of apoptosis), so the gradient gel (4–12%) was the most appropriate choice.
6. For evaluation of LC-MRM screening data and other data from publically available sources as part of assay development, single peptide hits have had a poor success rate for generating an assay (~25% of cases), but the best peptides selected from five or more detected sequences almost always (~95% of cases) did develop into useful assays. Data should also be considered with regard to the number of tryptic peptides that could be detected from a given protein. The apoptosis-regulating proteins in this study would generate only a few potential tryptic peptides amenable to LC-MRM, so the rationale for proceeding with assay development should remain flexible enough to accommodate these issues.
7. The choice to synthesize the proteolytic peptide from the endogenous protein (without stable isotope-labeling) provides more flexibility and more accurate assessment of the assay performance via calibration curves in buffer and matrix. However, reverse calibration curves (varying the amount of SIS peptide in a sample with a known amount of endogenous protein) can also be used for assay characterization. Note that interference may be more likely in the transitions for the

proteolytic peptide from the endogenous protein than in the transitions for the SIS peptide.

8. Appropriate criteria need to be applied for tissue sample selection. These choices are even more important for evaluation of samples collected across different sites or under different protocols; these projects can include potential confounding variables that are not observed under highly controlled single institution prospective collections under one unified protocol.
9. Prior to analysis of clinical samples, cleaning, tuning, and calibration of the mass spectrometer are recommended. For large-scale datasets, the UPLC columns should be replaced, conditioned, and tested prior to sample analysis. Routine injections of the cell lysates samples are ideal for these quality control steps.
10. If quality control indicates problems with the datasets, troubleshooting can include the steps outlined in Table 3.
11. The anticipated results include relative quantification of proteins across cell lines and tumor tissues as well as minimum estimates of protein expression levels in terms of femtomoles per milligram of total protein. The immediate impact of this assay platform and others like it can be in improving our understanding of tumor biology in situ and in ranking the most important proteins to develop as candidate biomarkers for prognosis or prediction of therapeutic response. In addition, these measurements can be included as correlates in clinical trials to examine the mechanisms of drug response and drug resistance. Ultimately, multiplexed LC-MRM has the potential and capability to provide biomarker measurements supporting selection of clinical care regimens.

Table 3
Steps for troubleshooting dataset quality control-related issues

Problem	Solution
Pathology report has high stroma/necrosis in tissue sample	Discard the sample as most target proteins are from tumor cellularity.
Loss of signal for SIS peptide	Discard this data or perform relative quantification by normalizing with an adjacent SIS peptide with similar signal intensity
Poor data for all proteins in a sample	Check histology of the patient sample and analyze another section
Batch-to-batch variability	Normalize data from the tissue samples by using the data from QC cell line samples

Acknowledgments

This work was supported in part by the following core facilities at Moffitt Cancer Center: Tissue Core, Analytic Microscopy, Proteomics, and Biostatistics; funding provided in part by the Cancer Center Support Grant, P30-CA076292, from the National Cancer Institute. Proteomics instruments are supported by funding from the Moffitt Foundation and shared instrument grants from the Bankhead-Coley Cancer Research program of the Florida Department of Health (06BS-02-9614 and 09BN-14). Project funding was received as subcontract from the Moffitt National Functional Genomics Center funded by the US Army Medical Research and Materiel Command under award DAMD17-02-2-0051. Amino acid analysis was performed by the Protein Chemistry Laboratory at Texas A&M University by Virginia Johnson, MS, and Larry Dangott, PhD.

References

1. Siegel RL, Miller KD, Jemal A (2016) Cancer statistics. CA Cancer J Clin 66:7–30
2. Taylor IW, Wrana JL (2012) Protein interaction networks in medicine and disease. Proteomics 12(10):1706–1716
3. Koomen JM, Haura EB, Bepler G et al (2008) Proteomic contributions to personalized cancer care. Mol Cell Proteomics 7:1780–1794
4. Myers MV, Manning HC, Coffey RJ, Liebler DC (2012) Protein expression signatures for inhibition of epidermal growth factor receptor-mediated signaling. Mol Cell Proteomics 11:M111.015222
5. Jendrossek V (2012) The intrinsic apoptosis pathways as a target in anticancer therapy. Curr Pharm Biotechnol 13:1426–1438
6. Network CGA (2012) Comprehensive molecular characterization of human colon and rectal cancer. Nature 487:330–337
7. De Sousa E, Melo F, Wang X et al (2013) Poor-prognosis colon cancer is defined by a molecularly distinct subtype and develops from serrated precursor lesions. Nat Med 19:614–618
8. Sadanandam A, Lyssiotis CA, Homicsko K et al (2013) A colorectal cancer classification system that associates cellular phenotype and responses to therapy. Nat Med 19:619–625
9. Ren Z, Wang W, Li J (2016) Identifying molecular subtypes in human colon cancer using gene expression and DNA methylation microarray data. Int J Oncol 48:690–702
10. Zhang B, Wang J, Wang X et al (2014) Proteogenomic characterization of human colon and rectal cancer. Nature 513:382–387
11. Turtoi A, Musmeci D, Wang Y et al (2011) Identification of novel accessible proteins bearing diagnostic and therapeutic potential in human pancreatic ductal adenocarcinoma. J Proteome Res 10:4302–4313
12. Whiteaker JR, Lin C, Kennedy J et al (2011) A targeted proteomics-based pipeline for verification of biomarkers in plasma. Nat Biotechnol 29:625–634
13. Agard NJ, Mahrus S, Trinidad JC et al (2012) Global kinetic analysis of proteolysis via quantitative targeted proteomics. Proc Natl Acad Sci U S A 109:1913–1918
14. Shi T, Su D, Liu T et al (2012) Advancing the sensitivity of selected reaction monitoring-based targeted quantitative proteomics. Proteomics 12:1074–1092
15. Collins BC, Miller CA, Sposny A et al (2012) Development of a pharmaceutical hepatotoxicity biomarker panel using a discovery to targeted proteomics approach. Mol Cell Proteomics 11:394–410
16. Boja ES, Rodriguez H (2012) Mass spectrometry-based targeted quantitative proteomics: achieving sensitive and reproducible detection of proteins. Proteomics 12:1093–1110
17. Hoofnagle AN, Becker JO, Oda MN et al (2012) Multiple-reaction monitoring-mass spectrometric assays can accurately measure the relative protein abundance in complex mixtures. Clin Chem 58:777–781
18. Sprung RW, Martinez MA, Carpenter KL et al (2012) Precision of multiple reaction

monitoring mass spectrometry analysis of formalin-fixed, paraffin-embedded tissue. J Proteome Res 11:3498–3505

19. Kuhn E, Whiteaker JR, Mani DR et al (2012) Inter-laboratory evaluation of automated, multiplexed peptide immunoaffinity enrichment coupled to multiple reaction monitoring mass spectrometry for quantifying proteins in plasma. Mol Cell Proteomics 11:M111.013854
20. Lam MP, Scruggs SB, Kim TY et al (2012) An MRM-based workflow for quantifying cardiac mitochondrial protein phosphorylation in murine and human tissue. J Proteome 75 (15):4602–4609
21. Fonseca-Sánchez MA, Rodríguez Cuevas S, Mendoza-Hernández G et al (2012) Breast cancer proteomics reveals a positive correlation between glyoxalase 1 expression and high tumor grade. Int J Oncol 41:670–680
22. Tang A, Li N, Li X et al (2012) Dynamic activation of the key pathways: linking colitis to colorectal cancer in a mouse model. Carcinogenesis 33:1375–1383
23. Casadonte R, Caprioli RM (2011) Proteomic analysis of formalin-fixed paraffin-embedded tissue by MALDI imaging mass spectrometry. Nat Protoc 6:1695–1709
24. Clemis EJ, Smith DS, Camenzind AG et al (2012) Quantitation of spatially-localized proteins in tissue samples using MALDI-MRM imaging. Anal Chem 84:3514–3522
25. Tang HY, Beer LA, Barnhart KT, Speicher DW (2011) Rapid verification of candidate serological biomarkers using gel-based, label-free multiple reaction monitoring. J Proteome Res 10:4005–4017
26. Wang M, Heo GY, Omarova S et al (2012) Sample pre-fractionation for mass spectrometry quantification of low-abundance membrane proteins. Anal Chem 84:5186–5191
27. Martin DB, Holzman T, May D et al (2008) MRMer, an interactive open source and cross-platform system for data extraction and visualization of multiple reaction monitoring experiments. Mol Cell Proteomics 7:2270–2278
28. Mead JA, Bianco L, Ottone V et al (2009) MRMaid, the web-based tool for designing multiple reaction monitoring (MRM) transitions. Mol Cell Proteomics 8:696–705
29. MacLean B, Tomazela DM, Shulman N et al (2010) Skyline: an open source document editor for creating and analyzing targeted proteomics experiments. Bioinformatics 26:966–968
30. Picotti P, Lam H, Campbell D et al (2008) Database of mass spectrometric assays for the yeast proteome. Nat Methods 5:913–914
31. Sharma V, Eckels J, Taylor GK et al (2014) Panorama: a targeted proteomics knowledge base. J Proteome Res 13:4205–4210
32. Cham JA, Bianco L, Barton C, Bessant C (2010) MRMaid-DB: a repository of published SRM transitions. J Proteome Res 9:620–625
33. Remily-Wood ER, Liu RZ, Xiang Y et al (2011) A database of reaction monitoring mass spectrometry assays for elucidating therapeutic response in cancer. Proteomics Clin Appl 5:383–396
34. Reiter L, Rinner O, Picotti P et al (2011) mProphet: automated data processing and statistical validation for large-scale SRM experiments. Nat Methods 8:430–435
35. Picotti P, Bodenmiller B, Mueller LN et al (2009) Full dynamic range proteome analysis of S. cerevisiae by targeted proteomics. Cell 138:795–806
36. Bluemlein K, Ralser M (2011) Monitoring protein expression in whole-cell extracts by targeted label- and standard-free LC-MS/MS. Nat Protoc 6:859–869
37. Picotti P, Rinner O, Stallmach R et al (2010) High-throughput generation of selected reaction-monitoring assays for proteins and proteomes. Nat Methods 7:43–46
38. Wu R, Haas W, Dephoure N et al (2011) A large-scale method to measure absolute protein phosphorylation stoichiometries. Nat Methods 8:677–683
39. Jovanovic M, Reiter L, Picotti P et al (2010) A quantitative targeted proteomics approach to validate predicted microRNA targets in C. elegans. Nat Methods 7:837–842
40. Kettenbach AN, Rush J, Gerber SA (2011) Absolute quantification of protein and post-translational modification abundance with stable isotope-labeled synthetic peptides. Nat Protoc 6:175–186
41. Kim YJ, Zaidi-Ainouch Z, Gallien S, Domon B (2012) Mass spectrometry-based detection and quantification of plasma glycoproteins using selective reaction monitoring. Nat Protoc 7:859–871
42. Addona TA, Abbatiello SE, Schilling B et al (2009) Multi-site assessment of the precision and reproducibility of multiple reaction monitoring-based measurements of proteins in plasma. Nat Biotechnol 27:633–641
43. Prakash A, Rezai T, Krastins B et al (2010) Platform for establishing interlaboratory reproducibility of selected reaction monitoring-based mass spectrometry peptide assays. J Proteome Res 9:6678–6688

44. Desiere F, Deutsch EW, King NL et al (2006) The PeptideAtlas project. Nucleic Acids Res 34:D655–D658
45. Vizcaíno JA, Deutsch EW, Wang R et al (2014) ProteomeXchange provides globally coordinated proteomics data submission and dissemination. Nat Biotechnol 32:223–226
46. Vizcaíno JA, Côté R, Reisinger F et al (2009) A guide to the proteomics identifications database proteomics data repository. Proteomics 9:4276–4283
47. Prakash A, Tomazela DM, Frewen B et al (2009) Expediting the development of targeted SRM assays: using data from shotgun proteomics to automate method development. J Proteome Res 8:2733–2739
48. Saldanha AJ (2004) Java Treeview – extensible visualization of microarray data. Bioinformatics 20:3246–3248

Chapter 3

Monitoring Dynamic Changes of the Cell Surface Glycoproteome by Quantitative Proteomics

Mathias Kalxdorf, Hans Christian Eberl, and Marcus Bantscheff

Abstract

The analysis of the cell surface accessible proteome provides invaluable information about cellular identity, cellular functions, and interactions. Cell surface labeling in combination with quantitative proteomics enables the unbiased identification and quantification of cell surface proteins. We describe a fast, efficient, and robust protocol for the enrichment of the N-linked plasma membrane glycoproteome and subsequent analysis by mass spectrometry. Precise and multiplexed quantification of relative changes of cell surface protein presentation is enabled by an isobaric labeling strategy.

Key words N-Glycoproteomics, Isobaric mass tags, Mass spectrometry, Plasma membrane proteome dynamics, Biotinylation

1 Introduction

The plasma membrane proteome, its composition and dynamics define morphology, function, and interaction of cells with their environment. Extracellular domain-containing proteins are involved in diverse functional aspects like intercellular interaction, cell adhesion as well as influx and efflux of molecules including nutrients, salts and also bioactive compounds. These proteins are further defining cellular identity and their specific interactions are involved in regulating the acquired and innate immune system while pathogens like viruses or bacteria target specific cell surface markers to infiltrate their host cells. Due to these crucial roles, this subproteome is also frequently targeted by pharmaceuticals. Approximately 60% of all FDA-approved drugs target transmembrane proteins with G-protein coupled receptors (GPCRs) (>25%) and ion channels (>10%) being the most frequently targeted protein classes [1, 2]. Among the top ten globally selling drugs in 2014, eight drugs were directly or indirectly targeting membrane proteins [3].

Iulia M. Lazar et al. (eds.), *Proteomics for Drug Discovery: Methods and Protocols*, Methods in Molecular Biology, vol. 1647, DOI 10.1007/978-1-4939-7201-2_3, © Springer Science+Business Media LLC 2017

The analysis of the plasma membrane proteome by mass spectrometry greatly improved our appreciation of the wealth of proteins in this cellular compartment [4, 5]. However, the relatively low abundance and the low solubilization efficiency of many plasma membrane proteins in commonly used lysis buffers still hamper the comprehensive analysis of this subproteome. To address these challenges, enrichment steps have been suggested. The most commonly used ones are: membrane isolation by gradient density centrifugation [6, 7] and selective labeling of proteins on the cell surface [8]. Sometimes, both the methods are applied in combination [9, 10]. For cell surface labeling, the side chains of accessible amino acids (especially the primary amino groups of lysine residues) are modified. However, these samples often suffer from unspecific labeling of intracellular proteins. More specific labeling can be achieved by exploiting the N-linked glycosylations on extracellular domains which are present on the majority of plasma membrane proteins. In a two-step procedure, these glycosylations are oxidized using sodium metaperiodate to generate aldehydes, which can be selectively reacted with commercially available biotinylation reagents like Alkoxyamine-PEG_4-Biotin (Fig. 1a). Subsequently, cells are lysed in 4% SDS for efficient solubilization of membrane proteins. Labeled glycoproteins are specifically enriched by affinity matrices like streptavidin-coated agarose beads. Due to the very strong interaction between biotin and streptavidin, harsh washing conditions can be applied during the following washing steps to reduce the amount of unspecifically bound proteins. In the last step, bound proteins are eluted by on-bead digestion and analyzed by mass spectrometry. In contrast to enrichment at the protein level which is described in this protocol, an alternative workflow in which the enrichment step is performed at the peptide level, followed by Peptide N-Glycosidase F (PNGase F) cleavage to liberate these peptides from the matrix, has been described [10, 11]. However, in our experience, this leads to lower sequence coverage per protein (resulting in reduced quantification accuracy) and overall lower protein coverage, as only a fraction of glycosylated peptides generated with standard proteases are easily identified by tandem mass spectrometry.

A strategy to monitor the cell surface proteome dynamics with high precision includes the chemical labeling of tryptic peptides with tandem mass tags (TMT) [12] that enable multiplexed relative quantification of up to ten different states in a single experiment (Fig. 1b). This high level of multiplexing allows the combination of multiple replicates or extensive time series without missing values. In contrast, alternative quantification methods, like label-free quantification in which every sample is analyzed separately in the mass spectrometer or stable isotope labeling with amino acids in cell culture (SILAC) [13, 14] which allows the combination of up to

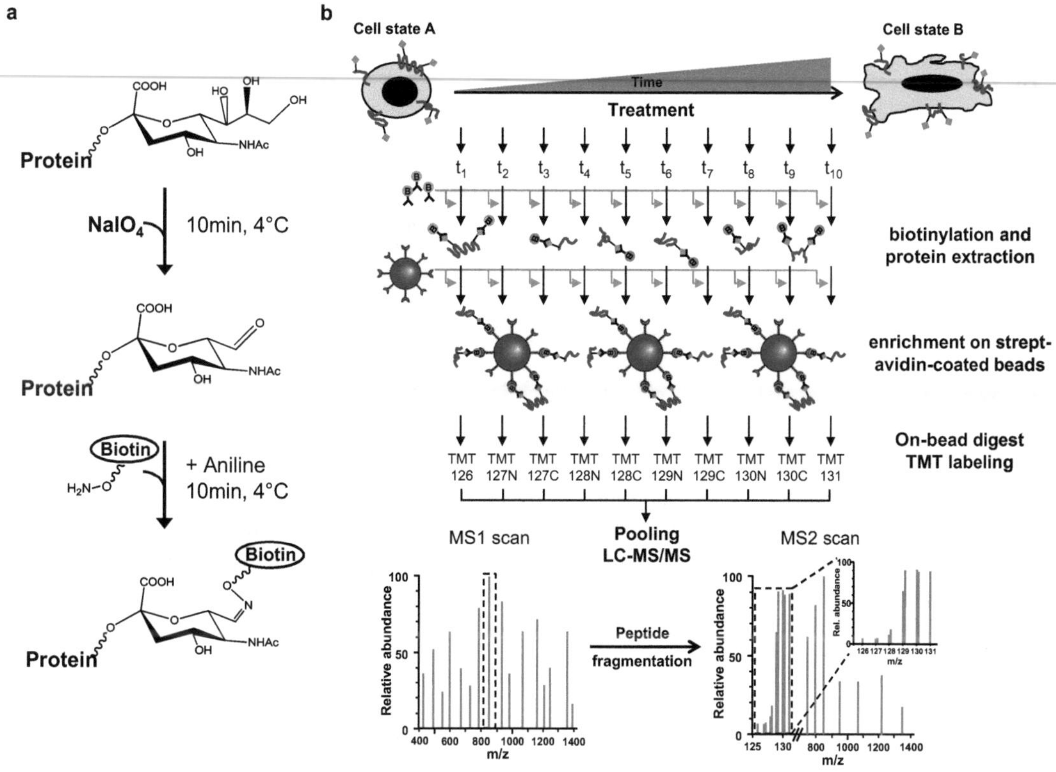

Fig. 1 Overview of the cell surface mapping workflow. (**a**) Cell surface proteins are labeled at their glycosylation site. Vicinal diols are oxidized with sodium metaperiodate to aldehydes. To covalently attach a biotin group to cell surface proteins, these aldehydes are reacted with alkoxyamine-PEG$_4$-biotin in the presence of aniline as a catalyst. (**b**) Example of a time course experiment: the cell surface proteins from ten different time points are biotinylated, enriched, digested with trypsin, and labeled with TMT reagents. After TMT labeling, samples are pooled and analyzed by LC-MS/MS. Reporter ions in the low mass region of the MS2 scan are used for relative quantification of the cell surface presentation of enriched proteins

three states per sample, often show missing values between separate runs, and, in addition, require substantially more instrument time. The accuracy of relative quantification by isobaric mass tags is often compromised by isobaric interference caused by co-isolation and co-fragmentation of multiple labeled peptide precursors in the same fragment ion scan. This artifact can be reduced by applying small isolation windows on the MS instrumentation [15] and by computational correction post data acquisition [16]. This protocol describes a simple and robust method for N-linked glycoproteome enrichment on protein level in combination with TMT-based quantification that enables deep coverage of the plasma membrane proteome for studies demanding a high level of multiplexing like time series or drug treatment experiments (Fig. 2).

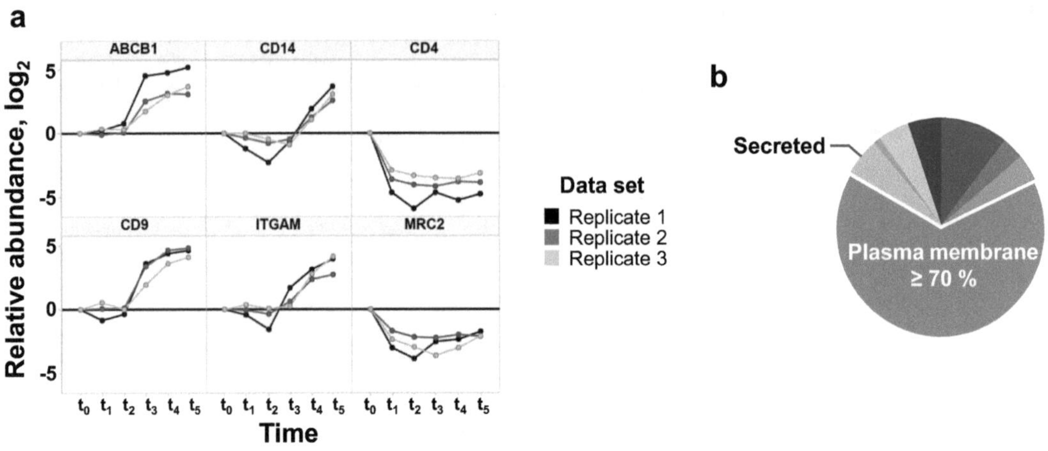

Fig. 2 Examples from cell surface labeling experiments. (**a**) Time-dependent cell surface abundance changes during monocyte to macrophage differentiation of selected transmembrane proteins. TMT labeling was used for multiplexed relative quantification. Relative abundance is calculated compared to undifferentiated cells. (**b**) Pie chart depicting the cell surface protein enrichment: estimated fractional abundance (as estimated by the intensity in the MS) of identified proteins classified by their localization. Usually, membrane proteins account for more than 70% of the observed protein intensities

2 Materials

Use only high purity reagents and HPLC grade organic solvents and water.

2.1 Cell Surface Labeling

1. Dulbecco's phosphate buffered saline (DPBS).
2. DPBS adjusted to pH 6.5 with concentrated HCl.
3. Sodium metaperiodate solution for suspension cells: 1 mM $NaIO_4$ in DPBS, pH 6.5.
4. DPBS, calcium, magnesium (+Ca^{2+} +Mg^{2+}).
5. Sodium metaperiodate solution for adherent cells: 1 mM $NaIO_4$ in DPBS (+Ca^{2+} +Mg^{2+}), pH 6.5.
6. Biotinylation solution for suspension cells: 1 mM EZ-Link™ Alkoxyamine-PEG_4-Biotin in DPBS, 10 mM aniline.
7. Biotinylation solution for adherent cells: 1 mM EZ-Link™ Alkoxyamine-PEG_4-Biotin in DPBS (+Ca^{2+} +Mg^{2+}), 10 mM aniline.
8. Rubber cell scraper.

2.2 Cell Lysis

1. SDS solution: 20% SDS in water.
2. Tris(hydroxymethyl)aminomethane hydrochloride (Tris–HCl) stock solution: 1 M Tris–HCl, pH 7.5.
3. Lysis buffer: 4% SDS, 60 mM Tris–HCl, pH 7.6.

4. Sonicator (e.g., Bandelin Sonopuls HD 2200).
5. BCA reagent for protein determination.

2.3 Enrichment of Biotinylated Proteins and Digestion

1. Dithiothreitol (DTT) solution: 45 mM DTT in water.
2. Iodoacetamide (IAA) solution: 100 mM IAA in water.
3. SDS Solution: 20% SDS in water.
4. Triethylammonium bicarbonate (TEAB) stock solution: 1 M TEAB in water.
5. 4-(2-hydroxyethyl)-1-piperazineethanesulfonic acid (HEPES) stock solution: 1 M HEPES, pH 7.5.
6. High capacity streptavidin agarose beads (ThermoFisher Scientific).
7. Trypsin (mass spectrometry grade) solution: 0.02 μg/μL trypsin in 50 mM TEAB.
8. Peptide N-Glycosidase F solution: 1 unit PNGase F in 50 μL 50 mM HEPES, pH 7.5.
9. Wash buffer 1: 0.4% SDS, 20 mM Tris–HCl (pH 7.6), 400 mM NaCl.
10. Wash buffer 2: 20 mM Tris–HCl (pH 7.6), 400 mM NaCl.
11. Wash buffer 3: 50 mM TEAB (pH 8.5), 2 M urea.
12. Wash buffer 4: 50 mM HEPES, pH 8.0.
13. Low-binding v-shaped 96-well plates, polypropylene (e.g., Nunc, Germany) and suitable lids for 96-well plates (e.g., Costar, Sigma-Aldrich).
14. 96-well Filtration Plate, 2 mL (e.g., Porvair Micrulute Combinatorial).
15. Microplate DeepWell 96-well Drain Cap (e.g., Porvair Micrulute Combinatorial).
16. Sealing mats square cap (e.g., ThermoFisher Scientific).
17. Vacuum pump.
18. MultiScreen®HTS vacuum manifold.
19. Over-head shaker.
20. Speedvac.

2.4 Labeling of Tryptic Digests with Isobaric Mass Tags

1. Dissolve solution: 200 mM TEAB, 10% acetonitrile (ACN), pH 8.5.
2. Stop solution: 2.5% hydroxylamine in 100 mM TEAB, pH 8.5.
3. 10-plex TMT reagents (TMT10, Thermo Fisher Scientific).

2.5 STAGE-Tip-Based Sample Desalting

1. 200 μL pipette tips.
2. Empore™ SPE Disks C18 (Sigma-Aldrich).

2.6 Mass Spectrometry

1. Trifluoroacetic acid (TFA) solution: 0.05% TFA in water.
2. Solution A: 0.1% formic acid, 2% acetonitrile (v/v).
3. Solution B: 0.1% formic acid, 35% acetonitrile (v/v).
4. Nano-flow HPLC system, e.g., RLSC nano (Thermofisher Scientific), or other systems of similar performance.
5. Pre-column, 100 μm ID x 2 cm, nano Viper C18, 5 μm, e.g., Thermo Scientific Acclaim PepMap® 100.
6. Nano-HPLC separation column, prepared in-house from 100 μm ID × 360 OD, 54 cm long fused silica capillary packed with C18 reversed phase particles (3 μm), e.g., Reprosil Pur C18-AQ 3 μm.
7. Mass spectrometer equipped with online nano-spray ion source, e.g., Q Exactive (Thermofisher Scientific). Other MS systems of similar performance can be used.

2.7 General Purpose Equipment

1. General purpose centrifuge with rotor suitable for 50 and 15 mL conical tubes.
2. Tabletop centrifuge for 2 mL reaction tubes.
3. Conical tubes (15 and 50 mL) and 2 mL reaction tubes.
4. Heat shaker.
5. Ultrasonic bath.

2.8 Data Analysis

1. Mascot™ 2.4 (Matrix Science).
2. Proteomics analysis suite, like MaxQuant [17] or isobarQuant package [18].
3. Statistical analysis software, e.g., R (www.r-project.org).

3 Methods

3.1 Cell Surface Protein Labeling of Suspension Cells (See Note 1)

This protocol describes the selective biotinylation of cell surface proteins on suspension cells.

1. Wash twice 10×10^6 cells with 5 mL cold DPBS, centrifuge at $314 \times g$ for 2 min, remove the supernatant.
2. Resuspend cells in 10 mL 1 mM Na-metaperiodate solution for suspension cells, incubate for 10 min at 4 °C in the dark, centrifuge at $314 \times g$ for 2 min, remove the supernatant.
3. Wash cells with 5 mL cold DPBS, centrifuge at $314 \times g$ for 2 min, remove the supernatant.
4. Resuspend cells in 0.4 mL biotinylation solution, transfer to a 2 mL reaction tube, incubate for 10 min at 4 °C in the dark, centrifuge at $1500 \times g$ for 2 min, remove the supernatant.

5. Wash twice the cells with 5 mL cold DPBS, centrifuge at 1500 × *g* for 2 min, remove the supernatant.
6. Freeze cell pellet in liquid nitrogen and store at −80 °C.

3.2 Cell Surface Protein Labeling of Adherent Cells (See Note 2)

This protocol describes the selective biotinylation of cell surface proteins on adherent cells.

1. Wash twice adherent cells on one 15 cm petri dish with ~80% confluency with 5 mL cold DPBS (+Ca^{2+} +Mg^{2+}), shake gently, remove the supernatant.
2. Add 15 mL 1 mM Na-metaperiodate solution for adherent cells, incubate for 10 min at 4 °C shaking in the dark, remove the supernatant.
3. Wash cells with 5 mL cold DPBS (+Ca^{2+} +Mg^{2+}), shake gently, remove the supernatant.
4. Add 3 mL biotinylation solution for adherent cells, incubate for 10 min at 4 °C while shaking in the dark, remove the supernatant.
5. Wash cells twice with 5 mL cold DPBS, shake gently, remove the supernatant.
6. Add 2 mL DPBS, gently detach adherent cells with a rubber cell scraper, transfer to a 2 mL reaction tube, centrifuge with 1500 × *g* for 2 min, remove the supernatant.
7. Freeze cell pellet in liquid nitrogen and store at −80 °C.

For comparative analysis, e.g., time series or drug treatment, process all samples that should be analyzed together in the same way. The samples will be labeled with isobaric mass tags and pooled later in the protocol.

3.3 Cell Lysis

This step describes the efficient solubilization of membrane proteins by cell lysis in 4% SDS.

1. Lyse cell pellet in 200 μL lysis buffer by pipetting ten times up and down.
2. Heat sample to 95 °C for 5 min shaking, cool on ice for 5 min.
3. Sonicate the sample at 50% power output with 1× sonication burst of 10 s on ice.
4. Centrifuge with 8000 × *g* for 5 min to remove air bubbles.
5. Dilute sample 1:10 in DPBS and store at −80 °C.
6. Determine protein concentration according to the BCA Protein Assay Kit manual.

3.4 Enrichment of Biotinylated Proteins and Digestion

This step describes the selective enrichment of cell surface proteins by the selective and strong interaction between biotin and streptavidin coated to beads. Wash steps are performed with a vacuum pump and MultiScreen®HTS vacuum manifold (*see* **Notes 3–5**).

1. Transfer 15 μL of streptavidin resin into a 2 mL 96-well filter plate (*see* **Note 3**).
2. Wash twice with 1 mL wash buffer 1.
3. Add cell lysate with biotinylated surface proteins (1 mg protein).
4. Close plate.
5. Incubate for 120 min at 25 °C on over-head shaker, centrifuge at 314 × *g* for 2 min to remove flow-through.
6. Wash four times with 1 mL wash buffer 1.
7. Wash eight times with 1 mL wash buffer 2.
8. Wash eight times with 1 mL wash buffer 3.
9. Centrifuge at 314 × *g* for 2 min to remove residual liquids.
10. Close the plate.
11. Add 30 μL of 45 mM DTT in water, incubate for 45 min at 25 °C in the dark.
12. Add 30 μL of 100 mM IAA in water, incubate for 30 min at 25 °C in the dark.
13. Centrifuge with 314 × *g* for 2 min to remove flow-through.
14. Wash five times with 1 mL wash buffer 3 (*see* **Note 5**).
15. Close the plate.
16. Add 40 μL of wash buffer 3 and 20 μL of trypsin solution, incubate overnight at 25 °C on an over-head shaker (*see* **Note 3**).
17. Add 20 μL of trypsin solution, incubate for an additional 4 h at 25 °C on an over-head shaker.
18. Place the filter plate on the top of a clean 96-well collector plate and centrifuge with 314 × *g* for 2 min.
19. Add 50 μL of 50 mM TEAB, incubate for 2 min shaking, centrifuge with 314 × *g* for 2 min.
20. Dry extracted peptides in collector plate in vacuo using a speedvac and store at −80 °C.

Optional further elution of biotinylated glyco-peptides by PNGase F (*see* **Note 4**).

21. Wash five times with 1 mL wash buffer 4.
22. Close the plate.
23. Add 50 μL of 1 unit PNGase F solution, and incubate for 3 h at 37 °C.
24. Place the filter plate on the top of a new 96-well collector plate and centrifuge with 314 × *g* for 2 min.

25. Add 50 μL of 50 mM HEPES, incubate for 2 min shaking, centrifuge with 314 × *g* for 2 min.
26. Dry extracted peptides in collector plate in vacuo using a speedvac.

3.5 Labeling of Peptides with Isobaric Mass Tags

To enable multiplexed quantification of different samples (e.g., in time series experiments), peptides are chemically labeled with isobaric mass tags. The following steps describe the labeling procedure of tryptic peptides with the TMT reagents (*see* **Note 6**).

1. Dissolve 5 mg TMT-label reagent in 600 μL ACN, vortex.
2. Dissolve samples in 100 μL dissolve solution on collection plate.
3. Add 50 μL of respective TMT-label to samples, incubate for 60 min at 25 °C while shaking.
4. Add 25 μL stop solution, incubate for 15 min at 25 °C while shaking.
5. Pool samples into a 2 mL reaction tube.
6. Wash respective wells of collection plate with 20 μL 100 mM TEAB/ACN (60:40, v/v), and add to a 2 mL reaction tube.
7. Dry TMT-labeled peptides in vacuo using a speedvac and store at −80 °C.

3.6 STAGE-Tip-Based Sample Desalting

This step describes the desalting of samples prior to MS analysis by C18 STAGE-Tips [19].

1. Punch 4 plugs of C18 material into a 200 μL pipette tip.
2. Activate STAGE-Tips by adding 100 μL methanol, centrifuge at 2000 × *g* for 2 min.
3. Wash twice the STAGE-Tips with 200 μL 0.1% TFA/ACN (98:2, v/v), centrifuge at 2000 × *g* for 2 min.
4. Resuspend samples in 100 μL 0.1% TFA/ACN (98:2, v/v).
5. Add samples to STAGE-Tips, centrifuge with 2000 × *g* for 2 min, collect flow-through in a clean reaction tube.
6. Wash twice with 200 μL 0.1% TFA/ACN (98:2, v/v), centrifuge with 2000 × *g* for 2 min.
7. Elute peptides with 50 μL 0.1% TFA/ACN (20:80, v/v) in a new reaction tube, centrifuge with 2000 × *g* for 2 min.
8. Repeat elution 1× with 50 μL 0.1% TFA/ACN (20:80, v/v), centrifuge with 2000 × *g* for 2 min.
9. Dry desalted peptides in vacuo using a speed vac and store at −80 °C.

3.7 LC-MS/MS Analysis

Appropriate instrumentation for mass spectrometric analysis of the samples generated in Subheading 3.5 is available at many institutions and core facilities. Many different instrument platforms are similarly well suited for this purpose. As an example, the following section describes the typical instrument settings for the commonly used Q-Exactive mass spectrometers.

1. Resuspend samples in 20 μL 0.05% TFA.
2. Inject 25% of the sample into a nano-LC system coupled online to the mass spectrometer.
3. Peptides are separated on a custom-made 50 cm reversed phase column (Reprosil) using a 2–4 h linear elution gradient from 100% solvent A to 100% solvent B (2–35% acetonitrile), at a flowrate of 350 nL/min.
4. Peptide masses are detected at 70,000 resolution with an ion target value of 3E6 in MS mode and at 35,000 resolution with an ion target of 2E5 in MS/MS mode.
5. Intact peptide ions are isolated with isolation width set to 1.0 Th and HCD fragmentation is performed using 25% normalized collision energy (NCE) for label-free or 35% NCE for TMT samples.

3.8 Data Analysis

In the last step, the enriched proteins are identified, accurately quantified, and grouped according to their annotated subcellular localization (*see* **Notes 7** and **8**).

1. For protein identification, tandem mass spectra are typically converted to peak lists and submitted to a data base search against a nonredundant sequence data base of the species of interest. A variety of free and commercial search engines such as Mascot [16, 17], Sequest [16], and Andromeda [18] are available for this purpose.
2. In case of label-free quantification, a top 3 method [20], in a slightly modified form [21], has proven to provide a reliable quantification. In brief: for each protein the XIC of the three most intense peptide sequences are log10 transformed and averaged. The derived value, ms1intensity, is a good proxy for protein abundance in the sample. Popular intensity-based label-free quantification approaches, e.g., the LFQ module in MaxQuant [22], can also be applied.

   For TMT-based quantification, analysis tools like MaxQuant or the isobarQuant package [18] are available. The isobarQuant package incorporates a previously described method [16] to computationally address isobaric interference by inferring and correcting for the extent of co-fragmentation for individual peptide MS/MS spectra (*see* **Note 8**). Briefly, reporter ion intensities are read from raw data and multiplied with ion accumulation times so as to yield a measure proportional to the number of ions; this measure is referred to as ion area [15]. Reporter ion containing MS/MS spectra originating from low intense precursor MS signals close to the noise level (low precursor-to-threshold ratio, P2T) or containing a substantial amount of co-fragmented signals (low signal-to-interference ratio, S2I) within the defined isolation window (P2T <4, S2I < 0.5) is discarded [16]. Remaining reporter ion area values are corrected by a simple algorithm utilizing the S2I measure which has been shown to strongly reduce isobaric interference due to co-fragmentation and to produce more accurate peptide and protein fold changes [16].

3. Annotations for subcellular localizations of proteins can be derived from the UniProtKB database. In brief, e.g., proteins annotated with a subcellular localization or Gene Ontology Cellular Component (GO-CC) related to the cell surface, plasma membrane, or extracellular space are considered plasma membrane-associated proteins while proteins annotated with a subcellular localization or GO-CC related to the cytoplasm, cytosol, or intracellular organelles are considered cytoplasmic proteins.

4 Notes

1. It is important to separately perform the oxidation and the biotinylation reaction. Periodate has been described to oxidize aliphatic and aromatic amines [23] and inactivates alkoxyamines by conversion to amines. This can lead to inefficient labeling of cell surface glycoproteins and might result in low plasma membrane proteome coverage. Efficient removal of periodate prior to the addition of the biotinylation reagent and the catalyst aniline is recommended.
2. When working with adherent cells, we recommend using PBS supplemented with Ca^{2+} and Mg^{2+} to avoid detachment from the plate.
3. Tryptic digestion on streptavidin beads leads to contamination of the final peptide sample with streptavidin peptides. Thus, the amount of streptavidin beads used should be kept low and adapted to the input amount of biotinylated proteins. If abundances of streptavidin-derived peptides are still limiting the dynamic range of the analysis, further reduction of the protease: bead ratio and shorter trypsinization should be considered.
4. While efficient and selective elution of biotinylated peptides by the N-glycosidase PNGase F can be applied for the enrichment at the peptide level [10], efficient elution of enriched proteins

by PNGase F has not been possible in our hands. This is most likely due to steric hindrance.

5. Efficient removal of residual DTT and IAA by additional washing steps is recommended. The presence of these reagents during TMT-labeling of peptides can lead to side reactions with the labeling reagent resulting in high amounts of interfering ions in the chromatogram.
6. When establishing TMT-based quantification, it is beneficial to thoroughly check digestion and labeling efficacy to reduce variability and avoid artifacts. Both the parameters should be checked on the individual samples before pooling. The digestion efficacy can be expressed by the number of identified peptides without missed cleavage sites to all the identified peptides. Ideally, more than 80% of the identified peptides should be without missed cleavage sites, and this percentage should be highly reproducible between the samples to be pooled. The TMT-labeling efficacy can be expressed by the number of identified chemically labeled peptides to all identified peptides and can be assessed by setting the TMT label as a variable modification in the Mascot search. Ideally close to 100% of the lysine residues and more than 90% of the N-termini should be labeled. Modification of the hydroxyl groups of serine, threonine, and tyrosine side chains can also occur. This over-labeling can be reversed by quenching the TMT-labeling reaction with hydroxylamine reversing O-acylation reactions while preserving acyl modifications on primary amines [24]. Ideally, less than 5% of the hydroxy amino acids should be TMT labeled.
7. Despite harsh washing conditions during the enrichment procedure, contaminations with intracellular proteins cannot be avoided. After enrichment, usually about 30% of all the identified proteins are annotated cell surface proteins representing 60–70% of the total protein abundance in the enriched samples (Fig. 2b).
8. Isobaric interference can be addressed by reduced precursor isolation windows [15], computational algorithms [16] or by the recently described Synchronous Precursor Selection (SPS) [25] analyzing reporter ion intensities of selected intense MS^2 fragments of a precursor ion after a second fragmentation.

Acknowledgments

We would like to thank all our colleagues at Cellzome for support and fruitful discussions.

References

1. Klabunde T, Hessler G (2002) Drug design strategies for targeting G-protein-coupled receptors. Chembiochem 3:928–944
2. Overington JP, Al-Lazikani B, Hopkins AL (2006) How many drug targets are there? Nat Rev Drug Discov 5:993–996
3. Lindsley CW (2015) 2014 global prescription medication statistics: strong growth and CNS well represented. ACS Chem Neurosci 6:505–506
4. Shin BK, Wang H, Yim AM et al (2003) Global profiling of the cell surface proteome of cancer cells uncovers an abundance of proteins with chaperone function. J Biol Chem 278:7607–7616
5. Pshezhetsky AV, Fedjaev M, Ashmarina L et al (2007) Subcellular proteomics of cell differentiation: quantitative analysis of the plasma membrane proteome of Caco-2 cells. Proteomics 7:2201–2215
6. Leonard RT, Vanderwoude WJ (1976) Isolation of plasma membranes from corn roots by sucrose density gradient centrifugation: an anomalous effect of ficoll. Plant Physiol 57:105–114
7. Jones DH, Matus AI (1974) Isolation of synaptic plasma membrane from brain by combined flotation-sedimentation density gradient centrifugation. Biochim Biophys Acta 356:276–287
8. Karhemo PR, Ravela S, Laakso M et al (2012) An optimized isolation of biotinylated cell surface proteins reveals novel players in cancer metastasis. J Proteome 77:87–100
9. Nunomura K, Nagano K, Itagaki C et al (2005) Cell surface labeling and mass spectrometry reveal diversity of cell surface markers and signaling molecules expressed in undifferentiated mouse embryonic stem cells. Mol Cell Proteomics 4:1968–1976
10. Wollscheid B, Bausch-Fluck D, Henderson C et al (2009) Mass-spectrometric identification and relative quantification of N-linked cell surface glycoproteins. Nat Biotechnol 27:378–386
11. Bausch-Fluck D, Hofmann A, Bock T et al (2015) A mass spectrometric-derived cell surface protein atlas. PLoS One 10:e0121314
12. Werner T, Sweetman G, Savitski MF et al (2014) Ion coalescence of neutron encoded TMT 10-plex reporter ions. Anal Chem 86:3594–3601
13. Oda Y, Huang K, Cross FR et al (1999) Accurate quantitation of protein expression and site-specific phosphorylation. Proc Natl Acad Sci U S A 96:6591–6596
14. Ong SE, Blagoev B, Kratchmarova I et al (2002) Stable isotope labeling by amino acids in cell culture, SILAC, as a simple and accurate approach to expression proteomics. Mol Cell Proteomics 1:376–386
15. Savitski MM, Sweetman G, Askenazi M et al (2011) Delayed fragmentation and optimized isolation width settings for improvement of protein identification and accuracy of isobaric mass tag quantification on Orbitrap-type mass spectrometers. Anal Chem 83:8959–8967
16. Savitski MM, Mathieson T, Zinn N et al (2013) Measuring and managing ratio compression for accurate iTRAQ/TMT quantification. J Proteome Res 12:3586–3598
17. Cox J, Mann M (2008) MaxQuant enables high peptide identification rates, individualized p.p.b.-range mass accuracies and proteome-wide protein quantification. Nat Biotechnol 26:1367–1372
18. Franken H, Mathieson T, Childs D et al (2015) Thermal proteome profiling for unbiased identification of direct and indirect drug targets using multiplexed quantitative mass spectrometry. Nat Protoc 10:1567–1593
19. Rappsilber J, Mann M, Ishihama Y (2007) Protocol for micro-purification, enrichment, pre-fractionation and storage of peptides for proteomics using StageTips. Nat Protoc 2:1896–1906
20. Silva JC, Gorenstein MV, Li GZ et al (2006) Absolute quantification of proteins by LCMSE: a virtue of parallel MS acquisition. Mol Cell Proteomics 5:144–156
21. Becher I, Savitski MM, Savitski MF et al (2013) Affinity profiling of the cellular kinome for the nucleotide cofactors ATP, ADP, and GTP. ACS Chem Biol 8:599–607
22. Cox J, Hein MY, Luber CA et al (2014) Accurate proteome-wide label-free quantification by delayed normalization and maximal peptide ratio extraction, termed MaxLFQ. Mol Cell Proteomics 13:2513–2526
23. Clamp JR, Hough L (1966) Some observations on the periodate oxidation of amino compounds. Biochem J 101:120–126
24. Rauniyar N, Yates JR III (2014) Isobaric labeling-based relative quantification in shotgun proteomics. J Proteome Res 13:5293–5309
25. McAlister GC, Nusinow DP, Jedrychowski MP et al (2014) MultiNotch MS3 enables accurate, sensitive, and multiplexed detection of differential expression across cancer cell line proteomes. Anal Chem 86:7150–7158

Chapter 4

High-Resolution Parallel Reaction Monitoring with Electron Transfer Dissociation for Middle-Down Proteomics: An Application to Study the Quantitative Changes Induced by Histone Modifying Enzyme Inhibitors and Activators

Michael J. Sweredoski, Annie Moradian, and Sonja Hess

Abstract

With the advent of new methodologies, proteomics-based assays are increasingly used to study the efficacy of drugs on a molecular level. For these studies to be meaningful, the proteomics assays need to be sensitive, selective, accurate, and reproducible. This is often accomplished through a targeted approach, either using single or multiple reaction monitoring (SRM/MRM) or, more recently, parallel reaction monitoring (PRM). In PRM, the parallel detection of all product ions in a high-resolution mass spectrometer affords higher selectivity than SRM/MRM. PRM is thus better suited to analyze peptides larger than 2 kDa. Similar to SRM/MRM, PRM provides sensitive, accurate, and reproducible quantitative data. Here, we present a specific PRM method to characterize the effects of histone modifying enzyme drugs such as histone deacetylase inhibitors (HDAC) on the posttranslational modifications of histones, in a quantitative manner. More specifically, we characterize the heavily modified N-terminal tail of histone H3 after treatment with the HDAC inhibitor butyric acid, and monitor the acetylation and methylation events after treatment. To take most advantage of the multiply charged N-terminal histone peptides that are generated by an endoproteinase GluC-digestion, we use electron transfer dissociation (ETD) as the method of MS/MS fragmentation. This provides high sequence coverage for the modified peptides. The methodology is not limited to HDAC inhibitors, and can be used for any modifying enzyme. In fact, it can even be expanded beyond histone analyses. To give guidance for the development of a PRM assay, we present here HDAC inhibited H3 histone N-terminal tails as an example.

Key words Parallel reaction monitoring (PRM), Quadrupole-Orbitrap mass spectrometry, Quantitation, Histones, Electron transfer dissociation (ETD)

1 Introduction

Cellular processes such as gene transcription, DNA repair, and DNA replication are, in part, regulated by histone posttranslational modifications (PTMs) [1, 2]. Histones exist in five major families: H1, H2A, H2B, H3, and H4 with the first being a linker histone

Iulia M. Lazar et al. (eds.), *Proteomics for Drug Discovery: Methods and Protocols*, Methods in Molecular Biology, vol. 1647, DOI 10.1007/978-1-4939-7201-2_4,

and the remaining being core histones. Four dimers of the core histones form an octameric nucleosome, which is wrapped around by DNA [2]. The regulation of gene expression by histones is mostly achieved through changes of the acetylation and methylation status of lysines of histone's N-terminal tails [3, 4]. Mainly, four groups of histone modifying enzymes are adding and removing acetyl and methyl groups, respectively: histone accetyltransferases and histone deacetylases, protein methyl transferases, and demethylases [2]. While methylation stabilizes the charge of the side chain amines of lysines, acetylation neutralizes it. This results in the following scenario: positively charged methylated histones are bound tightly with the negatively charged DNA that is wrapped around the histones, forming a relatively closed, inactive heterochromatin. In contrast, neutral acetylated histones are more loosely bound to DNA forming a transcriptionally active euchromatin by enabling RNA polymerase access to the DNA. The quantitative study of changes in histone PTMs is thus important for the understanding of gene regulation [1, 5].

Toward this goal, we have developed a parallel reaction monitoring method for the characterization of histones treated with a histone modifying enzyme [6]. The principles of PRM and single or multiple reaction monitoring (SRM/MRM) are shown in Fig. 1. PRM is similar to SRM/MRM in that it provides sensitive, accurate, and reproducible quantitative data. Due to the high-resolution mass spectrometers, the parallel detection of all product ions in PRM offers higher selectivity than SRM/MRM, making it the tool of choice for larger peptides (>2 kDa) that cannot be resolved in a typical SRM/MRM assay. While our specific example focuses on the analysis of H3 histone N-terminal tails from differentiating myeloerythroid (MEL) cells treated with the HDAC inhibitor butyric acid, in principle, this method can be applied to study quantitative changes in any histone tails treated with any histone modifying enzyme inhibitor or activator. In fact, it can be used to study other enzymatic processes in a similar fashion. Rather than using standard collision-induced dissociation, we take advantage of the fact that electron transfer dissociation tends to better fragment highly charged peptides [7]. Here, we present PRM using ETD as an effective method to analyze and quantify PTM changes of large and highly charged peptides. This method quantifies the global acetylation and methylation changes on all histone H3 N-terminal tails that were present in the cell at the time of the analysis, e.g., the majority of the histone tails have three acetylation sites and 5–6 methylation sites. This could be looked at as a unique "bird's eye" view. The use of MS/MS additionally enables us to determine where the modifications are located.

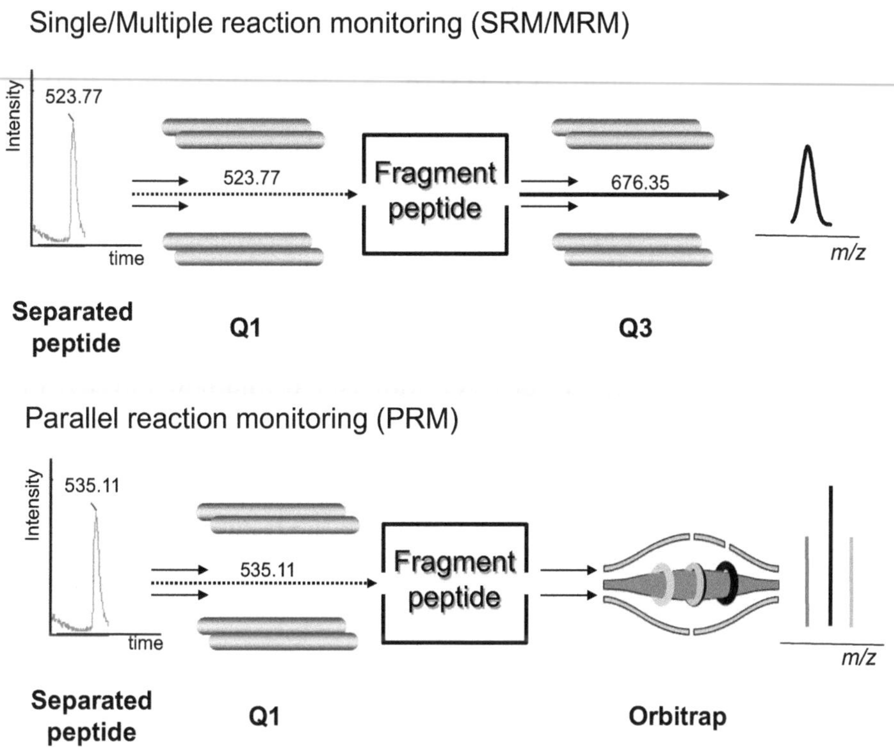

Fig. 1 Single and multiple reaction monitoring (SRM/MRM) vs. parallel reaction monitoring PRM. An LC-separated peptide is selected and filtered through Q1, fragmented in the collision cell and selected fragment ions are filtered through Q3 before MS analysis. When only one ion is monitored, it is referred to as SRM, when multiple ions are monitored, it is referred to as MRM. TripleQuad (QQQ) and Q-TRAP instruments are usually used for this analysis. Sensitivity is high, selectivity is usually achieved by monitoring multiple fragment ions per precursor ion. PRM: An LC-separated peptide is selected and filtered through Q1, and fragmented before multiple fragments are analyzed in parallel in the Orbitrap. The Orbitrap Fusion and Lumos instruments offer multiple fragmentation techniques: ETD, CID, and HCD. This enables a high degree of analytical flexibility. In case of larger peptides such as histone N-terminal tails, we recommend using ETD. Due to the high-resolution instruments, selectivity is high in this experiment

2 Materials

Prepare all solutions with LC-MS grade water and solvents. All reagents should be of the highest purity available. Standard lab equipment such as pipettes, Eppendorf tubes, sample vials, pH meter, sonicator, lyophilizer, incubator, and freezer is needed.

2.1 LC Samples

1. Formic acid: 0.2%. Add 20 μL of formic acid (puriss. p.a.) to 9.98 mL of water.
2. Histone extracts from differentially treated cells (e.g., MEL) (*see* **Note 1**).
3. Histone samples: Dissolve approximately 200–300 μg histones in 100 μL of 0.2% formic acid.
4. Sample vials for HPLC (*see* **Note 2**).

2.2 HPLC System for Histone Fractionation

1. Agilent 1100 LC-MS (Santa Clara, CA) system consisting of a degasser, binary pump system, autosampler, column compartment, UV or DAD detector and 1100 series LC/MSD SL mass selective detector with ChemStation software (*see* **Note 3**).
2. Zorbax SB-C3 column, 4.6 × 150 mm, particle size: 5 μm.
3. HPLC Solvent A: 949.5 mL of water, 50 mL of acetonitrile, 50 μL of trifluoroacetic acid (TFA; LC-MS Ultra).
4. HPLC Solvent B: 949.5 mL of acetonitrile, 50 mL of water, 50 μL of TFA (LC-MS Ultra).
5. A flow splitter is introduced after the UV detector and before the MS detector with 10% of the flow directed to the MS detector and 90% of the flow collected for further processing.

2.3 Glu-C Digestion of Histones

1. Endoproteinase Glu-C.
2. Ammonium bicarbonate (NH_4HCO_3) buffer: 100 mM, pH = 4.0. Dissolve 0.79 g NH_4HCO_3 (ppa) in 100 mL of water. Add 200 μL acetic acid to 10 mL of 100 mM NH_4HCO_3 to adjust the pH.

2.4 Weak Cation Exchange Liquid Chromatography—Mass Spectrometry (WCX-LC-MS/MS) System for Histone Characterization

1. Orbitrap Fusion (Thermo Fisher) with an EASY-nLC 1000 system and Xcalibur 3.1.66.10 software. The LC system is coupled to a nanospray Flex ion source (Thermo Fisher) (*see* **Note 4**).
2. WCX-HILIC column: Pack 16 cm (75 μm ID × 360 μm OD) fused silica capillary column in-house with PolyCat A resin (3 μm, 1500 Å pore size from PolyLC).
3. WCX LC-MS Solvent A: 70% acetonitrile, 30% water containing 20 mM propionic acid. Add 149.5 μL propionic acid (analytical standard; >99.88%) to 29.85 mL of water. Add 70 mL of acetonitrile. Add one to two drops of concentrated ammonia (analytical standard; >99.98%) to adjust to pH 6.0.
4. WCX LC-MS Solvent B: 25% acetonitrile, 75% water containing 0.3% formic acid. Add 300 μL formic acid (for mass spectrometry; 98%) to 74.7 mL of water. Add 25 mL of acetonitrile.
5. Sonicate WCX LC-MS solvents A and B prior to use (*see* **Note 5**).

2.5 Data Analysis

1. Computer with XCalibur 3.0.63 (Thermo Fisher), ProteoWizard v. 3.0.4006 and Deconv [8] programs installed.

3 Methods

3.1 HPLC Fractionation of Histones

1. For the fractionation of histone samples, an Agilent LC-MSD is used (*see* **Note 3**).
2. Histones are separated using a Zorbax SB-C3 column using HPLC solvent A and B. The flow rate is 1 mL/min. The

gradient is as follows: 0–30% B in 3 min, 30–38% B in 30 min, and 38–60% B in 8 min, 60–100% B in 3 min and 100% B for 10 min. The flow is split to collect ca. 90% in Eppendorf tubes and the remaining 10% is used for MS analysis to identify the histone subtypes (*see* **Note 6**).

3. Inject 100 μL of histone mixture.
4. Start LC-MS analysis.
5. Mass spectra are acquired in the mass range of 500–1700 m/z with a fragmentor set at 50. Capillary voltage is 4000 V and drying gas is set at 350 °C.
6. Collect fractions every minute, starting from min 5 (*see* **Note 6**).
7. Lyophilize all fractions.
8. Store at −20 °C.

3.2 Preparation of GluC Digestion of Histones

1. Histone fractions (10 μg) are digested with 1 μg endoproteinase Glu-C in 10 μL of 100 mM NH_4HCO_3 (pH 4.0) at 25 °C for 5 h.
2. Digestions are quenched by freezing at −80 °C and lyophilizing.

3.3 WCX-LC-MS/MS System for Histone Characterization

1. Sample: Redissolve lyophilized, digested histones in 50 μL of WCX-LC-MS Solvent A.
2. Inject 5 μL of the sample using the EASY-nLC coupled to an Orbitrap Fusion mass spectrometer, equipped with a nanospray Flex ion source.
3. Digested histones are separated using the WCX-HILIC column.
4. The column is heated with a butterfly column heater (Phoenix S&T), set at 50 °C.
5. Using WCX-LC-MS solvents A and B, and a flow rate of 350 nL/min, the following gradient is used: 0–50% Solvent B (10 min), 50–80% B (120 min), 80–100% B (1 min), and 100% B (15 min).
6. Data acquisition is controlled by Xcalibur 3.0.63 software. Orbitrap Fusion is operated in PRM acquisition mode: MS1 scan (m/z range 350–1000) is followed by ETD MS/MS PRM scans for the +10 charged ions of the 20 methyl equivalents of the histone N-terminal tail (for histone H3, i.e., m/z = 535.11, 536.52, 537.92, 539.32, 540.72, 542.12, 543.52, 544.92, 546.32, 547.72, 549.12, 550.53, 551.93, 553.33, 554.73, 556.13, 557.53, 558.93, 560.33, 561.73 (*see* **Note 7**). MS2 scans are acquired with an m/z range of 150–2000. Precursor and product ions are analyzed in the Orbitrap in profile mode.

Use the following instrument settings: MS1 resolution: 120,000 at m/z 200 with an automatic gain control (AGC) target value of 2 × 10^5 and maximum ion injection time (maxIIT) of 100 ms. Window for quadrupole isolation of MS2 ions is 1.4 m/z. The AGC target value is 2 × 10^5 and maxIIT at 25 ms. The maximum ETD reaction injection time is 10 ms at an ETD reagent target of 200,000 ions. Supplemental collision energy of 40% with EThcD is used. Three microscans per ETD spectrum. Resolution for ETD mass spectra is 30,000 at m/z 200.

7. Repeat this protocol with a minimum of three biological replicates.

3.4 MS1 Quantitation

1. Convert raw files into mzxml using MSConvert (ProteoWizard v. 3.0.4006). Either use an in-house script to identify the chromatographic peak boundaries for the different numbers of modified (here: acetylated and methylated) proteoforms, or do it manually (*see* **Note 8**).
2. To quantitate proteoforms, use Skyline [9] (*see* **Note 9**) to extract areas under the extracted ion chromatograms (XICs) of the top seven most abundant isotopes of each of the modified precursors using the calculated peak boundaries.
3. Normalize peak areas to the sum of all histone N-terminal peptide peak areas.
4. Use Student's t-test to assess statistical significance (*see* **Note 10**).

3.5 MS2 Identification

1. Sum and centroid MS2 spectra within the calculated peak boundaries.
2. Deconvolute all MS2 spectra using MS-Deconv [8].
3. Annotate deconvoluted spectra using MS-Product in Protein-Prospector [10] (*see* **Note 11**).

4 Notes

1. We follow the Shechter method for histone extraction [11] after treating MEL cells with HDAC inhibitor butyric acid for up to 4 days. This PRM protocol is suitable for other cells treated with histone modifying enzyme inhibitors or activators. In fact, it can even be used beyond the analysis of histones.
2. Due to losses observed with glass vials, we use SunSRI sample vials and caps.
3. LC systems from other manufacturers can be used. We recommend a system that allows for a flowrate of 1 mL/min.

4. This method can currently only be used with an Orbitrap Fusion or Orbitrap Lumos. If higher energy collision dissociation (HCD) or collision-induced dissociation (CID) is used instead of ETD, Q-Exactive or Q-TOF instruments, respectively, can be used in a similar fashion. The principle advantage of using ETD is related to the fact that larger peptides tend to have multiple charge states that do not fragment as well with collisional fragmentation techniques.
5. Sonication helps to degas the solvents and thus eliminate air bubbles in the LC system. This is particularly important when operating in the nanoflow regime as it is done here.
6. The MS chromatogram is used to determine elution profile and peak boundaries of the individual histone subtypes. Using the described conditions, histones elute in the following order: H1, H2B, H4, H2A, and H3 (Fig. 2).
7. After Glu-C digestion, the N-terminus of H3 has the following sequence: ARTKQTARKSTGGKAPRKQLATKAARKSA-PATGGVKKPHRYRPGTVALRE resulting in an average molecular weight of 5341.2. The average mass of the +10 charge state envelope of the unmodified peptide was 535.11. A methyl equivalent adds 14 amu to this value or 1.4 for each +10 charge state.
8. In the given example, the precursor mass differences between 1 acetyl group and 3 methyl groups are greater than 7 ppm. The instrument error is typically less than 2 ppm. Therefore, on the basis of MS1 mass error alone trimethylated and acetylated proteoforms can be distinguished. If other modifications are studied, this may or may not be the case.
9. This software tool is freely available at proteome.gs.washington.edu/software/skyline/. The website gives in-depth advice for usage, data analysis, and interpretation.

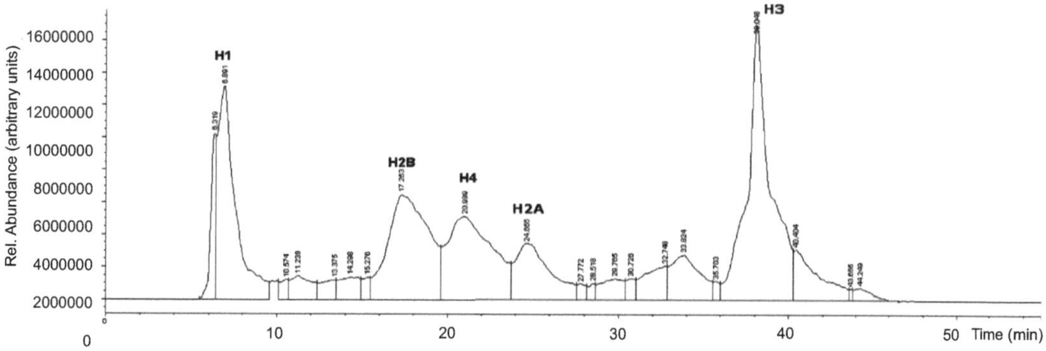

Fig. 2 MSD Chromatogram of histones using a Zorbax SB-C3 column. Elution profile is H1, H2B, H4, H2A, and H3

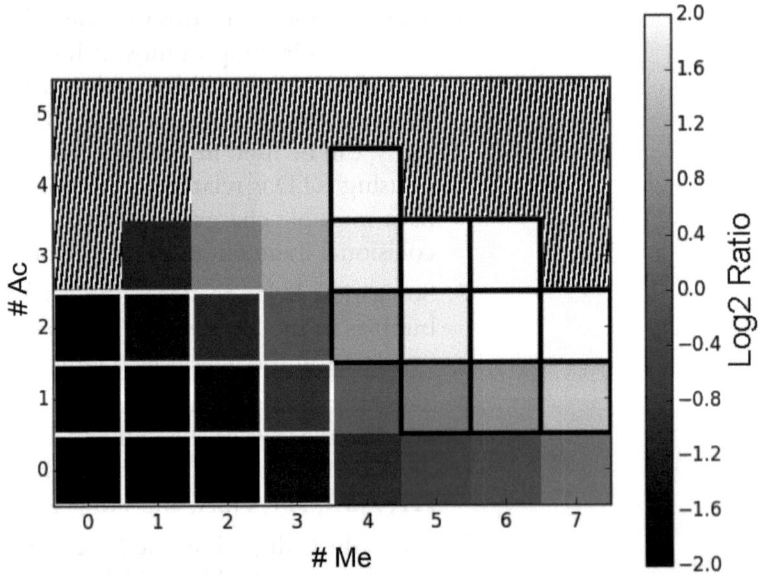

Fig. 3 Example of a "bird's eye" view on the quantitative acetylation and methylation status of histone H3 tails from differentiating MEL cells treated with HDAC inhibitor butyric acid (BA) for 4 days vs. untreated cells. As expected, HAD-treated MEL cells show a significant increase in the number of acetylations. Unexpectedly, HDAC-treated cells also showed an increase in the number of methylations. # Ac is the number of acetylations, # Me is the number of methylations. *Black boxed cells* indicate a significant up-regulation, *white boxed cells* indicate a significant down-regulation ($p < 0.05$)

10. The primary goal of this study is to determine the quantitative changes of histone N-termini in a "birds-eye" view (Fig. 3).
11. The use of Prospector (freely available at http://prospector.ucsf.edu/prospector/cgi-bin/msform.cgi?form=msproduct) enables the identification of the modification sites. Currently, manual interpretation is recommended to confirm assignment of peaks.

Acknowledgment

This work was supported by the Beckman Institute, the Gordon and Betty Moore Foundation through Grant GBMF775 and HHMI (Orbitrap Fusion).

References

1. Moradian A, Kalli A, Sweredoski MJ, Hess S (2014) The top-down, middle-down, and bottom-up mass spectrometry approaches for characterization of histone variants and their post-translational modifications. Proteomics 14 (4–5):489–497. doi:10.1002/pmic.201300256
2. Bannister AJ, Kouzarides T (2011) Regulation of chromatin by histone modifications. Cell Res 21(3):381–395. doi:10.1038/cr.2011.22
3. Loyola A, Almouzni G (2007) Marking histone H3 variants: how, when and why? Trends

Biochem Sci 32(9):425–433. doi:10.1016/j.tibs.2007.08.004

4. Voigt P, Reinberg D (2011) Histone tails: ideal motifs for probing epigenetics through chemical biology approaches. Chembiochem 12 (2):236–252. doi:10.1002/cbic.201000493
5. Onder O, Sidoli S, Carroll M, Garcia BA (2015) Progress in epigenetic histone modification analysis by mass spectrometry for clinical investigations. Expert Rev Proteomics 12(5):499–517. doi:10.1586/14789450.2015.1084231
6. Sweredoski MJ, Moradian A, Raedle M et al (2015) High resolution parallel reaction monitoring with electron transfer dissociation for middle-down proteomics. Anal Chem 87 (16):8360–8366. doi:10.1021/acs.analchem.5b01542
7. Mikesh LM, Ueberheide B, Chi A et al (2006) The utility of ETD mass spectrometry in proteomic analysis. Biochim Biophys Acta 1764 (12):1811–1822. doi:10.1016/j.bbapap.2006.10.003
8. Liu X, Inbar Y, Dorrestein PC et al (2010) Deconvolution and database search of complex tandem mass spectra of intact proteins: a combinatorial approach. Mol Cell Proteomics 9 (12):2772–2782. doi:10.1074/mcp.M110.002766
9. MacLean B, Tomazela DM, Shulman N et al (2010) Skyline: an open source document editor for creating and analyzing targeted proteomics experiments. Bioinformatics 26 (7):966–968. doi:10.1093/bioinformatics/btq054
10. Chalkley RJ, Baker PR, Huang L et al (2005) Comprehensive analysis of a multidimensional liquid chromatography mass spectrometry dataset acquired on a quadrupole selecting, quadrupole collision cell, time-of-flight mass spectrometer: II. New developments in protein prospector allow for reliable and comprehensive automatic analysis of large datasets. Mol Cell Proteomics 4(8):1194–1204. doi:10.1074/mcp.D500002-MCP200
11. Shechter D, Dormann HL, Allis CD, Hake SB (2007) Extraction, purification and analysis of histones. Nat Protoc 2(6):1445–1457. doi:10.1038/nprot.2007.202

Chapter 5

Preparation and Immunoaffinity Depletion of Fresh Frozen Tissue Homogenates for Mass Spectrometry-Based Proteomics in the Context of Drug Target/Biomarker Discovery

DaRue A. Prieto, King C. Chan, Donald J. Johann Jr, Xiaoying Ye, Gordon Whitely, and Josip Blonder

Abstract

The discovery of novel drug targets and biomarkers via mass spectrometry (MS)-based proteomic analysis of clinical specimens has proven to be challenging. The wide dynamic range of protein concentration in clinical specimens and the high background/noise originating from highly abundant proteins in tissue homogenates and serum/plasma encompass two major analytical obstacles. Immunoaffinity depletion of highly abundant blood-derived proteins from serum/plasma is a well-established approach adopted by numerous researchers; however, the utilization of this technique for immunodepletion of tissue homogenates obtained from fresh frozen clinical specimens is lacking. We first developed immunoaffinity depletion of highly abundant blood-derived proteins from tissue homogenates, using renal cell carcinoma as a model disease, and followed this study by applying it to different tissue types. Tissue homogenate immunoaffinity depletion of highly abundant proteins may be equally important as is the recognized need for depletion of serum/plasma, enabling more sensitive MS-based discovery of novel drug targets, and/or clinical biomarkers from complex clinical samples. Provided is a detailed protocol designed to guide the researcher through the preparation and immunoaffinity depletion of fresh frozen tissue homogenates for two-dimensional liquid chromatography, tandem mass spectrometry (2D-LC-MS/MS)-based molecular profiling of tissue specimens in the context of drug target and/or biomarker discovery.

Key words Immunoaffinity depletion, Fresh frozen tissue, Drug discovery, Biomarker discovery, Clinical proteomics, Mass spectrometry (MS)- based proteomics, 2D-LC-MS/MS

1 Introduction

Mass spectrometry-based profiling of clinical specimens (e.g., tissue, serum/plasma) has been increasingly used in drug target and biomarker discovery to elucidate changes in protein expression between diseased and healthy individuals [1]. Differentially expressed proteins not only serve as viable drug targets and/or important indicators of disease state, but also provide the basis for

Iulia M. Lazar et al. (eds.), *Proteomics for Drug Discovery: Methods and Protocols*, Methods in Molecular Biology, vol. 1647, DOI 10.1007/978-1-4939-7201-2_5, © Springer Science+Business Media LLC 2017

characterizing underlying pathologic mechanisms [2]. Ideally, newly discovered differentially expressed protein targets (e.g., tyrosine kinase-type cell surface receptor *HER2*) may serve as biomarkers for: (1) diagnosis and/or disease subtyping, (2) companion diagnostics, and/or (3) precision therapy targeting (i.e., trastuzumab for *HER2* positive cancers).

Despite recent advances in clinical proteomics, MS-based discovery of viable drug targets and/or relevant biomarkers has proven challenging. This is primarily due to the extensive dynamic range of protein concentration in tissue/blood, which exceeds six orders of magnitude in cells, and 10–12 orders of magnitude in plasma/serum [3]. Notably, an excess of peptides derived from highly abundant blood proteins (e.g., albumins, globulins) further complicates the dynamic range challenges and creates a background, which interferes with the identification of lower-abundant and biologically significant targets/biomarkers [4]. While the dynamic range issue has been dealt with by multidimensional separation/fractionation at the protein and/or peptide level, the proteolytic background from highly abundant proteins has been addressed by depletion of highly abundant blood derived proteins [5]. Typically, immunoaffinity depletion of highly abundant proteins from human plasma and/or serum specimens has been carried out as a mandatory step in a myriad of proteomic studies [6, 7], the assumption being, that if relevant drug targets and/or clinical biomarkers are to be uncovered/identified, they will be present in the moderate to low-abundant protein pools [8]. Currently, immunoaffinity depletion is an established technique that has been used for more than a decade for plasma/serum MS-based proteomic applications [9, 10]. It has gained wide acceptance, and has been adopted by many researchers [11–13] with immunoaffinity column formats available through numerous manufacturers [14–17].

Often overlooked, tissue specimens contain a formidable amount of blood/plasma, containing highly abundant blood-derived proteins found in interstitial fluid and blood or lymph capillaries [18, 19]. To our knowledge, we are the first to develop and optimize the immunodepletion of highly abundant blood-derived proteins from tumor and normal tissue homogenates (Fig. 1) in our proof of principle study using renal cell carcinoma (RCC) as a model disease [18]. We expanded upon that study by outlining the principles of tissue physiology and the rationale behind depletion of highly abundant blood-derived proteins from tissue homogenates and applied our technique to a variety of tissue types depicted in Figs. 2, 3, and 4 [19]. In our laboratory, we have found that the immunodepletion of abundant blood-derived proteins from fresh-frozen tissue homogenates is of equal importance as immunoaffinity depletion of serum and/or plasma. We apply this technique to the preparation of all tissue homogenates subjected to

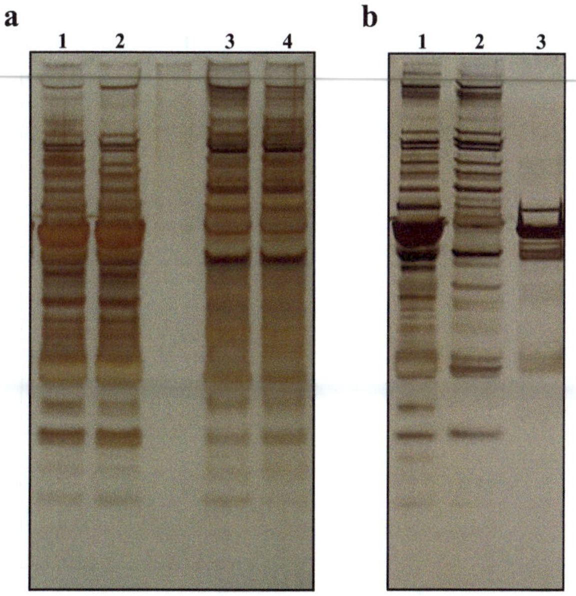

Fig. 1 SDS-PAGE analysis of immunoaffinity-depleted renal cell carcinoma (RCC) tissue homogenate (**a**) and matched patient serum (**b**). Samples were resolved on a 4–12% Bis-Tris gel and stained with SilverQuest Silver Staining Kit. Lane designations for gel (**a**) are as follows: Lane 1: normal kidney tissue homogenate (5 μg), Lane 2: RCC tissue homogenate (5 μg), Lane 3: immunoaffinity-depleted normal kidney tissue homogenate (5 μg), Lane 4: immunoaffinity-depleted RCC tissue homogenate (5 μg). Lane designations for gel (**b**) are as follows: Lane 1: RCC patient serum (5 μg), Lane 2: immunoaffinity-depleted matched patient serum (5 μg), Lane 3: high abundant fraction (15 μL)

global molecular profiling using 2D-LC-MS/MS analysis in the context of drug target and/or biomarker discovery [19].

Herein, we describe the preparation of fresh frozen tissue homogenates prior to strong cation exchange (SCX)-based peptide fractionation and high-resolution/accuracy LC-MS analysis [20]. Our workflow is depicted in Fig. 5. In the absence of access to off-line SCX-based fractionation, other forms of separation at the protein level (e.g., tube or slab gel electrophoresis) or at the peptide level (e.g., isoelectric focusing of peptides) can be considered an alternative [21, 22]. Furthermore, the highly abundant protein fraction may be further analyzed by standard in-gel digestion LC-MS analysis [23–25] to investigate the significance/rate of non-specifically-bound protein species to the immunoaffinity column.

The approximate amount of time to perform this protocol is generously broken down as follows: Tissue preparation: 1 h; Tissue homogenization: 1 h; Immunoaffinity depletion of the tissue homogenate: 2 h; Desalting/concentration of the tissue homogenate: 1.5 h; Protein concentration determination: 1 h; Tryptic digestion: 1 h plus overnight digestion; C18 cleanup of the proteolytic digest: 1 h; QC SDS-PAGE gels: 2.5 h.

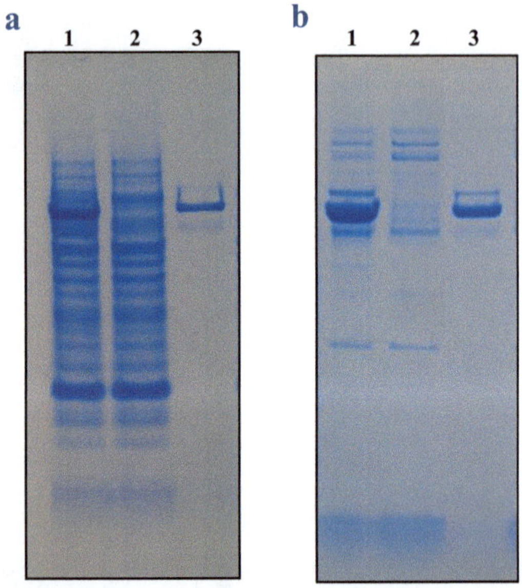

Fig. 2 SDS-PAGE analysis of immunoaffinity-depleted Ewing Sarcoma (EWS) tissue homogenate (**a**) and matched patient serum (**b**). Samples were resolved on a 4–12% Bis-Tris gel and stained with SimplyBlue Coomassie G-250 SafeStain. Lane designations for gel (**a**) are as follows: Lane 1: EWS tissue homogenate (10 μg), Lane 2: immunoaffinity-depleted EWS tissue homogenate (10 μg), Lane 4: high abundant fraction (20 μL). Lane designations for gel (**b**) are as follows: Lane 1: matched patient serum (10 μg), Lane 2: immunoaffinity-depleted patient serum (20 μL), Lane 4: high abundant fraction (20 μL)

2 Materials

2.1 General Reagents/Equipment

1. Single channel manual pipettes, 0.1–2 μL, 0.5–10 μL, 2–20 μL, 10–100 μL, 20–200 μL, and 100–1000 μL.
2. Racked pipette tips, 10 μL, 250 μL, and 1000 μL.
3. Safe-lock Eppendorf tubes, 0.5 mL, 1.5 mL, and 2.0 mL.
4. Drummond Portable Pipet-Aid.
5. Glass serological pipettes, 2 mL, 5 mL, 10 mL, 25 mL, and 50 mL.
6. Falcon tubes, 14 mL and 50 mL.
7. Syringes, polypropylene, 6 cc.
8. Vortex mixer.
9. Genie vortex mixer.
10. Refrigerated microcentrifuge 20R.
11. Allegra 6KR centrifuge.
12. Shaking 37 °C incubator.

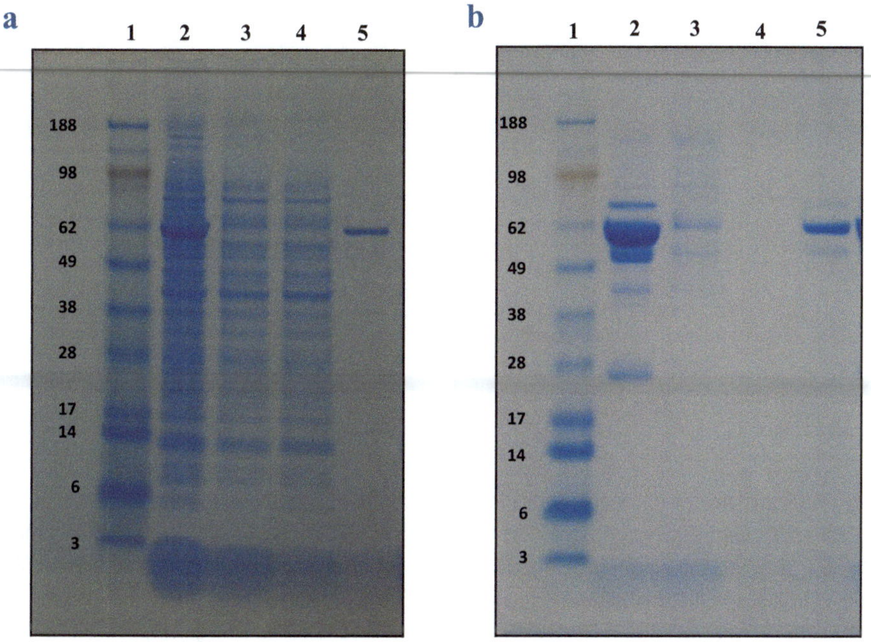

Fig. 3 SDS-PAGE analysis of immunoaffinity-depleted breast tissue homogenate (**a**) and matched patient serum (**b**). Samples were resolved on a 4–12% Bis-Tris gel and stained with SimplyBlue Coomassie G-250 SafeStain. Lane designations for gel (**a**) are as follows: Lane 1: SeeBlue Plus2 prestained protein MW standard (kDa), Lane 2: breast tissue homogenate (15 μg), Lane 3: immunoaffinity-depleted breast tissue homogenate, flow-through fraction #1 (10 μg), Lane 4: immunoaffinity-depleted breast tissue homogenate, flow-through fraction #2 (10 μg), Lane 5: high abundant fraction (20 μL). Lane designations for gel (**b**) are as follows: Lane 1: SeeBlue Plus2 prestained protein MW standard (kDa), Lane 2: matched patient serum (10 μg), Lane 3: immunoaffinity-depleted patient serum, flow-through fraction #1 (20 μL), Lane 4: immunoaffinity-depleted patient serum, flow-through fraction #2 (20 μL), Lane 5: high abundant fraction (20 μL)

2.2 Tissue Preparation

1. Snap fresh frozen tissue.
2. Optimal Cutting Temperature (OCT) compound.
3. Leica CM 1950 cryostat.

2.3 Tissue Homogenization

1. Snap fresh frozen tissue, 8-μm-thick slices in 2 mL Eppendorf tubes.
2. Ammonium bicarbonate (NH_4HCO_3) stock solution: 25 mM NH_4HCO_3, pH 8.0. Prepare 100 mL stock solution by dissolving 200 mg of NH_4HCO_3 in 80 mL of LC-MS CHROMASOLV® water. Mix and adjust pH to 8.0 with formic acid (HCOOH) or ammonium hydroxide (NH_4OH). Adjust the volume to 100 mL with LC-MS CHROMASOLV® water.
3. Protease inhibitor solution:
   (a) Phenylmethanesulfonyl fluoride (PMSF) stock solution: 100 mM in ethanol. Prepare a stock solution by dissolving

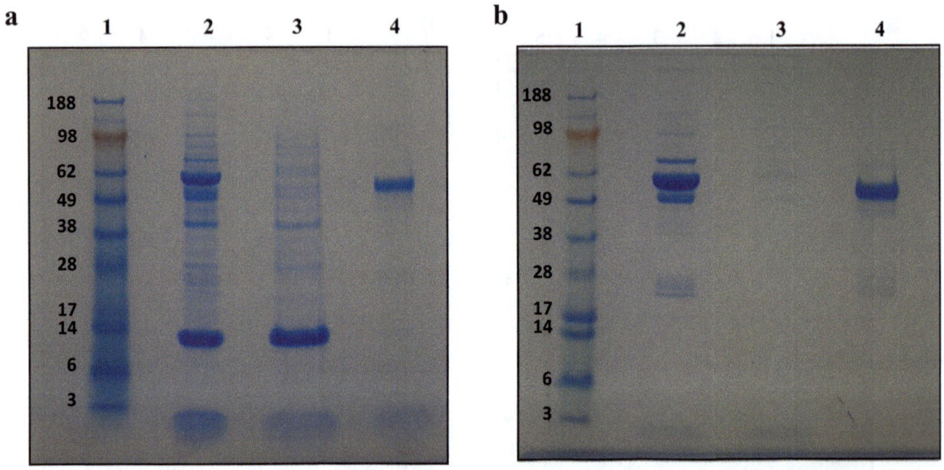

Fig. 4 SDS-PAGE analysis of immunoaffinity-depleted adipose tissue homogenate (**a**) and matched patient serum (**b**). Samples were resolved on a 4–12% Bis-Tris gel and stained with SimplyBlue Coomassie G-250 SafeStain. Lane designations for gel (**a**) are as follows: Lane 1: SeeBlue Plus2 prestained protein MW standard (kDa), Lane 2: adipose tissue homogenate (5 μg), Lane 3: immunoaffinity-depleted adipose tissue homogenate (5 μg), Lane 4: high abundant fraction (20 μL). Lane designations for gel (**b**) are as follows: Lane 1: SeeBlue Plus2 prestained protein MW standard (kDa), Lane 2: matched patient serum (5 μg), Lane 3: immunoaffinity-depleted matched patient serum (20 μL), Lane 4: high abundant fraction (20 μL)

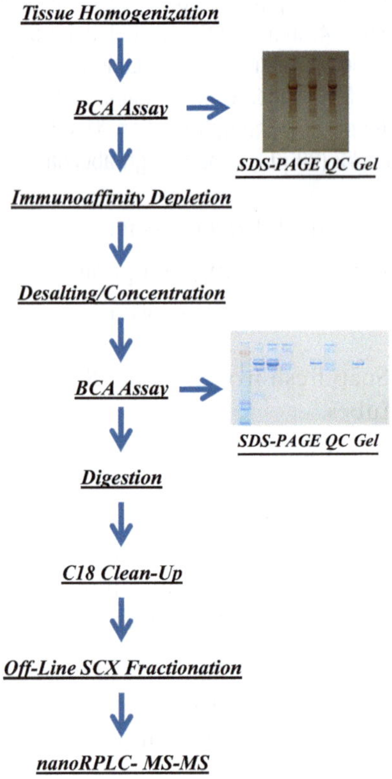

Fig. 5 Fresh frozen tissue preparation workflow: Tissue preparation for proteomic analysis

0.5226 g PMSF (Sigma Aldrich) in 30 mL of HPLC-grade ethanol, or,

(b) Complete protease inhibitor cocktail (25×): Prepare a 25× stock solution by adding 1 complete protease inhibitor cocktail tablet (Roche Applied Science) to 2 mL LC-MS CHROMASOLV® water. Mix until complete dissolution.

4. Lysis buffer:

   (a) Lysis buffer (a): 25 mM NH_4HCO_3, pH 8.0, 1 mM PMSF. Add 100 μL of 100 mM PMSF to 9.9 mL of 25 mM NH_4HCO_3 (pH 8.0) stock solution, or,

   (b) Lysis buffer (b): 25 mM NH_4HCO_3 (pH 8.0) containing 1× protease inhibitor cocktail. Add 400 μL of 25× complete protease inhibitor cocktail solution to 9.6 mL of 25 mM NH_4HCO_3 (pH 8.0) stock solution.

5. Branson digital sonifier.
6. Branson ultrasonic cleaner.

2.4 Immunoaffinity Depletion

1. Agilent Technologies MARS 0.45 mL affinity spin cartridge of choice: Human 14, Human 7, Human 6, Human Albumin/IgG, Human Albumin, or Mouse 3.
2. Protease inhibitor solution: stock solution prepared as described in Subheading 2.3, **item 3**.
3. Buffer A (Agilent Technologies)/Protease inhibitor solution: Prepare Buffer A/Protease inhibitor in 50 mL Falcon conical tubes:

   (a) Mix 39.6 mL of Buffer A and 400 μL of 100 mM PMSF, or,

   (b) Mix 38.4 mL Buffer A and 1.6 mL of 25× complete protease inhibitor cocktail.

4. Buffer B (Agilent Technologies)/Protease inhibitor solution: Prepare Buffer B/Protease inhibitor in 50 mL Falcon conical tubes:

   (a) Mix 39.6 mL of Buffer B and 400 μL of 100 mM PMSF, or,

   (b) Mix 38.4 mL Buffer B and 1.6 mL of 25× complete protease inhibitor cocktail.

5. Spin filters, 0.22 μm (pk/25) (Agilent Technologies).
6. Luer-lock adapters (pk/2) (Agilent Technologies).
7. Plastic luer-lock syringes, 5 mL (pk/2) (Agilent Technologies).
8. Screw-top Eppendorf-style microtubes, 1.5 mL (pk/100) (Agilent Technologies).
9. Spin cartridge screw caps and plugs (pk/6) (Agilent Technologies).
10. Reagent reservoirs.

2.5 SDS-PAGE

SDS-PAGE was performed in this work with equipment and reagents from Invitrogen Life Technologies. Other commercial products and equipment, of similar performance, can be used.

1. SureLock XCELL mini gel apparatus.
2. PowerEase 500 power programmable power supply.
3. Eppendorf Thermomixer R heating block.
4. NUPAGE 4–12% Bis-Tris gels.
5. NUPAGE sample reducing agent (10×).
6. NUPAGE LDS sample buffer (4×).
7. See Blue Plus 2 prestained protein standard.
8. NUPAGE antioxidant.
9. NUPAGE MES SDS running buffer (1×): Add 50 mL of 20× NUPAGE MES SDS running buffer to 950 mL of LC-MS CHROMASOLV® water and mix.
10. Gel staining trays.
11. Simply Blue SafeStain.
12. Platform rocker 200.

2.6 Desalt/Concentrate Tissue Homogenate

1. MWCO spin concentrators, 5 kDa.
2. Ammonium bicarbonate stock solution: Solution prepared as described in Subheading 2.3, **item 2**.

2.7 Protein Concentration Determination

1. Pierce BCA protein assay kit.
2. Corning Costar 96-well microplates.
3. Plate reader capable of measuring at an absorbance of 562 nm.

2.8 Tryptic Digestion

1. Ammonium bicarbonate stock solution: Solution prepared as described in Subheading 2.3, **item 2**.
2. Bond-breaker TCEP stock solution: 50 mM. Prepare a stock solution by mixing 10 μL of 500 mM Bond-breaker TCEP solution, neutral pH (Thermo Scientific), with 90 μL of LC-MS CHROMASOLV® water (*see* **Note 1**).
3. Trypsin Gold mass spectrometry grade trypsin solution: 1 μg/μL. Add 100 μL of 25 mM NH_4HCO_3 (pH 8.0) stock solution to 100 μg lyophilized Trypsin Gold (Promega). Vortex to mix (*see* **Note 1**).

2.9 C18 Cleanup of the Digest

1. Sep-Pak C18 Plus Light cartridges (Waters).
2. Ring stand with two double clamps.
3. Safe-lock Eppendorf tubes, 2 mL.
4. Syringes, 6 cc (polypropylene).

5. Trifluoroacetic acid (TFA) stock solution: 20% TFA. Add 8 mL of LC-MS grade TFA to 32 mL of LC-MS CHROMASOLV® water (*see* **Note 2**).
6. TFA dilute solution: 0.1%. Add 1 mL of 20% TFA stock solution to 199 mL of LC-MS CHROMASOLV® water (*see* **Note 2**).
7. Acetonitrile (ACN) solution: 70% ACN, 0.1% TFA. Add 140 mL of LC-MS grade ACN and 1 mL of 20% TFA stock solution to 59 mL of LC-MS CHROMASOLV® water (*see* **Note 2**).
8. pH paper.

3 Methods

3.1 Tissue Preparation

1. Set Leica CM 1950 cryostat at −15 °C.
2. Adhere snap frozen tissue to the cryostat specimen holder with a minimal amount of OCT.
3. Cut 8 μm frozen tissue slices, while avoiding the OCT-adhered tissue portion (*see* **Notes 3** and **4**).
4. Collect tissue slices in a labeled prechilled 2.0 mL Eppendorf tube to approximately the 0.75 mL mark.
5. Place the filled tube directly on dry ice. Process immediately or transfer to a −80 °C freezer.

3.2 Tissue Homogenization

This protocol is intended for use with 8 μm sectioned snap fresh-frozen tissue slices. We have simplified our lysis buffer to accommodate our immunoaffinity depletion protocol, SCX chromatography, and mass spectrometry. Additionally, MARS is an antibody-based immunoaffinity method; therefore, proteins must be in their native conformation. We have omitted from our lysis buffer any component that may denature proteins and/or interfere with further down-stream processing.

1. Prepare fresh lysis buffer as described in Subheading 2.3, **item 4**. Store on ice until needed.
2. Add a starting volume of 250–500 μL cold lysis buffer to the 2 mL Eppendorf tube containing the tissue slices (*see* **Notes 5** and **6**).
3. Using a Branson Digital Sonifier, homogenize the tissue on ice using a 10 s burst at 20% amplitude (*see* **Notes** 7 and **8**).
4. Incubate the sample on ice for 1 min (*see* **Note 9**).
5. Repeat **steps 3** and **4** until complete homogenization is achieved (*see* **Note 10**).
6. Sonicate the homogenate for 10 min in an ice-cold sonicating water bath (*see* **Note 11**).

7. Centrifuge the tissue homogenate 5–10 min at 10,000 × *g* in order to pellet the tissue debris.
8. Collect the supernatant and store on ice if BCA assay is to be performed immediately; otherwise, store the tissue homogenate at −80 °C.

3.3 Determination of Tissue Homogenate Protein Concentration: BCA Assay

1. Perform the BCA assay as per manufacturer's instructions (Pierce BCA Protein Assay Kit, ThermoFisher Scientific Manual).
2. Remove 10–20 μg of the tissue homogenate for QC gel analysis. Store on ice or at −80 °C if the gel is to be run at a later date.
3. Transfer 2000–3000 μg of tissue homogenate to a fresh 1.5 mL Eppendorf tube for immunoaffinity depletion (*see* **Note 12**). Store on ice if immunoaffinity depletion is to be performed immediately, otherwise, store at −80 °C.

3.4 QC SDS-PAGE Gel: Post-Tissue Homogenization

The purpose of running the tissue homogenate on a 4–12% Bis-Tris gel is two-fold. First, the gel serves to confirm that the homogenate preparation is of high quality. Second, as evidenced by the albumin band (human albumin isoform #1: 69,367 Da, UniProt database), the homogenate dilution may need to be adjusted for immunoaffinity depletion. The gel helps to establish this need.

1. Assemble the SureLock XCELL mini gel apparatus with a 4–12% Bis-Tris gel and 1× MES SDS running buffer as per manufacturer's instructions.
2. Transfer 5 μg of the tissue homogenate to a 0.5 mL Eppendorf tube.
3. Add LC-MS CHROMASOLV® water, 10× NUPAGE reducing agent and 4× NUPAGE sample buffer according to the manufacturer's recommendations (Table 1).

Table 1
Sample and reagent volumes for preparing the tissue homogenate for QC SDS-PAGE

Reagent	Reduced sample	Non-reduced sample
Sample (5 μg)	X μL	X μL
LC-MS CHROMASOLV® water	to 13 μL	to 15 μL
NUPAGE (10×) reducing agent	2 μL	–
NUPAGE (4×) LDS sample buffer	5 μL	5 μL
Total volume	**20 μL**	**20 μL**

4. Heat the tissue homogenate for 10 min at 70 °C.
5. Load the tissue homogenate in the well of a 4–12% Bis-Tris gel.
6. Load 15–20 μL of See Blue Plus 2 prestained protein standard in an adjacent well.
7. Perform gel electrophoresis at 125 V.
8. Remove the gel from the cassette, place the gel in a staining tray containing enough LC-MS CHROMASOLV® water to cover the gel, and incubate on a platform rocker for 5 min.
9. Decant the LC-MS CHROMASOLV® water.
10. Stain the gel with SimplyBlue™ SafeStain as per manufacturer's instructions (Life Technologies User Guide).

3.5 Immunoaffinity Depletion

Fresh frozen tissue samples are depleted of select highly abundant proteins using Agilent's Multiple Affinity Removal System (MARS) spin cartridges (Agilent Technologies). This technique utilizes 0.45 mL immunoaffinity cartridges packed with antibody-modified resin. MARS cartridges are specifically designed to remove up to 14 abundant proteins (albumin, IgG, IgA, antitrypsin, transferrin, haptoglobin, fibrinogen, alpha-2-macroglobulin, alpha-1-acid glycoprotein, IgM, apolipoprotein A1, apolipoprotein A2, complement C3, and transthyretin), depending upon the cartridge format. An enriched pool of low abundant proteins is then collected and combined from several flow-through fractions.

Human immunoaffinity cartridges have been successfully used in our laboratory for numerous primate and canine studies. Please note that transferrin is not depleted in canine models. Additionally, the mouse cartridge has also been effective in rat studies performed in our laboratory. Agilent Technologies has published comparative studies dealing with antibody species cross-reactivity (*see* Agilent Technologies application notes).

1. Prepare fresh Buffer A/Protease inhibitor as described in Subheading 2.4, **item 3**. Store on ice until needed (*see* **Note 13**).
2. Prepare fresh Buffer B/Protease inhibitor as described in Subheading 2.4, **item 4**. Store on ice until needed (*see* **Note 14**).
3. Transfer 2000–3000 μg of the tissue homogenate to a 1.5 mL Eppendorf tube. Adjust the volume of the homogenate to 200 μL (*see* **Note 15** for details regarding this adjustment).
4. Dilute the tissue homogenate 1:5 with Buffer A/Protease inhibitor: 200 μL homogenate, 800 μL Buffer A/Protease inhibitor (*see* **Note 16**).
5. Filter the diluted tissue homogenate through a 0.22 μm spin filter by centrifuging 3–5 min at 10,000 × *g*. Discard the inner filter and place the 1.5 mL Eppendorf tube containing the filtered homogenate on ice (*see* **Note 17**).

6. Prepare the 0.45 mL immunoaffinity cartridge as follows:
   (a) Remove the 0.45 mL immunoaffinity cartridge screw cap and bottom plug. Attach a luer-lock adapter to the top of the 0.45 mL immunoaffinity cartridge.
   (b) Draw 4 mL of Buffer A/Protease inhibitor into a 6 cc syringe.
   (c) Attach the syringe to the top of the 0.45 mL immunoaffinity cartridge via luer-lock adapter.
   (d) In a dropwise fashion, slowly dispense Buffer A/Protease inhibitor through the 0.45 mL immunoaffinity cartridge into a waste container.
   (e) Unscrew the 0.45 mL immunoaffinity cartridge from the syringe via luer-lock.
   (f) Remove the excess Buffer A/Protease inhibitor from the top of the 0.45 mL immunoaffinity cartridge resin bed with a pipette.
7. Place the 0.45 mL immunoaffinity cartridge into a 1.5 mL screw-top tube.
8. Add 200 μL of diluted tissue lysate to the top of the 0.45 mL immunoaffinity cartridge resin bed. Lightly tighten the screw cap, do not over tighten (this will be the flow-through #1 screw-top tube).
9. Centrifuge for 2 min at 100 × *g*.
10. Leaving the 0.45 mL immunoaffinity cartridge in the flow-through #1 screw-top tube, remove the screw-cap, and add 350 μL of Buffer A/Protease inhibitor to the top of the 0.45 mL immunoaffinity cartridge resin bed. Lightly tighten the cap, do not over tighten.
11. Centrifuge for 3 min at 100 × *g*.
12. Remove the screw-cap. Remove the 0.45 mL immunoaffinity cartridge from the flow-through #1 tube and place it into a fresh 1.5 mL screw-top tube (this will be the flow-through #2 tube). Cap the flow-through #1 screw-top tube and store it on ice. Add 350 μL of Buffer A/Protease inhibitor to the top of the 0.45 mL immunoaffinity cartridge resin bed and lightly tighten the screw cap, making sure not to over tighten.
13. Centrifuge for 3 min at 100 × *g*.
14. Remove the screw-cap, remove the 0.45 mL immunoaffinity cartridge from the flow-through #2 collection tube, and set aside. Store flow-through #2 collection tube (capped) on ice.
15. Elute the high abundant protein fraction as follows:
   (a) Attach a luer-lock adapter to the top of 0.45 mL immunoaffinity cartridge.

   (b) Fill a second syringe with 2 mL of Buffer B/Protease inhibitor.
   (c) Attach the syringe to the 0.45 mL immunoaffinity cartridge via luer-lock.
   (d) In a dropwise fashion, slowly elute the bound high abundant protein fraction by dispensing 2 mL of Buffer B/Protease inhibitor through the 0.45 mL immunoaffinity cartridge into a 2 mL Eppendorf tube.
   (e) Cap the 2 mL Eppendorf tube and place the high abundant protein fraction on ice.
16. Re-equilibrate the immunoaffinity cartridge as follows:
   (a) Remove the 0.45 mL immunoaffinity cartridge from the syringe via luer-lock adapter.
   (b) Attach the luer-lock adapter to the 0.45 mL immunoaffinity cartridge.
   (c) Draw 4 mL of Buffer A/Protease inhibitor into the syringe.
   (d) Attach the syringe to the 0.45 mL immunoaffinity cartridge via the luer-lock.
   (e) Slowly dispense Buffer A/Protease inhibitor through the immunoaffinity cartridge into a waste container.
   (f) Remove excess buffer from the top of resin bed with a pipette.
17. Repeat **steps 8–16** four times (total # depletions = 5, total volume depleted = 1 mL).
18. Pool flow-through fractions #1 and #2 from all depletions. Remove 20 μL and store on ice or at −80 °C to be used for a *QC SDS-PAGE Gel.* Store the remaining low abundant pool on ice for desalting/concentration or at −80 °C if desalting/concentration is to be performed at a later date.
19. Pool the high abundant fractions. Remove 20 μL and store on ice or at −80 °C to be used for a *QC SDS-PAGE Gel.* Store the remaining pooled high abundant fraction on ice or at −80 °C if it is to be used for further analysis.

3.6 QC SDS-PAGE Gel: Post Immunoaffinity Depletion

QC measures are employed to assess depletion efficiency by removing an aliquot of both the low and high abundant pools and subsequently resolving the proteins on a 4–12% Bis-Tris SDS-PAGE gel. Proteins are visualized by staining with Coomassie blue.

1. Assemble the SureLock XCELL Mini gel apparatus with a 4–12% Bis-Tris gel and MES SDS running buffer per manufacturer's instructions.
2. Transfer 5 μg of undepleted tissue homogenate to a 0.5 mL Eppendorf tube and prepare as outlined in Subheading 3.4, **step 3**.

3. Prepare 20 μL of the pooled low abundant flow-through fraction by mixing 20 μL fraction, 2 μL of 10× NUPAGE reducing agent, and 5 μL of 4× NUPAGE sample buffer in a 0.5 mL Eppendorf tube.
4. Prepare 20 μL of the pooled high abundant fraction by mixing 20 μL fraction, 2 μL of 10× NUPAGE reducing agent, and 5 μL of 4× NUPAGE sample buffer in a 0.5 mL Eppendorf tube.
5. Heat the undepleted tissue homogenate, low abundant fraction, and high abundant fraction for 10 min at 70 °C (*see* **Note 18**).
6. Load the samples in the wells of the 4–12% Bis-Tris gel.
7. Load 15–20 μL of See Blue Plus 2 prestained protein standard in an adjacent well.
8. Perform gel electrophoresis at 125 V.
9. Remove the gel from the cassette, place the gel in a staining tray containing enough LC-MS CHROMASOLV® water to cover the gel and incubate on a platform rocker for 5 min.
10. Decant the LC-MS CHROMASOLV® water.
11. Stain the gel with Simply Blue SafeStain as per manufacturer's instructions (ThermoFisher Scientific Manual).

3.7 Desalt/Concentrate Tissue Homogenate

1. Transfer the pooled low abundant flow-through fraction (approximately 4.5 mL) to a 5 kDa MWCO spin concentrator.
2. Centrifuge the homogenate at 5000 × *g* for approximately 30–45 min at 4 °C using an Allegra 6R centrifuge (reduce the sample volume to approximately 500 μL).
3. Decant the flow-through into a waste container.
4. Fill the concentrator with 25 mM NH_4HCO_3, pH 8.0 (approximately 4 mL).
5. Centrifuge until the sample volume has been reduced to approximately 500 μL (Allegra 6R centrifuge, 5000 × *g*, 4 °C).
6. Decant the flow-through into a waste container.
7. Repeat **steps 4–6** twice (*see* **Note 19**).
8. Transfer the desalted/concentrated homogenate to a fresh 1.5 mL Eppendorf tube.
9. Wash the walls of concentrator three times with 100 μL of 25 mM NH_4HCO_3 (pH 8.0). Add the washes to the 1.5 mL Eppendorf tube.
10. Lyophilize the desalted/concentrated tissue homogenate to approximately 200–250 μL and store on ice or at −80 °C if the BCA protein assay is to be performed on a later date.

3.8 Determination of Immunoaffinity-Depleted Homogenate Protein Concentration: BCA Assay

We typically utilize 100–200 μg of immunoaffinity-depleted protein for each study. The BCA assay is a critical step because standardization is always performed at the protein level. This is especially critical if relying upon spectral count. By standardizing at the protein level, we assure that every sample is consistent and the reproducibility of the study is preserved.

1. Perform a BCA assay as per manufacturer's instructions (Pierce BCA Protein Assay Kit, ThermoFisher Scientific Manual).
2. Transfer 100–200 μg of immunoaffinity-depleted, desalted, and concentrated tissue homogenate to a fresh 1.5 mL Eppendorf tube.
3. Store on ice if a tryptic digestion is to be performed next, otherwise, store at −80 °C.

3.9 Protein Digestion: In-Solution Tryptic Digestion

1. Adjust the sample volume of 100–200 μg of immunoaffinity-depleted, desalted, and concentrated tissue homogenate to 312 μL with 25 mM NH_4HCO_3 (pH 8.0).
2. Heat 10 min a 95 °C.
3. Cool on ice 5 min.
4. Add 8 μL of the 50 mM TCEP stock solution (*see* **Note 20**).
5. Incubate for 30 min at 37 °C with shaking.
6. Cool on ice 10 min.
7. Add 80 μL of 100% HPLC-grade methanol (*see* **Note 21**).
8. Add porcine-grade modified trypsin at a 1:20 (trypsin: protein) ratio.
9. Digest overnight at 37 °C.

3.10 C18 Cleanup of the Digestate

1. Set up the Sep-Pak C18 Plus Light cartridge as depicted in Fig. 6.
2. Acidify the digestate with 5–20 μL of 20% TFA.
3. Check digestate acidity using pH paper (*see* **Note 22**).
4. Connect a 6 cc syringe (without plunger) to the top of the Sep-Pak C18 Plus Light cartridge.
5. Prewet the Sep-Pak C18 Plus Light Cartridge by adding 5 mL of LC-MS-grade ACN to the syringe, making sure that there is no air at the syringe/column interface. Replace the plunger and slowly expel the solvent through the Sep-Pak C18 Plus Light cartridge in a dropwise fashion, discarding the flow-through.
6. Remove the syringe from the Sep-Pak C18 Plus Light cartridge, remove the plunger from the syringe, and reattach the syringe (without plunger) to the Sep-Pak C18 Plus Light Cartridge.

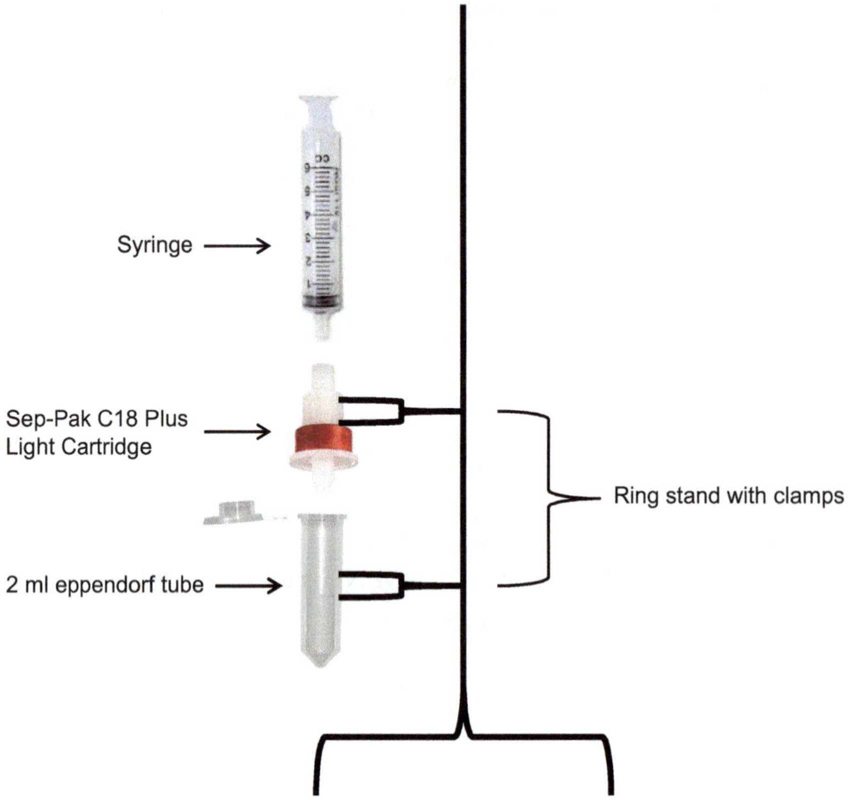

Fig. 6 Tissue digestate cleanup using Sep-Pak C18 Plus Light cartridges

7. Equilibrate the Sep-Pak C18 Plus Light Cartridge with 5 mL of 0.1% TFA as in **steps 4–6**.
8. Transfer the digestate to the syringe, replace the plunger, and slowly expel the digestate through the cartridge in a dropwise fashion, making sure to recover it in a clean 2 mL Eppendorf tube.
9. Remove the syringe from the Sep-Pak C18 Plus Light cartridge, remove the plunger from the syringe, and reattach the syringe (without plunger) to the Sep-Pak C18 Plus Light Cartridge.
10. Repeat **steps 8** and **9** twice.
11. Wash the Sep-Pak C18 Plus Light cartridge twice with 5 mL of 0.1% TFA as in **steps 4–6**.
12. Elute the bound digestate into a fresh 2 mL Eppendorf tube with 2 mL of 70% ACN/0.1% TFA.
13. Lyophilize to dryness.
14. The C18-purified immunoaffinity-depleted tissue digestate is now ready for fractionation via off-line SCX chromatography (*see* **Notes 23** and **24**) followed by MS analysis or by standard in-gel digestive MS analysis in lieu of SCX chromatographic capability.

4 Notes

1. Prepare a fresh stock solution directly before trypsin digestion.
2. Do not use plastic containers when preparing this stock solution. Use only glass.
3. Tissue slices **must not** contain OCT.
4. It is very important to obtain the tissue specimen from an experienced histo-technician, one who is able to use a very minimal amount of OCT to attach the tissue to the specimen holder thus avoiding OCT contamination.
5. The lysis buffer starting volume is dependent upon the amount of tissue to be homogenized.
6. Keep in mind that the cartridges used for immunoaffinity depletion are 0.45 mL cartridges; therefore, tissue homogenization in the lowest possible lysis buffer volume needed is recommended.
7. An easy way to homogenize the tissue on ice is to place the 2 mL Eppendorf tube in a small beaker packed with ice.
8. The sample can rapidly over-heat. Perform all the steps of the protocol on ice in order to avoid overheating, thereby minimizing proteolysis.
9. You may need to increase the one-min incubation on ice. This will be dependent upon the temperature of your homogenate. Do not allow the homogenate to over-heat.
10. We typically start with four or five cycles of homogenize/cool on ice.
11. We perform this step to ensure that total homogenization of the tissue has occurred. The sonicating water bath can become very warm. Adding ice to the bath is a good way to reduce the chance of the homogenate over-heating, thereby minimizing the chance of proteolysis.
12. Depletion efficiency is dependent upon several factors, i.e., tissue type, high abundant protein content of the tissue, type of cartridge used (human 6, 7, 14, albumin/IgG, albumin, mouse, etc.), therefore, low-abundant protein yield will vary. The lowest protein yields typically occur with Human 14 immunoaffinity cartridges. For our laboratory, the ultimate goal is to obtain a final concentration of 100–200 μg of immunoaffinity-depleted tissue homogenate for off-line SCX chromatography. We have found that a starting yield of 2–3 mg of tissue homogenate is sufficient for most studies and cartridge types. It is best to start with the highest protein yield possible.

13. We typically prepare Buffer A/Protease inhibitor in small aliquots and store on ice until needed. Prepare 2–50 mL Falcon tubes containing 40 mL of Buffer A/Protease inhibitor for each sample to be immunoaffinity depleted. Store on ice until ready to use.
14. We typically prepare Buffer B/Protease inhibitor in small aliquots and store on ice until needed. Prepare 1–50 mL Falcon tubes containing 40 mL of Buffer B/Protease inhibitor for each sample to be immunoaffinity depleted. Store on ice until ready to use.
15. Adjusting the volume of the tissue homogenate typically involves lyophilizing to 200 μL or adding 25 mM NH_4HCO_3 (pH 8.0) to 200 μL. Whether the homogenate is to be lyophilized to 200 μL or adjusted to 200 μL with 25 mM NH_4HCO_3 (pH 8.0) is dependent upon the tissue homogenization lysis buffer starting volume.
16. Dilutions are dependent upon the tissue type and the amount of high abundant protein contamination, especially albumin. We have found typical dilutions to be 1:5 or 1:10. This step of the protocol is highly dependent upon tissue type and may vary.
17. Filtration of the tissue homogenate is extremely important as particulate matter can damage the immunoaffinity cartridge.
18. The formulation of Buffer A and Buffer B is proprietary; however, Buffer B does contain urea. Do not become alarmed if the high abundant fraction containing reducing agent and sample buffer turns a blue/green-ish hue upon heating.
19. Check progress every 30 min. If flow-through approaches the level of the bottom of the concentrator, remove the concentrator and set aside. Decant flow-through, re-assemble, and continue centrifugation.
20. We prefer to use TCEP in place of DTT as our reducing agent. The neutral pH of the formulation minimizes the exposure of proteins to carbohydrate modification and acid hydrolysis. (https://tools.thermofisher.com/content/sfs/manuals/MAN0011394_BondBreaker_TCEPSolution_NeutralpH_UG.pdf)
21. There are numerous protein enzymatic digestion enzymes and protocols [23–26]. Each proteomic laboratory has its preferred method. We have found our methanol-assisted digestion to be simple, clean, and effective [20]. Our goal is to minimize components that may interfere with down-stream mass spectrometric analysis. Our methanol-assisted digestion accomplishes this goal.

22. Sample acidity should be approximately pH 3.0. This pH accommodates peptide binding to the Sep-Pak C18 Plus Light cartridge sorbent.
23. Our SCX chromatography and our immunoaffinity depletion are two novel aspects of our tissue processing workflow for LC-MS-MS analysis allowing us to delve deep into the proteome to identify potential biomarker candidates.
24. Details of our SCX chromatography technique are beyond the scope of this manuscript and have been previously published [20]; however, certain aspects of our SCX chromatography system deserve to be highlighted as follows:
    (a) Our SCX chromatography system is custom built.
    (b) Its novel features include a fluorescent detection system and solid state UV laser technology.
    (c) Our detector is the first in the world to utilize solid state UV laser technology.
    (d) The system can accommodate a wide range of column sizes.
    (e) The SCX separation is visualized as it occurs.
    (f) Volatile and low background producing buffers are used: (1) 25% ACN and (2) 25% ACN/0.5 M ammonium formate, pH 3. Our volatile ammonium formate buffer minimizes downstream sample manipulation.
    (g) We consistently see minimal carryover (<0.5%).

Acknowledgment

"This project has been funded in whole or in part with Federal funds from the National Cancer Institute, National Institutes of Health, under Contract No. HHSN261200800001E. The content of this publication does not necessarily reflect the views or policies of the Department of Health and Human Services, nor does mention of trade names, commercial products, or organizations imply endorsement by the U.S. Government."

References

1. Johann DJ Jr, Blonder J (2007) Biomarker discovery: tissues versus fluids versus both. Expert Rev Mol Diagn 7(5):473–475
2. Puangpila C, Mayadunne E, Rassi SE (2015) Liquid-phase-based separation systems for depletion, pre-fractionation and enrichment of proteins in biological fluids and matrices for in-depth proteomics analysis–an update covering the period 2011–2014. Electrophoresis 36(1):238–252
3. Bjorhall K, Miliotis T, Davidson P (2005) Comparison of different depletion strategies for improved resolution in proteomic analysis of human serum samples. Proteomics 5 (1):307–317
4. Faca V, Pitteri SJ, Newcomb L et al (2007) Contribution of protein fractionation to depth of analysis of the serum and plasma proteomes. J Proteome Res 6(9):3558–3565

5. Yadav AK, Bhardwaj G, Basak T et al (2011) A systematic analysis of eluted fraction of plasma post immunoaffinity depletion: implications in biomarker discovery. PLoS One 6(9):e24442
6. Huang Z, Yan G, Gao M et al (2016) Array-based online two dimensional liquid chromatography system applied to effective depletion of high-abundance proteins in human plasma. Anal Chem 88(4):2440–2445
7. Borg J, Campos A, Diema C et al (2011) Spectral counting assessment of protein dynamic range in cerebrospinal fluid following depletion with plasma-designed immunoaffinity columns. Clin Proteomics 8:6–14
8. Cheon DH, Nam EJ, Park KH et al (2016) Comprehensive analysis of low-molecular-weight human plasma proteome using top-down mass spectrometry. J Proteome Res 5 (1):229–244
9. Fountoulakis M, Juranville JF, Jiang L et al (2004) Depletion of the high-abundance plasma proteins. Amino Acids 27 (3–4):249–259
10. Martosella J, Zolotarjova N, Liu H et al (2005) Reversed-phase high-performance liquid chromatographic pre-fractionation of immunodepleted human serum proteins to enhance mass spectrometry identification of lower-abundant proteins. J Proteome Res 4(5):1522–1537
11. Hyung SW, Piehowski PD, Moore RJ et al (2014) Microscale depletion of high abundance proteins in human biofluids using IgY14 immunoaffinity resin: analysis of human plasma and cerebrospinal fluid. Anal Bioanal Chem 406:7117–7125
12. Galindo AN, Kussman M, Dayon L (2015) Proteomics of cerebrospinal fluid: throughput and robustness using a scalable automated analysis pipeline for biomarker discovery. Anal Chem 87(21):10755–10761
13. Cellar NA, Karnoup AS, Albers DR (2009) Immunodepletion of high abundance proteins coupled on-line with reversed-phase liquid chromatography: a two-dimensional LC sample enrichment and fractionation technique for mammalian proteomics. J Chromatogr B Analyt Technol Biomed Life Sci 877(1–2):79–85
14. Bellei E, Bergamini S, Monari E et al (2011) High-abundance protein depletion for serum proteomic analysis: concomitant removal of non-targeted proteins. Amino Acids 40:145–156
15. Millioni R, Tolin S, Puricelli L et al (2011) High abundance proteins depletion vs low abundance proteins enrichment: comparison of methods to reduce the plasma proteome complexity. PLoS One 6(5):e19603
16. Janecki DJ, Pomerantz SC, Beil EJ et al (2012) A fully integrated multi-column system for abundant protein depletion from serum/plasma. J Chromatogr B Analyt Technol Biomed Life Sci 902:35–41
17. Kullolli M, Warren J, Arampatzidou M et al (2013) Performance evaluation of affinity ligands for depletion of abundant plasma proteins. J Chromatogr B Analyt Technol Biomed Life Sci 939:10–16
18. Johann DJ, Wei BR, Prieto DA et al (2010) Combined blood/tissue analysis for cancer biomarker discovery: application to renal cell carcinoma. Anal Chem 82(5):1584–1588
19. Prieto DA, Johann DJ Jr, Wei BR et al (2014) Mass spectrometry in cancer biomarker research: a case for immunodepletion of abundant blood-derived proteins from clinical tissue specimens. Biomark Med 8(2):269–286
20. Blonder J, Chan KC, Issaq HJ et al (2006) Identification of membrane proteins from mammalian cell/tissue using methanol-facilitated solubilization and tryptic digestion coupled with 2D-LC-MS/MS. Nat Protoc 1 (6):2784–2790
21. Tran JC, Doucette A (2008) Gel-eluted liquid fraction entrapment electrophoresis: an electrophoretic method for broad molecular weight range proteome separation. Anal Chem 80:1568–1573
22. Slebos RJ, Brock JW, Winters NF et al (2008) Evaluation of strong cation exchange versus isoelectric focusing of peptides for multidimensional liquid chromatography-tandem mass spectrometry. J Proteome Res 12:5286–5294
23. Dzieciatkowska M, Hill R, Hansen KC (2014) GeLC-MS/MS analysis of complex protein mixtures. Methods Mol Biol 1156:53–66
24. Gundry RL, White MY, Murray CI et al (2010) Preparation of proteins and peptides for mass spectrometry analysis in a bottom-up proteomics workflow. Curr Protoc Mol Biol Chapter 10: Unit 10.25
25. Shevchenko A, Tomas H, Havlis J et al (2006) In-gel digestion for mass spectrometric characterization of proteins and proteomes. Nat Protoc 1(6):2856–2860
26. Giansanti P, Tsiatsiani L, Low TY et al (2016) Six alternative proteases for mass spectrometry-based proteomics beyond trypsin. Nat Protoc 11(5):993–1006

Chapter 6

Target Identification Using Cell Permeable and Cleavable Chloroalkane Derivatized Small Molecules

Jacqui L. Mendez-Johnson, Danette L. Daniels, Marjeta Urh, and Rachel Friedman Ohana

Abstract

An important aspect for gaining functional insight into the activity of small molecules revealed through phenotypic screening is the identification of their interacting proteins. Yet, isolating and validating these interacting proteins remains difficult. Here, we present a new approach utilizing a chloroalkane (CA) moiety capture handle, which can be chemically attached to small molecules to isolate their respective protein targets. Derivatization of small molecules with the CA moiety has been shown to not significantly impact their cell permeability or potency, allowing for phenotypic validation of the derivatized small molecule prior to capture. The retention of cell permeability also allows for treatment of live cells with the derivatized small molecule and the CA moiety enables rapid covalent capture onto HaloTag coated magnetic beads. Additionally, several options are available for the elution of interacting proteins, including chemical cleavage of the CA moiety, competitive elution using excess unmodified small molecule, or sodium dodecyl sulfate (SDS) elution. These features taken together yield a highly robust and efficient process for target identification, including capture of weak or low abundance interactors.

Key words Target identification, Chemoproteomics, Phenotypic screening, Small molecule, Derivatized small molecule, Chloroalkane, HaloTag, Chemical cleavage, Palladium catalyst, Mass spectrometry

1 Introduction

Among many challenges facing pharmaceutical industry, the selection of novel high potency drug candidates with good selectivity and low toxicity early in the development process, is certainly one of the key challenges. Over the last several decades development of small molecule drugs largely depended on assessing their activity against purified recombinant proteins. However, in recent years, we have observed a comeback of a more traditional, pharmacology-driven discovery, also known as phenotypic screening, where small molecules are screened for their ability to alter the phenotype of a cell or organism in a desired manner. The promise of this approach

Iulia M. Lazar et al. (eds.), *Proteomics for Drug Discovery: Methods and Protocols*, Methods in Molecular Biology, vol. 1647, DOI 10.1007/978-1-4939-7201-2_6,

is that it has the potential to accelerate the discovery of first-in-class small molecule drugs with novel mechanisms of action [1]. It has the additional advantage in that the binding of the small molecule occurs under more physiological conditions allowing us to gain knowledge, not only of the principle drug target, but also of the unforeseen "off targets." The success of phenotypic screening has in the past been encumbered by difficulties in identifying cellular targets eliciting the desired phenotype. The recent development of new tools for the discovery of targets for small molecule drugs has greatly aided the phenotypic screening approach [2–5]. The success afforded by these new tools is based on the integration of biology, chemistry, quantitative mass spectrometry, labeling technologies, and computational data analysis. This new approach has been given the name of chemoproteomics and it is rapidly becoming a default approach for the deconvolution of drug targets, assessment of the "off target" effects, and evaluation of the selectivity of small molecules [6, 7].

Chemoproteomics strategies include several different approaches that commonly rely on linking the small molecule to either an affinity tag or a surface. The key aim of all these approaches is the unbiased determination of drug targets in the complex cellular environment, and thus it is critical that the modification of the bioactive small molecule does not significantly perturb its biological activity. Furthermore, it is desirable that the elicited phenotype could be recapitulated with the derivatized small molecule, which among other parameters requires that the modification does not impede cellular permeability. This condition turns out to be quite challenging as many of the affinity moieties such as biotin, often negatively impact cell permeability. To address this limitation, different methods have been designed to enable cellular penetration of the derivatized small molecule and its binding to the targets within the cellular environment [8–11]. Typically, a small bi-functional tag comprised of a cross-linking moiety and an alkyne or an azide enabling click-chemistry ligation of an affinity tag (i.e., biotin) is appended onto the small molecule. This approach has been shown to allow the small molecule to penetrate cells where it can engage its targets. Subsequent photo-cross-linking freezes the transient interactions, allowing for isolation of low affinity targets. To address some of the undesirable side effects, such as low efficiency and nonspecificity of these reactions (i.e. photo-cross-linking and click ligation), we designed a new approach based on a chloroalkane capture moiety [12, 13]. Unlike the above-mentioned methods, the chloroalkane-mediated capture does not entail photo- cross-linking or click chemistry, but rather, it relies on rapid covalent binding of a chloroalkane derivatized small molecule onto a surface-immobilized HaloTag protein. The HaloTag protein is a protein fusion tag that was designed for highly specific, rapid and covalent binding of chloroalkane moieties to which different functional groups such as fluorophores or bioactive small molecules can be attached [14–16]. For the application of target identification, it is important to note that the chloroalkane capture moiety has been shown to have excellent cellular permeability, enabling testing of the derivatized small molecule for its ability to elicit the desired phenotype. In the systems tested to date, minimal influence on drug potency was observed [12, 13].

Other parameters critical for the success of the methodology relate to the ability to isolate low affinity, low abundance, short residence time protein targets as well as reducing nonspecific recovery of proteins. The fundamental aspect of the chloroalkane-driven enrichment and target identification is the covalent, rapid, and very specific binding of the chloroalkane to the HaloTag protein. These features have significant impact on the protocol and effectiveness of the method. The rapid binding allows for fast sample processing, thus minimizing the dissociation of the protein targets bound to the chloroalkane derivatized drug candidates. In addition, the covalent binding of the chloroalkane moiety to the immobilized HaloTag creates a high local concentration of the small molecule, which tips the equilibrium toward the bound state and thus increases the retention of bound targets. The highly specific nature of chloroalkane binding to the HaloTag protein, and the fact that the chloroalkane is not found in native cellular environments, significantly enhances the capture specificity of this method. The specificity of capture and elution is of utmost importance, since it can facilitate the identification of targets with low abundance or affinity over a nonspecific background. Furthermore, it can significantly reduce the time needed to further characterize the identified proteins and distinguish between the true protein targets and the proteins that co-purify due to nonspecific binding to the surface used for isolation. To this end, elution with excess of unmodified small molecule is frequently employed.

To improve the specificity of elution even further we designed a special linker molecule positioned between the chloroalkane and small molecule which carries a small chemically cleavable site. This approach allows for a selective release of protein targets captured to the HaloTag surface through cleavage of linker and leaves proteins nonspecifically bound to the resin behind. The cleavable linker comprises a small allyl-carbamate linkage, which can be efficiently cleaved by mild conditions utilizing a water-soluble palladium catalyst [13, 17]. This newly designed cleavable chloroalkane was shown to be stable in cellular environments and common buffers and furthermore, shows minimal impact on cellular permeability and efficacy of small molecule drugs [13].

Collectively, all the developments described above allowed us to develop a highly efficient and easy protocol for target deconvolution following a phenotypic screen (Fig. 1). As discussed above, it is important to test whether the derivatized small molecule

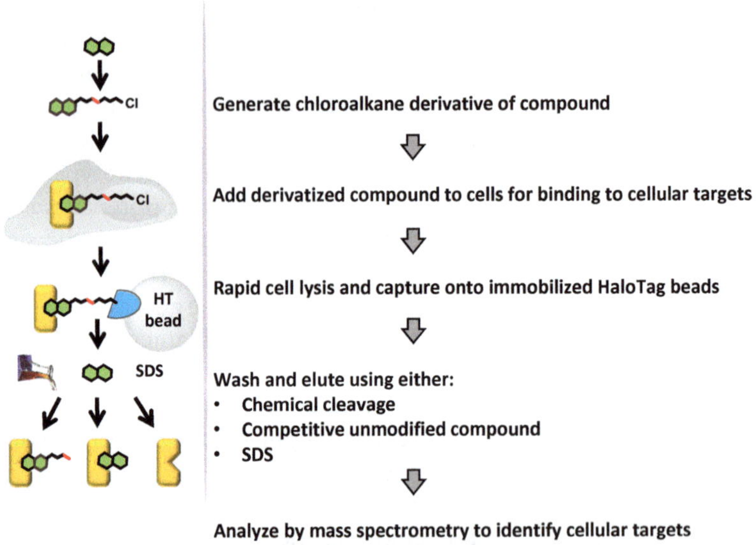

Fig. 1 Enrichment workflow utilizing a chloroalkane capture tag, which can be chemically attached to small molecules to isolate their respective cellular targets. Following target engagement inside cells, the cells are lysed and the derivatized small molecule together with the bound targets are captured onto HaloTag coated beads. Captured targets can then be released by either competitive elution with excess unmodified small molecule, chemical cleavage of the tag utilizing an optimized palladium catalyst, or SDS. The isolated proteins can be identified by mass spectrometry or Western blot analysis if the target is already known or suspected

candidate can elicit the same phenotype as the original molecule. It is also important to ascertain that appending the small molecule onto the chloroalkane does not negatively impact the binding of the chloroalkane to the HaloTag protein. This can be accomplished by evaluating the binding efficiency of the chloroalkane derivatized small molecule to a HaloTag protein expressed in live cells. This evaluation also provides an estimation of the molecule ability to penetrate cells. In the process, cells expressing the HaloTag protein are first treated with the chloroalkane derivatized small molecule for different extents of time, followed by treatment with a permeable fluorescent chloroalkane. Only HaloTag protein not bound by the chloroalkane derivatized small molecule will be accessible for binding by the fluorescent ligand and will therefore be fluorescently labeled. The comparison of fluorescent signal from control cells (not treated by chloroalkane derivatized small molecule) and samples pretreated with chloroalkane derivatized small molecule for different extents of time provides an estimation of binding kinetics to the HaloTag. Once the phenotype of the derivatized small molecule and its ability to bind to the HaloTag protein is established, one can proceed to the pull-down portion of the process.

The pulldown is designed for rapid processing and efficient isolation of cellular protein targets bound to the chloroalkane derivatized small molecule using magnetic beads carrying immobilized HaloTag protein. The isolated proteins are then analyzed by mass spectrometry or Western blotting in case the targets are known.

2 Materials

Certain materials and reagents are exclusively available from Promega Corp. as indicated.

2.1 Binding Efficiency of Chloroalkane Derivatized Small Molecule to HaloTag Protein

1. Reagents and equipment necessary for tissue culture including 10 cm dishes.
2. HeLa cells (preferred, *see* **Note 1**) or HEK293 cells.
3. Dulbecco's Modified Eagle Medium (DMEM) supplemented with 10% Fetal Bovine Serum (FBS).
4. Dulbecco's phosphate-buffered saline (DPBS).
5. Transfection reagent.
6. DNA expressing HaloTag protein (HaloTag Control Vector; Promega).
7. Cell striper (Corning) or 0.25% trypsin solution.
8. Lysis buffer **1**: Prepare 400 μL lysis buffer **1** by mixing 344 μL Mammalian Lysis buffer (Promega), 8 μL RQ1 RNase-free DNAse (Promega), 40 μL 10× RQ1 RNase-free DNase buffer (Promega), and 8 μL Protease Inhibitor Cocktail (Promega) (*see* **Note 2**).
9. HaloTag TMR (tetramethylrhodamine) ligand solution (4 μM). Prepare 1 mL solution by adding 0.8 μL HaloTag TMR ligand (Promega) to 1 mL DPBS.
10. Chloroalkane derivatized small molecule stock solution: 10 mM in dimethyl sulfoxide (DMSO).
11. Chloroalkane derivatized small molecule solution (1 mM). Prepare 10 μL solution by adding 1 μL of 10 mM chloroalkane derivatized small molecule stock solution to 9 μL DPBS.
12. Sodium dodecyl sulfate-polyacrylamide gel electrophoresis (SDS-PAGE) loading dye buffer (4×): 240 mM Tris–HCl (pH 6.8), 3 mM bromophenol blue, 50% glycerol, 400 mM dithiothreitol, 2% SDS.
13. Reagents and equipment necessary for SDS-PAGE.
14. Fluorescent imager.

2.2 Cellular Permeability of Chloroalkane Derivatized Small Molecule and Binding Efficiency to HaloTag in Intact Cells

1. Use or prepare reagent solutions and equipment as described in Subheading 2.1, **items 2–6, 10** and **12–14**.
2. Reagents and equipment necessary for tissue culture including 24-well plates.
3. Chloroalkane derivatized small molecule solution (200 μM). Prepare 500 μL solution by adding 10 μL of 10 mM chloroalkane derivatized small molecule stock solution to 490 μL DPBS.
4. HaloTag TMR (tetramethylrhodamine) ligand solution (50 μM): Prepare 1 mL solution by adding 10 μL HaloTag TMR ligand (Promega) to 990 μL DPBS.
5. Lysis buffer **2**: Prepare 1 mL lysis buffer **2** by mixing 860 μL Mammalian Lysis buffer (Promega), 20 μL RQ1 RNase-free DNAse (Promega), 100 μL 10× RQ1 RNase-free DNase buffer (Promega), and 20 μL Protease Inhibitor Cocktail (Promega) (*see* **Note 2**).

2.3 Pulldown for the Isolation of Cellular Targets Using the Chloroalkane Derivatized Small Molecules

1. Reagents and equipment necessary for tissue culture including 150 mm dishes.
2. Relevant culture cells and appropriate growth medium.
3. Unmodified small molecule stock solution: 10 mM in DMSO (50 mM in DMSO is needed for the competitive elution option *see* **Note 3**).
4. Chloroalkane derivatized small molecule solution (200 μM/growth medium): Prepare 9 mL solution by mixing 180 μL of 10 mM chloroalkane derivatized small molecule stock solution with 9 mL growth medium.
5. *Unmodified* small molecule solution (200 μM/growth medium): Prepare 9 mL solution by mixing 180 μL of 10 mM unmodified small molecule stock solution with 9 mL growth medium (*see* **Note 4**).
6. DMSO/growth medium solution: Prepare 9 mL solution by mixing 180 μL DMSO with 9 mL growth medium (*see* **Note 4**). (Optional)
7. HaloTag Coated Magnetic Beads (Promega).
8. IGEPAL-CA 630 (Sigma-Aldrich).
9. IGEPAL-CA 630 solution (10%): Prepare 5 mL solution by carefully pipetting 500 μL IGEPAL-CA 630 (viscous reagent) in 4.5 mL water, and mixing well by vortexing or end-to-end mixing. This solution must be prepared fresh (*see* **Note 5**).
10. Pull-down buffer **1**: 50 mM MOPS pH 7.5, 150 mM NaCl, 0.01% IGEPAL CA-630 final concentration (1000 fold dilution of the 10% IGEPAL-CA 630 solution). Prepare 150 mL of this reagent.

11. Pull-down buffer **2**: This formulation is only needed when performing chemical cleavage elution. Prepare 50 mL of 50 mM MOPS pH 7.5, 150 mM NaCl (no IGEPAL-CA 630) and degas under Argon or a similar degassing method for at least 30 min. After degassing is complete, add IGEPAL CA-630 to a final concentration of 0.01% (1000 fold dilution of the 10% IGEPAL-CA 630 solution).
12. Heater Shaker Magnet (HSM, Promega) or equivalent shaker for mixing 50 mL conical tubes and an appropriate magnet (*see* **Note 6**).
13. Chloroalkane derivatized small molecule solution (20 μM): Prepare 15 mL solution by adding 30 μL of 10 mM chloroalkane derivatized small molecule stock solution to 15 mL DPBS. This solution will be used with the *experimental samples* only.
14. Lysis buffer **3**: Prepare 10 mL lysis buffer **3** solution by mixing 8.6 mL Mammalian Lysis buffer (Promega), 200 μL RQ1 RNase-free DNAse (Promega), 1 mL 10× RQ1 RNase-free DNase buffer (Promega), 200 μL Protease Inhibitor Cocktail (Promega, *see* **Note 2**), and 1 μL of 10 mM chloroalkane derivatized small molecule stock solution (1 μM final). This solution will be used with the *experimental samples* only.
15. Lysis buffer **4**: Prepare 10 mL of lysis buffer **4** solution by mixing 8.6 mL Mammalian Lysis buffer (Promega), 200 μL RQ1 RNase-free DNAse (Promega), 1 mL 10× RQ1 RNase-free DNase buffer (Promega), and 200 μL Protease Inhibitor Cocktail (Promega, *see* **Note 2**). This solution will be used with the *control samples* only.
16. Prepare one of the following elution solutions to release captured protein targets. Selective elution methods include competitive elution with excess unmodified small molecule; or by chemical cleavage elution with a palladium catalyst. Targets may also be eluted nonselectively with SDS.
    (a) Competitive elution solution: 400 μM unmodified small molecule in pull-down buffer **1** (*see* **Note 3**).
    (b) Chemical cleavage elution with palladium catalyst solution: allow one vial of Palladium Cleavage Reagent (Promega) to reach room temperature and resuspend in 1 mL pull-down buffer **2** that has been degassed and supplemented with 0.01% IGEPAL-CA 630.
    (c) SDS elution solution: 1% SDS in 50 mM Tris–HCl (pH 7.5). Prepare 1 mL solution (*see* **Note 7**).
17. Magnetic stand for 1.5 mL micro-centrifuge tubes.
18. Rotator or shaker for 15 mL and Eppendorf tubes.

3 Methods

3.1 Validation of the Chloroalkane Derivatized Small Molecule for Bioactivity

Following derivatization of the small molecule with the chloroalkane capture tag (Promega) as recommended by the manufacturer (*see* **Note 8**), the bioactivity of the derivatized small molecule should be verified by recapitulating the phenotype of the unmodified small molecule (*see* **Note 9**).

3.2 Binding Efficiency of Chloroalkane Derivatized Small Molecule to HaloTag Protein

This protocol will evaluate the capability of the chloroalkane derivatized small molecule to bind the HaloTag protein in solution within the time frame the derivatized small molecule and its bound targets are typically captured onto the HaloTag Coated Magnetic Beads. The chloroalkane derivatized small molecule binds covalently to the HaloTag protein. To evaluate the binding kinetics of the chloroalkane derivatized small molecule to the HaloTag protein in solution, a fluorescent HaloTag ligand can be added to the binding reaction at different time points to label and detect any HaloTag protein not previously bound to the chloroalkane derivatized small molecule.

3.2.1 Cell Plating and Transfection

1. Plate 10 mL of HeLa cells in a 10 cm dish at a cell density of 1×10^5/mL.
2. Incubate at 37 °C, 5% CO_2 for 18–24 h.
3. Next day, transfect the cells with HaloTag Control Vector using the preferred transfection reagent as recommended by the manufacturer.
4. Incubate the plate at 37 °C, 5% CO_2 for 18–24 h.
5. Aspirate the medium and wash the cells with 5 mL DPBS.
6. Aspirate the DPBS, add 5 mL cell striper, and incubate at 37 °C, 5% CO_2 for 5–10 min (*see* **Note 10**).
7. Collect the cells and harvest by spinning for 5 min at $200 \times g$.
8. Aspirate the cell striper solution, resuspend the cells with 2 mL DPBS, and aliquot in two vials, 1 mL each, and harvest again by spinning for 5 min at $200 \times g$.
9. Aspirate the DPBS and freeze the cell pellets at −80 °C for at least 15 min to aid in passive lysis.

3.2.2 Binding Analysis

1. Thaw one of the cells pellets and resuspend in 400 μL lysis buffer 1.
2. Incubate for 15 min at room temperature with constant mixing on a plate shaker or rotating mixer.
3. Dilute 1:2 with 800 μL DPBS for a final volume of 1.2 mL diluted cell lysate.
4. Remove 60 μL of the lysate and label this as Time 0.

5. Add 20 μL of the 4 μM HaloTag TMR ligand solution to the Time 0 aliquot for a final concentration of 1 μM and incubate for 15 min at room temperature. This represents 100% available HaloTag protein not treated with the chloroalkane derivatized small molecule and will be used as reference in the calculations.
6. To the remainder of the lysate, 1.14 mL, add 1.14 μL of the 1 mM chloroalkane derivatized small molecule solution for a final concentration of 1 μM. Incubate at room temperature with constant mixing on a plate shaker or rotating mixer.
7. Sequentially, after 5, 10, 15, 30, 45, and 60 min incubation time, remove 60 μL lysate and immediately add to it 20 μL of the 4 μM HaloTag TMR ligand solution, incubating each at room temperature for 15 min. The HaloTag TMR ligand will bind to the HaloTag which was not already bound to the chloroalkane derivatized small molecule.
8. After labeling of each time point fraction, add 26 μL of 4× SDS loading dye, denature samples at 70 °C for 5 min or at 95 °C for 2 min, and analyze 10–15 μL of each fraction by SDS-PAGE.
9. Scan the gel on a fluorescence imager and quantitate each band intensity (Fig. 2a).
10. To calculate the % binding efficiency for each time point, first calculate the % of HaloTag protein labeled with the HaloTag TMR ligand relative to Time 0 (% T = 0 on Fig. 2a). This amount represents the HaloTag protein not previously labeled by the chloroalkane derivatized small molecule. Then subtract those values from 100% to infer the % bound to the chloroalkane derivatized small molecule (% Bound on Fig. 2a).
11. Plot the % binding values vs. time in min to determine the binding kinetics of the chloroalkane derivatized small molecule to the HaloTag (Fig. 2b). Ideally, at least 50%–60% of the HaloTag protein should be bound within the 15 min time point (*see* **Note 11**).

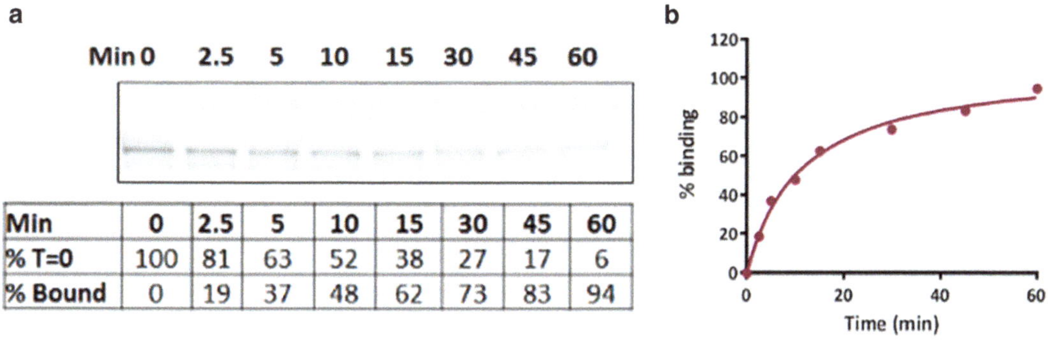

Min	0	2.5	5	10	15	30	45	60
% T=0	100	81	63	52	38	27	17	6
% Bound	0	19	37	48	62	73	83	94

Fig. 2 Binding kinetics of dasatinib-chloroalkane to the HaloTag in the cell lysate. (**a**) SDS-PAGE analysis of binding to the HaloTag. (**b**) % Binding with time. Results indicates sufficient binding to the HaloTag within the 15 min capture time frame

3.3 Cellular Permeability of Chloroalkane Derivatized Small Molecule and Binding Efficiency to HaloTag Protein in Intact Cells

This protocol will evaluate the cellular permeability of the chloroalkane derivatized small molecule within the time frame the small molecule is typically allowed to engage with the cellular targets. HaloTag protein expressed within intact cells is used to determine the chloroalkane derivatized small molecule permeability. Upon entering the cell, the chloroalkane derivatized small molecule binds covalently to the HaloTag protein. At a series of time points, a permeable fluorescent HaloTag ligand is added to label and detect any HaloTag protein not previously bound to the chloroalkane derivatized small molecule.

3.3.1 Cell Plating and Transfection

1. Seed 10 wells of a 24-well plate with 0.45 mL HeLa cells/well at a cell density of 1.1×10^5/mL.
2. Incubate at 37 °C, 5% CO_2 for 18–24 h.
3. Next day, transfect the cells with the HaloTag Control Vector using the preferred transfection reagent as recommended by the manufacturer.
4. Incubate the plate at 37 °C, 5% CO_2 for 18–24 h.

3.3.2 Binding Analysis

1. Label each well on the plate with the corresponding time points as shown in Table 1.
2. Sequentially add 50 μL of the 200 μM chloroalkane derivatized small molecule solution to each well starting with the longest

Table 1
Time points for the addition of the chloroalkane derivatized small molecule

Time Point	Time in minutes	Time interval for addition of chloroalkane derivatized small molecule from previous time point
1	180	
2	150	30 min after T = 1
3	120	30 min after T = 2
4	90	30 min after T = 3
5	60	30 min after T = 4
6	45	15 min after T = 5
7	30	15 min after T = 6
8	15	15 min after T = 7
9	5	10 min after T = 8
10	0	No chloroalkane derivatized small molecule addition

time point for the course of 180 min. Return the plate to 37 °C, 5% CO_2.

3. Follow each remaining time point, adding 50 μL of the 200 μM chloroalkane derivatized small molecule solution to all wells except for the 0 min well (no binding of derivatized small molecule). Return the plate to 37 °C, 5% CO_2 in between time point additions.
4. Once all time points are completed, add 60 μL of the 50 μM HaloTag TMR ligand solution to each well (5 μM final concentration) and incubate the plate for 15 min at 37 °C, 5% CO_2. The HaloTag TMR ligand will bind to any HaloTag protein which was not already bound to the chloroalkane derivatized small molecule.
5. Being careful to not disturb the cells (*see* **Note 1**), aspirate the medium from each well, wash with 200 μL PBS, and then add 90 μL lysis buffer **2** to each well.
6. Incubate the plate on a plate shaker for 30 min.
7. To each well add 30 μL of 4× SDS loading buffer, mix and transfer samples to individual tubes. Heat denature samples at 70 °C for 5 min or at 95 °C for 2 min.
8. Analyze 10–15 μL of each time point by SDS-PAGE.
9. Scan the gel on a fluorescence imager and quantitate the band intensities (Fig. 3a).
10. To calculate the % binding efficiency for each time point, first calculate the % of HaloTag protein labeled with the HaloTag TMR ligand relative to Time 0 (% T = 0 on Fig. 3a). This amount represents the HaloTag protein not previously labeled by the chloroalkane derivatized small molecule. Then subtract those values from 100% to infer the % bound to the chloroalkane derivatized small molecule (% Bound on Fig. 3a).

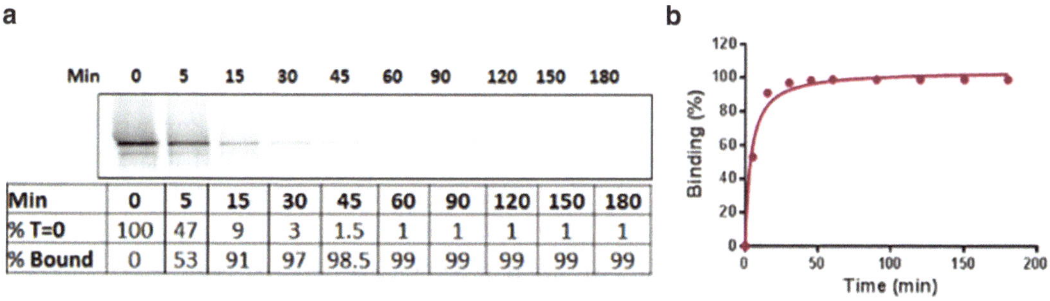

Min	0	5	15	30	45	60	90	120	150	180
% T=0	100	47	9	3	1.5	1	1	1	1	1
% Bound	0	53	91	97	98.5	99	99	99	99	99

Fig. 3 Binding kinetics of dasatinib-chloroalkane to the HaloTag inside the cells. (**a**) SDS-PAGE analysis of binding to the HaloTag. (**b**) % Binding with time. Results indicates sufficient cellular permeability within the target engagement time frame

11. Plot the % binding values vs. time in min to determine the binding kinetics of the chloroalkane derivatized small molecule to the HaloTag (Fig. 3b) inside the cells. In most cases, nearly 100% of the HaloTag protein is usually bound within the 150 min (2.5 h) time point which is the incubation time recommended for cellular treatment with the derivatized small molecule (*see* **Note 12**).

3.4 Pulldown for the Isolation of Cellular Targets Using the Chloroalkane Derivatized Small Molecules

This protocol is for the enrichment of cellular targets for a bioactive small molecule derivatized with the chloroalkane capture tag, which was already validated for bioactivity, cellular permeability, and binding efficiency to the HaloTag. Captured protein targets released by either selective or nonselective elution can be subjected to mass spectrometry analysis for identification. Selective elution methods include competitive release with excess unmodified small molecule or chemical cleavage of the tag by treatment with an optimized palladium catalyst. Nonselective elution includes treatment with 1% SDS.

3.4.1 Cell Plating and Treatment with Chloroalkane Derivatized Small Molecule

When attempting to identify new targets by mass spectrometry it is recommended to perform these enrichment experiments in triplicate, i.e., three experimental and three control samples, *see* **Note 13**. For the first step in this protocol please follow the seeding conditions for adherent cells (Subheading 3.4.1, **step 1**) or for suspension cells (Subheading 3.4.1, **step 2**). The remaining steps in the protocol are identical.

1. For adherent cells seed six 150 mm dishes with 27 mL of appropriate cells at 0.8–1 $\times$ 10^6 cells/mL (total of 2–3 $\times$ 10^7 cells /dish) and incubate for 12–16 h at 37 °C, 5% CO_2 to allow cell adhesion.
2. For suspension cells seed six 150 mm dishes with a total of 3–5 $\times$ 10^7 cells in 27 mL of medium; there is no need to incubate further before proceeding with the experiment.
3. Label 3 of the dishes experimental samples and 3 of the dishes control samples.
4. To each *experimental dish* add 3 mL of 200 μM/growth medium chloroalkane derivatized small molecule solution (20 μM final concentration).
5. To each *control dish* add 3 mL of 200 μM/growth medium unmodified small molecule solution (20 μM final concentration) or 3 mL DMSO/growth medium solution if using as negative control (*see* **Note 4**).
6. Incubate for 2.5 h at 37 °C, 5% CO_2 (*see* **Note 12**).

3.4.2 Equilibration of HaloTag Coated Magnetic Beads During Cell Treatment

This protocol is for use with a Heater Shaker Magnet (HSM) instrument. If not using the HSM, you will need a shaker for mixing 50 mL conical tubes and an appropriate magnet (*see* **Note 6**).

1. Allow HaloTag Coated Magnetic Beads to equilibrate to room temperature for at least 15 min.
2. Resuspend the beads thoroughly before dispensing for 5 min at 750 RPM (*see* **Note 14**).
3. Dispense 75 μL settled beads (i.e., 300 μL of 25% beads slurry) into each of 6 × 50 mL conical tubes, three labeled *experimental* and three *control*.
4. Add to each tube 2 mL of pull-down buffer **1**.
5. Place on the HSM and following the instrument prompts wash with the pull-down buffer **1** with constant mixing for 5 min (*see* **Note 14**).
6. Magnetize, aspirate, and remove the buffer, and resuspend again in 2 mL pull-down buffer **1**.
7. Repeat the previous **steps 5** and **6** for a total of five washes of the beads with pull-down buffer **1**.
8. After the last wash, magnetize, aspirate the buffer, and resuspend the beads in 4.8 mL pull-down buffer **1**.

3.4.3 Cellular Lysis

To minimize the dissociation of the targets from the chloroalkane derivatized small molecule during the lysis and capture steps, *the experimental samples only* are supplemented with additional chloroalkane derivatized small molecule during processing (*see* **Note 15**). For the first step in this protocol please follow the harvesting conditions for adherent cells (Subheading 3.4.3, **step 1**) or for suspension cells (Subheading 3.4.3, **step 2**). The remaining steps in the protocol are identical.

1. For adherent cells gently aspirate the medium from each dish. To each *experimental dish* add 5 mL 20 μM chloroalkane derivatized small molecule DPBS solution; and to each *control dish* add 5 mL DPBS without additives. Collect cells by scraping with a disposable cell lifter, transfer to 15 mL conical tubes and harvest by centrifugation at 200 × *g* for 5 min.
2. For suspension cells first transfer the content of each dish to a 50 mL conical tube and harvest by spinning at 200 × *g* for 5 min. Aspirate the medium from each cell pellet. To each *experimental cell pellet* add 5 mL 20 μM chloroalkane derivatized small molecule DPBS solution; and to each *control cell pellet* add 5 mL DPBS without additives. Harvest cells again by centrifugation at 200 × *g* for 5 min.

3. Aspirate the DPBS from harvested cells.
4. Resuspend each *experimental cell pellet* in 2.4 mL of lysis buffer **3** (supplemented with 1 μM chloroalkane derivatized small molecule) and each *control cell pellet* with 2.4 mL of lysis buffer **4** (no additives).
5. Incubate for 10 min with constant mixing at 750 RPM.
6. Harvest lysates by centrifugation at 3600 × *g* for 1 min.

3.4.4 Capture, Wash, and Elution

1. Carefully, without touching the cell debris pellet, transfer the supernatant lysates to the 50 mL tubes containing the previously washed HaloTag Coated Magnetic Beads.
2. Bind with constant mixing for 15 min on the HSM (*see* **Notes 5** and **11**). During this binding time prepare the desired elution solution as described in Subheading 2.3.
3. After 15 min of binding, magnetize and remove the supernatant.
4. Add 2 mL pull-down buffer **1** and mix with shaking for 3 min on the HSM.
5. Repeats the wash steps with pull-down buffer **1** for a total of three washes.
6. After the final wash, remove the buffer and resuspend the beads in 1 mL of pull-down buffer **1**.
7. Transfer the beads to a 1.5 mL micro-centrifuge tube, magnetize on a 1.5 mL magnetic stand, remove the buffer, and resuspend the beads in 150 μL of one of the elution solutions prepared: competitive elution; chemical cleavage palladium catalyst elution; or SDS elution.
8. Mix with shaking at 1000–1400 RPM for the following duration depending on elution method chosen: competitive elution for 60 min; chemical cleavage palladium catalyst elution for 30 min; or SDS elution for 30 min.
9. Magnetize and collect the supernatant into a 1.5 mL micro-centrifuge tube.
10. Repeat the magnetization of the supernatant and collect the supernatant into a new 1.5 mL micro-centrifuge tube to ensure complete removal of the beads.
11. Analyze the supernatant by mass spectrometry to identify the interacting cellular targets (Fig. 4). If a protein target is known or suspected, then Western blot analysis and confirmation can be performed (Fig. 5).

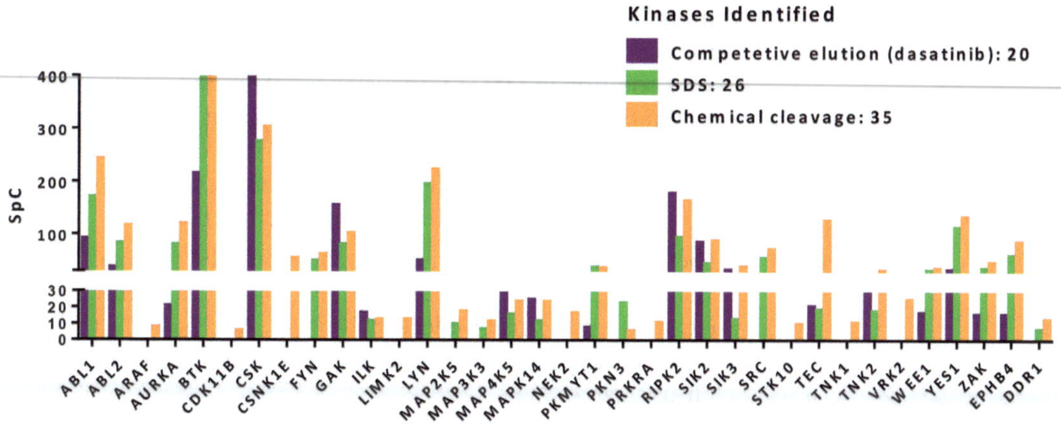

Fig. 4 Identification of dasatinib targets by LC-MS/MS analysis. Kinases enriched from K-562 cells by dasatinib-chloroalkane and eluted by either competitive elution (400 μM dasatinib), SDS, or chemical cleavage of the tag

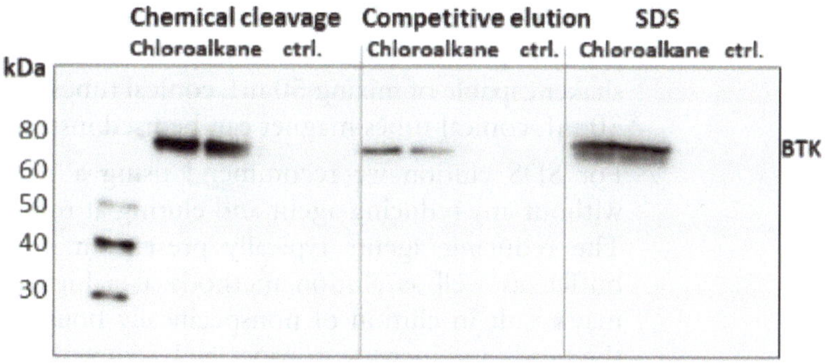

Fig. 5 Western analysis of a known dasatinib target. Bruton's Tyrosine Kinase (BTK) enriched from K-562 cells by dasatinib-chloroalkane and eluted with either competitive elution (400 μM dasatinib), SDS, or chemical cleavage of the tag

4 Notes

1. HeLa cells that do not detach easily from plastic are preferred for the binding efficiency and cell permeability assays. HEK293 cells could be substituted for HeLa cells, but as they detach easily from plates, extra attention should be paid during handling to prevent cell loss.
2. If not using the Protease Inhibitor Cocktail from Promega make sure the substituting reagent does not contain AEBSF which can reduce the binding efficiency of the chloroalkane to HaloTag.
3. For competitive elution with excess unmodified small molecule, we recommend using a concentration of 400 μM. To

minimize the amount of DMSO added to the elution buffer to no more than 1%, we recommend starting with a 40–50 mM stock solution in DMSO. In instances of limited small molecule solubility where a 40–50 mM stock solution is not available, the elution buffer can be made at a lower concentration (100–300 μM) to keep the amount of DMSO at 1% maximum. However, this may reduce the elution efficiency.

4. The most relevant negative control should be the unmodified small molecule, in case any specific biological pathways are activated upon small molecule addition. Alternatively, the control cells can be treated with an equivalent amount of DMSO or left untreated.
5. Use all the solutions of IGEPAL-CA 630 within a week. Because of its viscosity we recommend first preparing a 10% stock solution in water and using a final concentration of 0.01% in the pull-down buffers **1** and **2**.
6. The HSM instrument allows convenient programming and optimal mixing and magnetization in the various steps of the isolation protocol, but it is not required to perform the assay. A shaker capable of mixing 50 mL conical tubes at 750 RPM and 50 mL conical tubes magnet can be used instead.
7. For SDS elution we recommend using a 1% SDS solution without any reducing agent and eluting at room temperature. The reducing agents typically present in 4× SDS loading buffer, as well as elution methods at a higher temperature, may result in elution of nonspecifically bound proteins from the beads and increase nonspecific background.
8. Contact Promega in regard to materials and protocols to derivatize your small molecule with the chloroalkane moiety. In most cases, a reactive chloroalkane containing a nitrophenyl carbonate can be reacted with primary or secondary amines to generate a stable carbamate linkage.
9. In our experience most chloroalkane derivatized small molecules retain their phenotypic response with a shift in potency of no more than two to ten fold over the unmodified small molecule. If observing a greater reduction in potency, consider a different attachment point for the chloroalkane tag.
10. Cell striper reagent to lift adherent cells from the plate surface can be substituted with a trypsin solution following standard protocols for incubation times and neutralization.
11. For most cases we recommend a standard binding time of 15 min. This time should be sufficient to capture the chloroalkane derivatized small molecule with its bound targets on the HaloTag Coated Magnetic Beads while minimizing target dissociation. Do not reduce the binding time to less than 15 min.

If the binding kinetics of the derivatized small molecule to the HaloTag in lysate are deemed to be too slow (i.e., less than 50% at 15 min time point) the binding time for this step can be increased to up to 30 min. Keep in mind that some targets may dissociate from the small molecule with extended incubation time.

12. Although extremely rare, if the binding kinetics are slower inside intact cells (not close to 100% near the 150 min time point), repeat the experiment using longer time points. For example, drop the 5 and 15 min time points and add 210 min and 240 min time points to determine the optimal incubation time for cellular treatment. Use this new optimal incubation time for cellular treatment with the chloroalkane derivatized small molecule when performing the isolation of cellular targets protocol.
13. A minimum set of samples will include one experimental and one control sample. To increase statistical significance we strongly recommend performing the isolation protocol from at least duplicate or triplicate sets of samples.
14. The HaloTag Coated Magnetic Beads need to be completely resuspended prior to dispensing and during the binding and wash steps throughout the protocol. If the beads are allowed to settle, the binding and washing efficiency will be affected.
15. Ensure that the buffers and solutions supplemented with the chloroalkane derivatized small molecule are used only for the *experimental samples* during the lysis and binding steps. Set aside buffers without additives for the control samples.

References

1. Swinney DC, Anthony J (2011) How were new medicines discovered? Nat Rev Drug Discov 10:507–519
2. Moellering RE, Cravatt BF (2012) How chemoproteomics can enable drug discovery and development. Chem Biol 19:11–22
3. Cravatt BF, Wright AT, Kozarich JW (2008) Activity-based protein profiling: from enzyme chemistry to proteomic chemistry. Annu Rev Biochem 77:383–414
4. Fischer JJ, Michaelis S, Schrey AK et al (2010) Capture small molecule mass spectrometry sheds light on the molecular mechanisms of liver toxicity of two Parkinson drugs. Toxicol Sci 113:243
5. Bantscheff M, Eberhard D, Abraham Y et al (2007) Quantitative chemical proteomics reveals mechanisms of action of clinical ABL kinase inhibitors. Nat Biotechnol 25:1035–1044
6. Patterson SD, Aebersold RH (2003) Proteomics: the first decade and beyond. Nat Genet 33:R311–R323
7. Flory MR, Griffin RJ, Martin D, Aebersold R (2002) Advances in quantitative proteomics using stable isotope tags. Trends Biotechnol 20:S23–S29
8. Sato S, Murata A, Shirakawa T, Uesugi M (2010) Biochemical target isolation for novices: affinity-based strategies. Chem Biol 17:616–623
9. Salisbury CM, Cravatt BF (2008) Optimization of activity-based probes for proteomic profiling of histone deacetylase complexes. J Am Chem Soc 130:2184–2194
10. Su Y, Ge J, Zhu B et al (2013) Target identification of biologically active small molecules via in situ methods. Curr Opin Chem Biol 17:768–775

11. Salisbury CM, Cravatt BF (2007) Activity-based probes for proteomic profiling of histone deacetylase complexes. Proc Natl Acad Sci U S A 104:1171–1176
12. Ohana RF, Kirkland TA, Woodroofe CC et al (2015) Deciphering the cellular targets of bioactive small molecules using a chloroalkane capture tag. ACS Chem Biol 10:2316–2324
13. Ohana RF, Levin S, Wood MG et al (2016) Improved deconvolution of protein targets for bioactive small molecules using a palladium cleavable chloroalkane capture tag. ACS Chem Biol 11(9):2608–2617. Submitted
14. Los GV, Encell LP, McDougall MG et al (2008) HaloTag: a novel protein labeling technology for cell imaging and protein analysis. ACS Chem Biol 3:373–382
15. Urh M, Rosenberg M (2012) HaloTag, a platform technology for protein analysis. Curr Chem Genomics 6:72–78
16. Encell LP, Ohana FR, Zimmerman K et al (2012) Development of a dehalogenase-based protein fusion tag capable of rapid, selective and covalent attachment to customizable ligands. Curr Chem Genomics 6:55–71
17. Trost BM, Lee C (2000) Asymmetric allylic alkylation reaction in catalytic asymmetric synthesis, 2nd edn. Wiley-VCH, New York

Chapter 7

Microfluidics-Mass Spectrometry of Protein-Carbohydrate Interactions: Applications to the Development of Therapeutics and Biomarker Discovery

Alina D. Zamfir

Abstract

The functional interactions of carbohydrates and their protein receptors are the basis of biological events critical to the evolution of pathological states. Hence, for the past years, such interactions have become the focus of research for the development of therapeutics and discovery of novel glycan biomarkers based on their binding affinity. Due to the high sensitivity, throughput, reproducibility, and capability to ionize minor species in heterogeneous mixtures, microfluidics-mass spectrometry (MS) has recently emerged as a method of choice in protein-glycan interactomics. In this chapter, a straightforward microfluidics-based MS methodology for the assessment of protein-glycan interactions is presented. The general protocol encompasses: (1) submission of the interacting partners to a binding assay under conditions mimicking the in vivo environment; and (2) screening of the reaction products and their structural characterization by fully automated chip-nanoelectrospray (nanoESI) MS and multistage MS. The first section of the chapter is devoted to describing a method that enables the study of protein-oligosaccharide interactions by chip-nanoESI quadrupole time-of-flight (QTOF) MS and top-down complex analysis by collision-induced dissociation (CID). This section provides the protocol for the determination of the complex formed by standard β-lactoglobulin (BLG) with maltohexaose (Glc_6) and recommends as a concrete application the study of the interaction between BLG extracted from human milk with Glc_6, considered a ligand able to reduce the allergenicity of this protein. The second part is dedicated to presenting the protocols for the binding assay followed by chip-nanoESI ion trap (ITMS) and electron transfer dissociation (ETD) in combination with CID for protein-ganglioside interactions, using as an example the B subunit of cholera toxin (Ctb5) in interaction with comercially available GM1 species. The methodology described may be successfully applied to native ganglioside mixtures from human brain, in particular for discovery of biomarkers on the basis of their binding affinity.

Key words Chip-nanoelectrospray, Multistage mass spectrometry, Noncovalent interactions, Proteins, Glycans, Gangliosides, Top-down fragmentation

1 Introduction

Carbohydrates are ubiquitous biopolymers of high structural and functional complexity. They may occur in nature as free oligo- and polysaccharides or as glycoconjugates, in which the glycan is

Iulia M. Lazar et al. (eds.), *Proteomics for Drug Discovery: Methods and Protocols*, Methods in Molecular Biology, vol. 1647, DOI 10.1007/978-1-4939-7201-2_7, © Springer Science+Business Media LLC 2017

covalently linked to a protein or a lipid forming the known classes of glycoproteins, proteoglycans, and glycolipids. Carbohydrates are found mostly displayed on the surface of the cells participating in essential biological processes among which protein folding, stability, and trafficking [1–3]. The noncovalent interactions between carbohydrates and their protein receptors lie at the basis of the antigen recognition machinery and other biological functions [4, 5] such as signal transduction and cellular adhesion of bacteria and viruses. Protein-carbohydrate interactions mediate also the specific host-pathogen recognitions, which are of critical importance in the evolution of various diseases [4]. As a consequence, in the past decade, these interactions came into the focus of research for the development of therapeutics, and, in particular, for designing carbohydrate-based or protein-based inhibitors [6]. Knowledge of the structure of the interacting glycan-protein partners and of the interaction mechanism, which very often encompasses a cascade of complex processes, is a fundamental requirement for the development of such therapeutics.

On the other hand, particular carbohydrate structures were shown to represent valuable biomarkers of severe pathologies [7–9]. However, there is a large discrepancy between the need for natural glycoforms of extreme structural variability representative of biological information, the limited availability of samples from natural sources, and the ability to analyze and characterize such complex glycoform extract mixtures. These challenges resulted in substantial efforts dedicated to the development of sensitive and specific analytical methods for mapping the glycan mixtures, identifying biomarker or disease-associated species, and characterizing the detailed structure of a marker in a multicomponent sample. An efficient approach in this regard targets the discovery of novel structures based on their high binding affinity to specific protein receptors [10]. Hence, protein-glycan interactions may be efficiently exploited also for biomarker discovery in complex extracts.

Mass spectrometry (MS) with either electrospray ionization (ESI) or matrix-assisted laser desorption/ionization (MALDI) has developed over the past two decades into one of the most proficient methods, which greatly advanced the fields of genomics, proteomics, and glycomics [11–13]. The continuous technical improvements in the MS instrumentation in terms of resolution, mass accuracy, analysis throughput, and sensitivity have expanded the applicability of this technique from simple molecular mass determinations of individual compounds, to challenging de novo identifications, biomarker discovery, drug development, and structural analyses of heterogeneous mixtures, molecular aggregates, whole cells and tissues [14–16]. In the same context of performance augmentation, in the past years, modern achievements in chip technology and robotics were introduced also in MS [17, 18], in particular for replacing the classical ion sources with microfluidic devices [18–20]. Among these technologies, chip-based nanoESI with automatic sample infusion (NanoMate robot) exhibits multiple advantages, making this approach highly suitable for the characterization of protein-glycan complexes: a several-fold increase in sensitivity; capability to ionize minor components in heterogeneous mixtures; compatibility with a wide range of solvents; reported in-run and run-to-run reproducibility of the experiments between 98% and 100%; elevated signal-to-noise ratio; high-throughput; low flow rates; steady spray beneficial for fragmentation analysis using either collision-induced dissociations (CID), electron transfer dissociation (ETD) or their combination; and, potential to discover, on the basis of their binding affinity [10, 21], novel species, undetected before because of their reduced expression.

In this chapter, a straightforward MS-based methodology for the assessment of protein-glycan functional and structural interactions is described. The general protocol requires the conjunction of biochemical tools for purification and binding assays, with a modern analytical system based on fully automated chip-nanoESI (NanoMate robot) successively coupled to a high-resolution quadrupole time-of-flight (QTOF) MS and a high performance ion trap (IT) MS. The described methodologies are divided into two parts (Fig. 1): (1) a binding assay under conditions mimicking the in vivo environment; and (2) a detailed chip-nanoESI MS and multistage MS compositional and structural analysis protocol of the reaction products for confirming the noncovalent complex formation and its

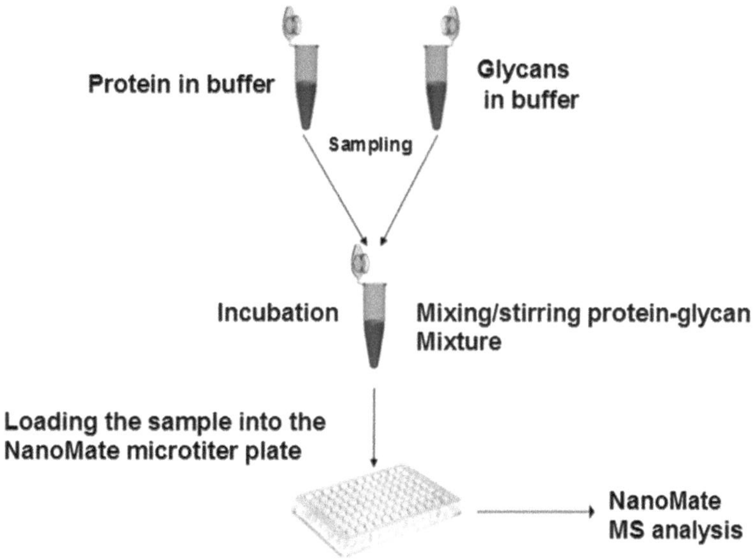

Fig. 1 Schematic of the workflow for protein-glycan noncovalent interaction assay and chip-nanoESI analysis of the reaction products (adapted from ref. 10 with permission from Taylor and Francis)

detailed characterization in terms of binding site(s) and structure of the bound glycan/glycoconjugate species. The chapter describes in detail the procedures to be followed for the assessment of protein-oligosaccharide interactions by chip-nanoESI QTOF MS and CID MS/MS, and of protein-ganglioside interactions by chip-nanoESI ITMS and ETD/CID MS^n. The standards chosen for exemplification are β-lactoglobulin (BLG) in interaction with maltohexaose (Glc_6) and the B subunit of cholera toxin (Ctb5) in interaction with GM1 ganglioside. Concrete applications, to be carried out following the described procedures, are also suggested. The first proposed application is related to the complex formed by BLG extracted from human milk with Glc_6, considered among the ligands able to reduce the allergenicity of this protein. The second recommended application is the interaction of Ctb5 with a native mixture of human brain gangliosides, which is able to offer data on novel species, of potential biomarker value, on the basis of their binding affinity.

2 Materials

2.1 Binding Assays

Prepare all solutions using HPLC or MS grade solvents and deionized water from a Milli-Q or other similar water purification systems. Follow the manufacturer's recommendations for storing reagents.

1. Ammonium acetate buffer: 10 mM, pH 6.0. Prepare 1 mL solution and adjust the pH with either 98% formic acid or acetic acid (*see* **Notes 1–3**). Filter the buffer solution using Millipack disposable filter units, 0.22 μm pore size (or other similar performance products).
2. Maltohexaose (Glc_6): ≥65% by HPLC (MW 990.86). The tested product was from Sigma-Aldrich, but other suppliers may be used.
3. Maltohexaose Glc_6 solution: 10 pmol/μL. Prepare the Glc_6 solution in filtered ammonium acetate buffer and vortex. Amount needed for analysis is 20 μL.
4. β-Lactoglobulin (BLG): ≥90% by PAGE. The tested product was from Sigma-Aldrich, but other suppliers may be used.
5. BLG from human milk: Extracted and purified as described in ref. [21].
6. BLG solution: 5 pmol/μL. Prepare the BLG protein solution in filtered ammonium acetate buffer and vortex. Amount needed for analysis is 50 μL.

7. Subunit B (Ctb5) of cholera toxin from *V. cholerae* (*see* **Note 4**). The tested product was from Sigma-Aldrich, but other suppliers may be used.
8. Ctb5 protein solution: 1 pmoL/μL. Prepare the ganglioside solution in filtered ammonium acetate buffer and vortex. Amount needed for analysis is 120 μL.
9. Monosialoganglioside GM1 fraction from bovine brain: ≥95%, lyophilized powder (*see* **Note 5**). The tested product was from Sigma-Aldrich, but other suppliers may be used.
10. Native ganglioside mixture from human brain: Extracted and purified as described in ref. 22.
11. Ganglioside solution: 10 pmoL/μL (*see* **Note 6**). Prepare the ganglioside solution in filtered ammonium acetate buffer and vortex. Amount needed for analysis is 120 μL.
12. SpeedVac concentrator for drying the samples coupled to a vacuum pump.
13. Thermomixer Eppendorf Comfort (or other similar devices).
14. pH meter.
15. Digital microbalance.
16. Micro DispoDialyzer system with 500 Da cutoff dialysis membrane (5–100 μL).
17. Dialysis membranes for protein purification, 10,000 Da cutoff.

2.2 Chip-Based nanoESI MS and MS/MS Analysis

1. NanoMate™ 400 or NanoMate™ TriVersa 400 robot from Advion BioSciences with commercial bracket from Advion BioSciences for NanoMate robot coupling to QTOF mass spectrometer.
2. Silicon nanoESI Chip™ with 400 nozzles from Advion BioSciences.
3. Microtiter sample plates with 96 wells, V-bottom, 200 μL, thermal cycler-compatible, polypropylene, or glass-coated from Beckman Coulter Inc. or Eppendorf AG.
4. Rack of 96 disposable conductive pipette tips from Advion BioSciences.
5. ChipSoft 8.1.0 software operating under Windows system from Advion BioSciences for controlling and manipulating the NanoMate robot.
6. Orthogonal hybrid quadrupole time-of-flight (QTOF) mass spectrometer (Micromass or Waters instruments) equipped with electrospray ion source in Z-spray geometry. Instrument characteristics: (1) resolution 5000 (full width at the half maximum, FWHM) for m/z 500; (2) average mass accuracy 20 ppm for m/z (150–4000); (3) scan rate from 2 to 10 scan/s; (4)

mass range: *m/z* 250–4500 for MS, *m/z* 150–4000 for CID MS/MS (*see* **Note 7**).

7. MassLynx software v. 4.1. from Waters/Micromass running on a PC under Windows to control the QTOF MS instrument, acquire the signal, and process the MS and MS/MS data.
8. Ion trap or Orbitrap mass spectrometer equipped with electrospray ionization source and with an ETD module using fluoranthene as anionic reagent and methane as the supporting gas. For example, high capacity ion trap ultra (HCT Ultra, PTM discovery) mass spectrometer (Bruker Daltonics) or amaZon speed ETD ion trap (Bruker Daltonics).
9. Compass™ Software (Bruker Daltonics) for HCT Ultra mass spectrometer.
10. Commercially available bracket from Advion BioSciences, for Bruker ion trap coupling with the NanoMate robot.
11. MS calibration solution: standard G2421A electrospray "tuning mix" from Agilent Technologies, 1:100 dilution in acetonitrile; or, alternatively, sodium iodide (2 μg/μL) or cesium iodide (50 ng/μL) in 2-propanol/water, 50:50 (v/v). Prepare fresh before analysis.
12. Nitrogen for QTOF mass spectrometer and NanoMate robot, purity $\geq$99.999%.
13. Argon for CID MS/MS experiments, purity $\geq$99.9995%.
14. Helium for ion trap mass spectrometer, purity $\geq$99.999% vol.
15. Methane as the supporting gas for anion production in ETD experiments, purity $\geq$99.995% vol.

3 Methods

3.1 Protein-Glycan Interactions. Application: β-Lactoglobulin in Interaction with Maltooligosaccharides

β-Lactoglobulin is a lipocalin protein highly expressed in cow's milk as the major whey proteic component. This protein is known as one of the most aggressive allergens for newborns, triggering a complex symptomatology and possibly life-threatening complications. BLG may incidentally occur in human milk by cow's milk intake. As no prophylactic treatment is presently available, finding ligands able to reduce the BLG allergenicity is currently the only possible option to address this problem. Earlier work conducted in the field [23] has shown that some classes of oligosaccharides represent ligands with such properties. A 2016 report [21] on chip-nanoESI MS and MS/MS application, with significant biomedical impact, revealed that maltohexaose binds BLG under conditions mimicking the in vivo environment and therefore might be considered among the ligands able to reduce the allergenicity of this protein.

3.1.1 Binding Assay

1. Incubate 30 μL standard BLG solution with 6 μL Glc_6 solution in the Eppendorf thermomixer under constant stirring at 37 °C (*see* **Note 8**).
2. Collect 5 μL of the reaction products after 1, 5, 10, 15, 20, 25, and 30 min incubation time and submit the aliquots immediately after collection to MS analysis by loading the solution directly into the vial of the NanoMate robot (*see* **Notes 9** and **10**).
3. For the binding assay using BLG extracted from human milk, follow the same procedure as described above in **steps 1** and **2** for the standard BLG solution.

3.1.2 Chip-Based nanoESI MS Detection of the Noncovalent Complex

1. Remove the original ESI source of the QTOF mass spectrometer and couple the NanoMate robot via the Advion commercial bracket.
2. Turn on the gas supplies, the mass spectrometer, and the NanoMate robot.
3. Tune the QTOF MS instrument for operating in positive ESI mode and set the parameters in the Mass Lynx software 4.1 as follows: acquisition in MS mode; cone voltage: 40 V; nitrogen nebulizer: 60 psi; source block temperature: 80 °C; mass range: m/z 250–4500.
4. Tune the NanoMate robot in the positive ion mode by setting on the ChipSoft software the following values for the operating parameters: sample aspiration volume 5 μL; air aspiration volume: 2 μL (*see* **Note 11**); nanoESI voltage on the pipette tip: 1.8 kV; nitrogen back pressure: 0.40 psi.
5. Slide the 400 ESI Chip in the chip holder of the robot with the frame notches oriented down and the chip frame tag upward, whereas the serial number faces the robot.
6. Select in the ChipSoft the chip with the corresponding serial number.
7. Place the rack of the 96 disposable conductive pipette tips and the sample microtiter plate in their corresponding holders.
8. Load the collected aliquot of the reaction products into a well of the microtiter plate and select in the ChipSoft program the corresponding well position (*see* **Note 12**).
9. Set on the ChipSoft the option "return to the initial well the unused sample."
10. Select "deliver sample" on the ChipSoft software of the NanoMate robot and "Acquire" in the MS mode on the Mass Lynx software of the QTOF MS (*see* **Note 13**).
11. After initiation of the electrospray, optimize the signal intensity and stability by fine adjustment of the chip position with respect to the mass spectrometer inlet. Select the back-forward,

right-left options in the "spray optimization" of the ChipSoft software.

12. Acquire the signal and switch to View Chromatogram (Total Ion Chromatogram, TIC). Generate the spectrum by "combining in progress" the TIC scans. Accumulate scans until a fair signal-to-noise ratio is obtained. Recommended acquisition time is above 2 min.
13. Load the calibration solution in a new well of the microtiter plate and infuse it by chip-based nanoESI. Generate the calibration spectrum. Apply the calibration file to the analyte spectrum (*see* **Note 14**).
14. Subtract, smooth, and center the mass spectrum using these options in the "Process" portal of Mass Lynx. Deconvolute the protein-related signals and determine the protein molecular weight by applying Process/Component/Find Auto or Find Manual.
15. Keep the same parameters and apply an identical procedure for the measurement of all seven reaction product aliquots collected.
16. Apply the **steps 1–15** described above in this section for detecting the noncovalent complexes formed by BLG extracted from human milk and maltooligosaccharides (*see* Fig. 2).

3.1.3 Top-Down Analysis of the Formed Noncovalent Complex by CID MS/MS

1. Identify in the screening mass spectrum, the ions that, according to the calculated mass, correspond to the protein-glycan noncovalent complex (*see* **Notes 15** and **16**).
2. Leave the chip nanoESI settings unaltered, turn on the Ar and set the LM Res and HM Res parameters for precursor ion isolation at 10 and 10 respectively (*see* **Note 17**).
3. Select "Acquire," set Function MS/MS and introduce the *m/z* of the chosen precursor ion.
4. Select the mass range with "start mass" 150 and "end mass" 4000, "scan time" 10 min.
5. Start the signal acquisition and acquire the scans while increasing gradually the collision energy from 40 to 100 eV (*see* **Note 18**).
6. Accumulate scans until a fair signal-to-noise ratio is obtained and vary the collision energy within the (40–100) eV range to produce fragment ions confirming the complex formation and the structure of the bound oligosaccharide (*see* **Notes 18–20**).
7. Stop the acquisition, switch to View Chromatogram, generate the MS/MS by combining over the entire TIC, and further calibrate the CID tandem mass spectrum as described in Subheading 3.1.2.

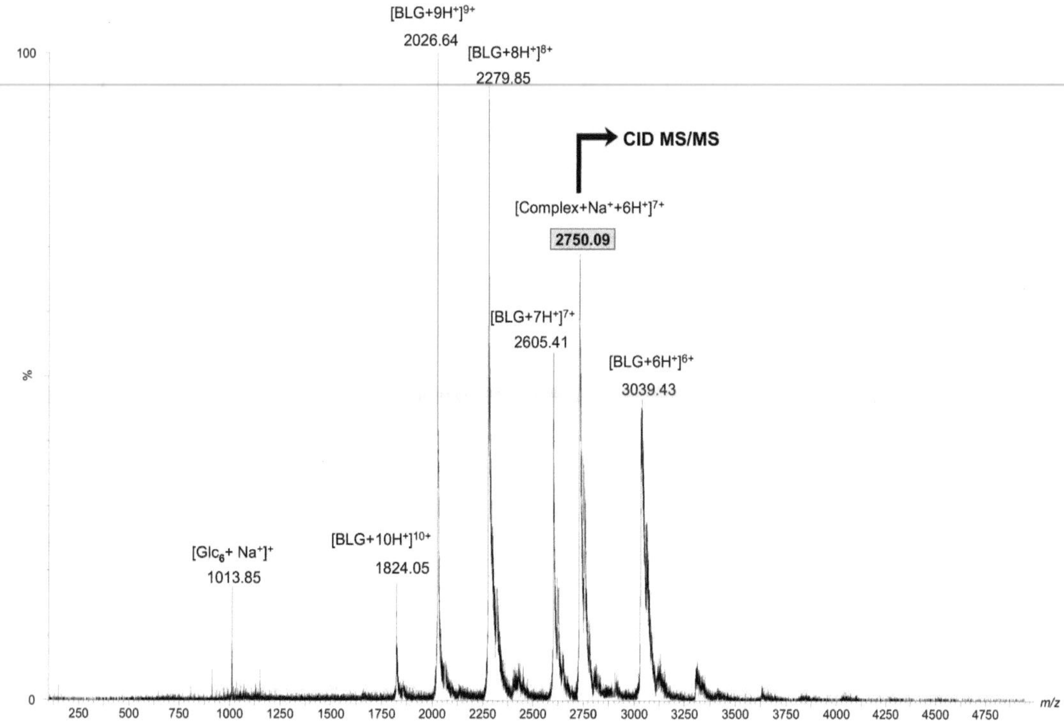

Fig. 2 (+) Chip-nanoESI QTOF MS of BLG-Glc_6 reaction products after 10 min incubation in buffer at 37 °C. Interaction partners: BLG-extracted from human milk upon deliberate intake of 300 mL cow's milk and standard Glc_6. Buffer: 10 mM ammonium acetate/formic acid pH 6.0. (+) Chip-based nanoESI QTOF MS conditions: nanoESI voltage 1.8 kV. Cone voltage 40 V. Nitrogen back pressure 0.40 psi. Nitrogen nebulizer pressure 60 psi. Acquisition time 2 min (reproduced from ref. 21 with permission from Springer)

8. Apply the same protocol for top-down analysis of the complexes formed by BLG extracted from human milk and maltooligosaccharides (*see* Fig. 3).

3.2 Protein-Ganglioside Interaction. Application: Cholera Toxin in Interaction with Human Brain Gangliosides

Gangliosides, sialylated glycosphingolipids highly expressed in the central nervous system (CNS), are valuable biochemical markers in early diagnosis of CNS pathologies [25–28], being in the focus of research as potential therapeutic targets [28–30] in particular for designing anticancer vaccines. While the hydrophobic ceramide (Cer) moiety anchors the molecule into the cell membrane, the glycan chain of gangliosides may interact freely with the soluble extracellular molecules and with the hydrophilic segments of other membrane components [31, 32]. Being enriched in specialized microdomains, gangliosides interact with signal transducers. They mediate carbohydrate-dependent cell adhesion, induce cell activation, motility and growth and participate in cell–cell and cell–matrix interactions. The interest in the study of ganglioside interactions with proteins was lately stimulated by findings indicating that the formed complexes might play a role in: (1) the molecular

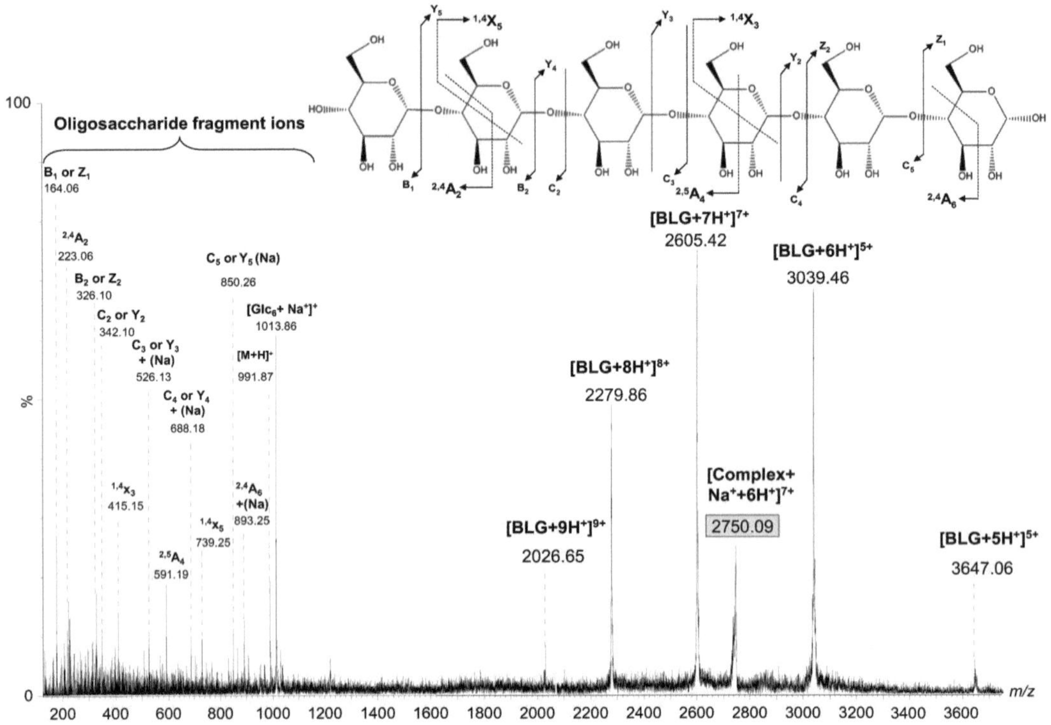

Fig. 3 Top-down by (+) chip-nanoESI QTOF CID MS/MS of the [Complex + Na + $6H^+]^{7+}$ detected at *m/z* 2750.09, corresponding to formed BLG-Glc_6 noncovalent complex. Variable collision energy within (40–100) eV. Acquisition time 5 min. Other MS conditions as in Fig. 1. Inset: ion fragmentation scheme of Glc_6 in top-down experiment. Glc_6 fragment ions are designated following the nomenclature introduced by Domon and Costello [24] (reproduced from ref. 21 with permission from Springer)

mechanisms of Alzheimer disease (AD), with an emphasis on the functions of amyloid β-protein-ganglioside complexes [33]; (2) the action of bacterial toxins, for which several gangliosides were found as specific receptors [23]; (3) the progression of malignant brain tumors [34]; and, (4) the discovery of novel species in complex mixtures based on their high binding affinity [10].

Since gangliosides are readily embedded on hydrophobic surfaces, their interactions with water-soluble proteins were studied by specific methods such as high performance thin layer chromatography, enzyme-linked immunosorbent assay, and surface plasmon resonance [25]. Flow cytometry [35] and carbohydrate microarray systems [36] were also implemented for the determination of ganglioside-cholera toxin complexes, which hitherto is the most studied interaction of gangliosides. The major drawbacks of these methods relate to reduced sensitivity and the type of environment in which the interactions are studied. To overcome these shortcomings, the lipid domain of gangliosides was used for integrating the molecule into liposomes [37], which allowed the interaction under conditions mimicking better the in vivo environment.

Nevertheless, the sensitivity issue and the impossibility to structurally characterize the complex still represent the limitations of all approaches. Due to its sensitivity, reproducibility, data accuracy, wealth of structural information and compatibility with aqueous solutions, ESI MS became a method of choice in ganglioside interactomics. A few valuable ESI MS approaches developed for glycolipid-protein interactions, and introduced to some extent also to ganglioside analysis, are available [38–41]. However, besides the significant gain in sensitivity and other benefits described, in the case of gangliosides, chip-nanoESI was shown to reduce the *in-source* fragmentation with particular loss of the labile Neu5Ac residues. Also, the steady spray provided by chip-nanoESI is beneficial to ETD/CID MS^n experiments on the protein-ganglioside complex since multiple fragmentations require stable signal of relatively high intensity.

3.2.1 Binding Assay

1. Incubate 100 μL Ctb5 solution with 100 μL standard GM1 ganglioside solution in the Eppendorf thermomixer under constant stirring at 37 °C (*see* **Note 21**).
2. Collect directly into the 96-well plate of the NanoMate 15–20 μL aliquots of the reaction products after 10, 30 min, and 60 min of incubation and proceed with the MS analysis (*see* **Notes 9** and **10**).
3. Proceed with the **steps 1** and **2** described above for the binding assay of Ctb5 and the native human brain ganglioside extract.

3.2.2 Chip-nanoESI MS Analysis

1. Dismantle the original ESI source of the ITMS and the connections to the nitrogen supply and couple the mass spectrometer and NanoMate robot via the Bruker commercial interface. Follow the HCT-NanoMate coupling principles (*see* ref. 22).
2. Connect the nitrogen supply to the NanoMate, turn on all gas supplies, the mass spectrometer and the NanoMate robot.
3. Tune the ion trap mass spectrometer for operating in the positive ion mode. Set on the Compass™ software the following parameters: "positive ion mode"; HCT capillary exit: 50 V; source block temperature: 80 °C; skimmer voltage: 50 V; nebulizer nitrogen pressure: 40 psi; detection in a range of (100–3000) *m/z*, scan speed: 8000 *m/z* per sec.
4. Tune the NanoMate robot in positive ion mode. Set on the ChipSoft program the following values of the electrospray: pipette tip voltage: 0.8 kV; nitrogen back pressure: 0.30 psi; sample aspiration volume: 10 μL; air aspiration volume: 2 μL (*see* **Notes 11** and **12**).
5. Proceed with **steps 5–9** described in Subheading 3.1.2.

6. Select "deliver sample" on the ChipSoft software of the NanoMate robot and "start acquisition" on the Compass software of the ion trap MS (*see* **Note 13**).
7. Proceed with **step 11** described in Subheading 3.1.2.
8. Acquire the spectrum until a fair *signal-to-noise* ratio is obtained. Recommended acquisition time is above 1 min.
9. After stopping the acquisition, switch to DataAnalysis portal of Compass™ software, import the TIC and generate the mass spectrum by combining it over all TIC scans.
10. Proceed with **step 13** described in Subheading 3.1.2.
11. Deconvolute the signals related to the protein and the protein-ganglioside complex and determine their MW and the standard deviation using the Compound Mass Spectra and Charge State Ruler options available in the Data Analysis program (*see* **Note 22**).
12. Keep the same parameters and apply an identical procedure for the measurement of the other aliquots collected.
13. Apply the **steps 1–12** described above in this section for detecting the noncovalent complexes formed by Ctb5 and native gangliosides extracted from human brain.

3.2.3 Top-Down Noncovalent Complex Analysis by ETD/CID MS^n

The ETD method is highly suitable for the characterization of noncovalent protein-ganglioside complexes in top-down experiments. By ETD, c and z type fragment ions are produced; N–C$_\alpha$ bonds within the protein are cleaved [42] with the preservation of other linkages [43, 44], among which, certainly, the noncovalent bonds. Currently, the most efficient and modern procedure for structural elucidation of noncovalent protein–carbohydrate complexes is based on multistage fragmentation analysis with alternate ETD and CID fragmentation stages. ETD provides information on the protein structure and site(s) of the noncovalent carbohydrate attachment, while CID generates data on the structure of the linked glycan. In the case of gangliosides, the ETD/CID methodology encompasses the following steps: (1) the noncovalent complex is first subjected to ETD MS^2 which will produce fragment ions corresponding to peptides and ganglioside-linked peptides; and (2) following ETD sequencing, the fragment ions related to the noncovalent peptide-ganglioside complex are isolated and further submitted to CID MS^n; CID analysis will offer valuable data on the structure of the bound ganglioside species in terms of carbohydrate sequence, sialylation, possible identification of the Neu5Ac positional isomers and on the ceramide composition (*see* Figs. 4a, b, and 5). In some instances CID is also able to discover or confirm the noncovalent attachment of the ganglioside species to a certain aminoacid of the peptide [10]. Hence, in a single run an exhaustive structural analysis of the molecules forming the noncovalent complex can be achieved.

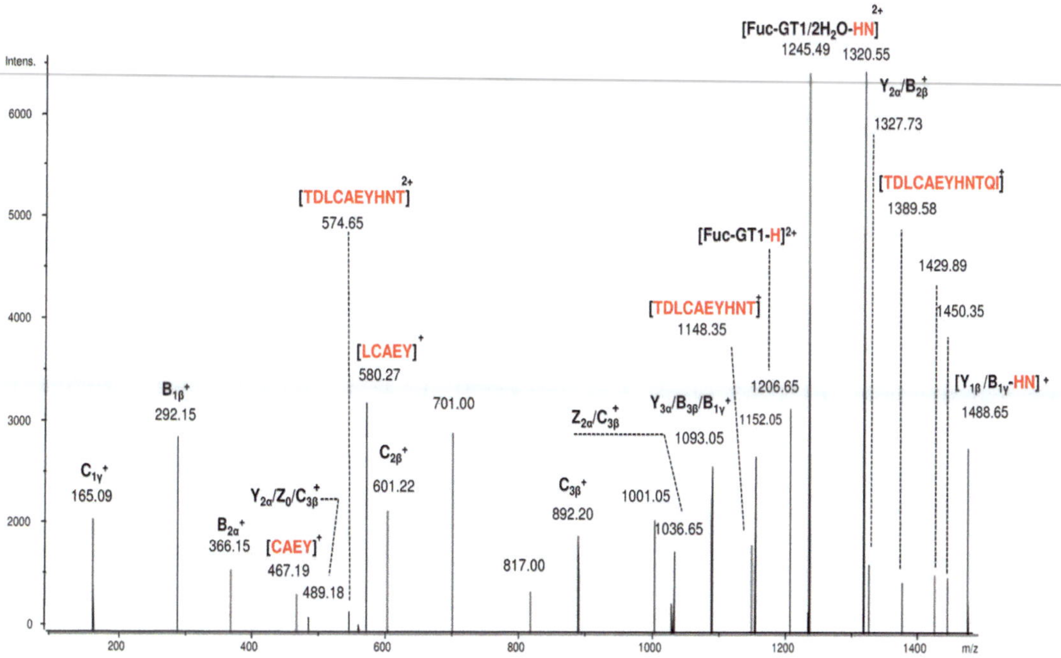

Fig. 4 ETD/CID MS^3 analysis of the $[M + 2H^+]^{2+}$ at m/z 1832.55 having the composition (Fuc-GT1) TDLCAEYHNTQI. (**a**) m/z range (100–1500); (**b**) m/z range (1500–3000). ETD: 75 eV ionization energy; 20 ms interaction time. CID: RF amplitude: 0.4–0.8 V. The signal at m/z 1206.65 corresponding to $[Fuc\text{-}GT1\text{-}His]^{2+}$ indicates the noncovalent attachment of the ganglioside to His. Assignment of the glycan fragment ions follows the nomenclature of Domon and Costello [24]. Assignment of the peptide fragment ions follows the nomenclature of Roepstorff and Fohlman [45] (reproduced from ref. 10 with permission from Taylor and Francis)

1. Identify in the screening mass spectrum, the ions that, according to the calculated mass, correspond to the protein-ganglioside noncovalent complex (*see* **Note 15**).
2. Leave the MS settings unaltered, switch to ETD, set the isolation window to 2u, introduce the m/z value of the precursor ion and restart the acquisition. ETD parameters: 600 °C reactant temperature; 2.5 μA emission current; – 4.5 V ionization chamber; 75 eV ionization energy; 20 ms interaction time between the fluoranthene anions and analyte cations (*see* **Note 23**).
3. Acquire the ETD MS^2 signal for at least 2 min.
4. For CID MS^3, identify in the ETD MS^2 the ion of interest for additional fragmentation (the GM1-linked peptide).
5. Do not stop the acquisition, leave unchanged the settings for MS and ETD MS^2, select CID MS^3, and introduce the m/z value of the fragment ion chosen for further fragmentation.
6. Acquire the CID MS^3 signal for at least 2 min while increasing gradually the fragmentation amplitude from 0.4 V to 0.8 V

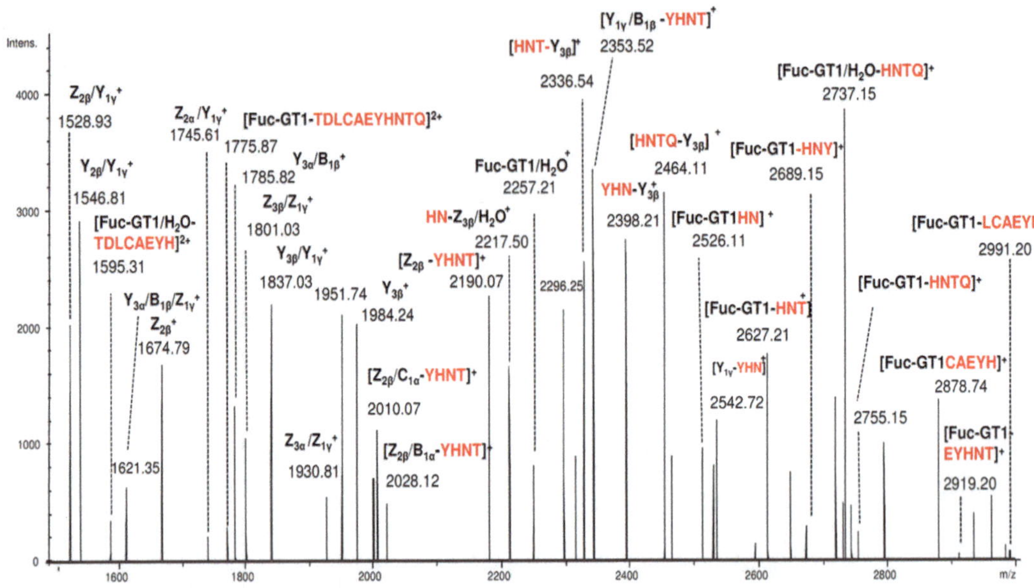

Fig. 4 (continued)

with a ramp from 30% to 200% within 40 ms per single spectrum and a fragmentation cutoff default of 27% of the precursor ion m/z (*see* **Note 24**).

7. For superior fragmentation stages, keep unaltered the settings and selections related to the MS screening and the previous fragmentation stages and proceed further as indicated above in **steps 1–3** for ETD and **steps 4–6** for CID (*see* **Note 25**).
8. Stop the acquisition, switch to DataAnalysis portal of Compass™ software, import the chromatogram, and generate the ETD and separately the CID mass spectra.
9. Calibrate the fragmentation mass spectra as described in Subheading 3.1.2.
10. Proceed with the **steps 1–9** described above in this section for the top-down analysis of the nocovalent complexes formed by Ctb5 and the native ganglioside mixture extracted from human brain (*see* Fig. 4a, b).

4 Notes

1. A higher ionic strength influences positively the interaction; however, it impedes the chip-nanoESI process. Therefore, 10 mM ammonium acetate concentration at pH 6.0 is a fair compromise between the complex formation and the solvent system requirements for chip-electrospray.

2. Perform this step under properly ventilated chemical fume hood to avoid formic or acetic acid inhalation.
3. Although of low toxicity, formic acid in 98% concentration is harmful for skin and eyes. Concentrated formic acid may irritate, even burn the skin and is dangerous for the eyes. Therefore, it is recommended to wear chemical splash goggles, laboratory coat and gloves (nitrile, or natural rubber). Avoid wearing shorts or open-toed shoes.
4. According to the specification of the producer, Ctb5 exhibits a purity above 95%, as determined by SDS-PAGE. However, prior to the interaction assay followed by chip-nanoESI, the protein should be desalted by overnight dialysis at 4 °C using a 10,000 Da cutoff membrane.
5. Prior to the interaction assay desalt the gangliosides by overnight dialysis at 4 °C against water using Micro DispoDialyzer systems with 500 Da cutoff dialysis membrane.
6. In the case of the native human brain ganglioside mixture an average molecular weight of 2200 g/mol should be considered for establishing the solution concentration.
7. Other high-resolution instruments with ESI and CID capabilities such as Orbitrap or FTMS may be used.
8. Other volumes of the solutions containing the interaction partners dissolved in buffer may be prepared and considered if the proper stoichiometry is chosen. The stoichiometry of 1:2.5 Glc_6:BLG was so far found to be the best suited for both complex formation and detection by chip-based nanoESI MS.
9. To reduce the interference in the reaction process during the collection procedure, do not remove the vial from the Thermomixer. Collect the aliquot with the pipette, close the vial and leave it for further incubation.
10. In some cases the noncovalent protein-glycan complex is formed rapidly so that after only 1 min incubation time the ions of the complex might be already visible in the mass spectrum. Nevertheless, as the reaction might be still ongoing, the aliquots of the reaction products should be further monitored at different time intervals.
11. In order to prevent sample dripping during manipulation, the robot should always be programmed to aspirate a few μL of air after sample aspiration.
12. As a general rule, prior to the binding assay, collection of the reaction products and their MS analysis, measure always the interaction partners separately in buffer for finding the conditions for their optimal ionization and detection and for assessing their individual spectra. The glycan/glycoconjugate

molecular ion should be also sequenced by CID. Knowing the spectrum of the protein in buffer and the induced charge states will allow a straightforward identification of those signals corresponding to the protein-glycan/glycoconjugate complex. On the other hand, prior knowledge of the glycan/glycoconjugate screening mass spectrum and the fragmentation pattern by CID will greatly facilitate the evaluation of the top-down experiment results.

13. If no ESI signal is detected, or the ESI process is unstable or stops immediately, proceed with the following steps: (1) at first, change the chip position with respect to the MS entrance orifice by fine adjustment "forward/back," "right/left" in the ChipSoft; (2) increase slightly the nitrogen back pressure; (3) alter slightly the ESI voltage.
14. On a QTOF mass spectrometer, under appropriate calibration conditions, a mass accuracy of at least 30 ppm can be achieved for BLG or proteins of similar MW.
15. The most efficient fragmentation by either CID or ETD will be obtained for the precursor ion exhibiting the highest charge state. On the flip side, usually, the highly charged ions have a reduced intensity signal, which results in a low yield of fragment ions. For this reason, in selecting the precursor ion, the best compromise between the charge state and the signal intensity is to be considered.
16. In the case of oligosaccharides, the ions may also be formed as adducts with alkali metals; hence the formed noncovalent complex could be detected as an alkali adduct ion.
17. HM Res and LM Res define the resolving DC on the quadrupoles and are set to give constant resolution across the mass range. Though a broader isolation window (lower HM and LM values) appears initially beneficial to the total ion current intensity, it provides a diminished selectivity that results in simultaneous isolation of other ions (exhibiting *m/z* values close to those of the chosen precursor).
18. Values of the collision energy lower than 40 eV (E_{lab}) are insufficient for any fragmentation, while values above 100 eV (E_{lab}) induce multiple ring cleavages of the oligosaccharide, which complicates the spectrum and makes difficult its interpretation.
19. Under properly optimized fragmentation conditions, a top-down CID MS/MS experiment, carried out using as precursor the ion corresponding to the protein-glycan complex, yields two events (*see* Fig. 5): (1) breakdown of the protein–glycan complex; this event, crucial for the confirmation of the complex formation, is observed through the occurrence in the mass spectrum of the signals corresponding to the *m/z* of the

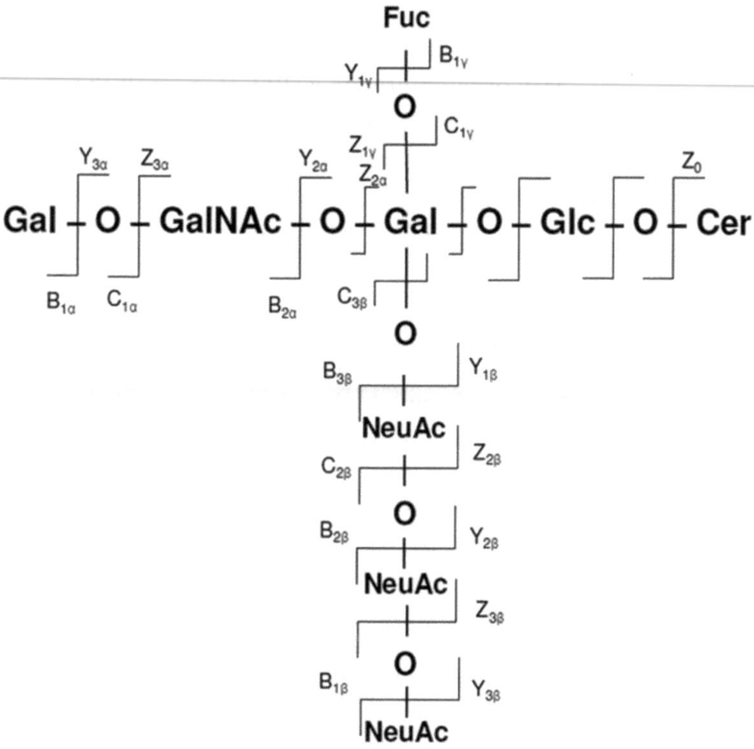

Fig. 5 CID fragmentation pathway and the structure of the bound Fuc-GT1 species. $C_{3\beta}$ corresponding to $(NeuAc)_3$, $Z_{2\alpha}/Y_{1\gamma}$ having the composition $(Neu5Ac)_3$GalGlcCer, and $Z_{3\alpha}/Z_{1\gamma}$ corresponding to GalNAc $(Neu5Ac)_3$GalGlcCer are diagnostic for GT1c isomer. Assignment of the glycan fragment ions follows the nomenclature of Costello et al. [46] (reproduced from ref. 10 with permission from Taylor and Francis)

protein molecular ion and the *m/z* of the bound glycan molecular ion; and (2) the sequencing of the released glycan ligand, indicated by the detection of glycan fragment ions in the low *m/z* region.

20. If instruments with ETD or ECD capabilities such as Orbitrap, FTMS, or advanced ion traps are available, a more structurally informative top-down analysis can be performed by the ETD/CID protocol described in Subheading 3.2.3.

21. Other solution volumes containing the interaction partners dissolved in buffer may be prepared, with the preservation of the stoichiometry. With the exception of GM3 [47], gangliosides form micelles in aqueous solutions. Therefore, higher concentrations are not recommended.

22. On a HCT mass spectrometer, under appropriate calibration conditions, a mass accuracy of at least 100 ppm can be achieved for Ctb5 or proteins of similar MW (58 kDa).

23. Prior to ETD fragmentation, always check the production of the fluoranthene anions. For this purpose generate the

spectrum of the fluoranthene by selecting "reactant only" option. Check the presence of the corresponding ion. The MW of fluoranthene is 202.26 g/mol; the signal of the fluoranthene radical anion is at *m/z* 202.26.

24. RF amplitudes below 0.40 V will result in an excitation energy insufficient for fragmentation. RF amplitudes above 0.80 V will induce extensive desialylation of the ganglioside glycan core which will impede the identification of the ganglioside isomer bound to the protein.
25. In MS^n experiments the intensity of the signal corresponding to the fragment ions decreases after each sequencing event. No further fragmentation is recommended after the intensity dropped below 100 counts/s.

Acknowledgments

Dedicated to the inspiring Diana Soviero, a life of outstanding artistry and teaching.
This work was supported by the Romanian National Authority for Scientific Research through the projects PN-II-PT-PCCA-2011-142 and PN-II-PT-PCCA-2013-0191.

References

1. Xu C, Ng DT (2015) Glycosylation-directed quality control of protein folding. Nat Rev Mol Cell Biol 16:742–752
2. Swiech K, de Freitas MC, Covas DT, Picanço-Castro V (2015) Recombinant glycoprotein production in human cell lines. Methods Mol Biol 1258:223–240
3. Zurzolo C, Simons K (2016) Glycosylphosphatidylinositol-anchored proteins: membrane organization and transport. Biochim Biophys Acta 1858:632–639
4. Cohen M (2015) Notable aspects of glycan-protein interactions. Biomol Ther 5:2056–2072
5. Cederkvist FH, Zamfir AD, Bahrke S et al (2006) Ication of a high-affinity-binding oligosaccharide by (+) nanoelectrospray quadrupole time-of-flight tandem mass spectrometry of a noncovalent enzyme-ligand complex. Agew Chem Int Ed Engl 45:2429–2434
6. Holgersson J, Gustafsson A, Breimer ME (2005) Characteristics of protein–carbohydrate interactions as a basis for developing novel carbohydrate-based antirejection therapies. Immunol Cell Biol 83:694–708
7. Krishn SR, Kaur S, Smith LM et al (2016) Mucins and associated glycan signatures in colon adenoma-carcinoma sequence: prospective pathological implication(s) for early diagnosis of colon cancer. Cancer Lett 374:304–314
8. Li X, Wang X, Tan Z et al (2016) Role of glycans in cancer cells undergoing epithelial-mesenchymal transition. Front Oncol 17:6–33
9. Zamfir AD (2014) Neurological analyses: focus on gangliosides and mass spectrometry. Adv Exp Med Biol 806:153–204
10. Căpitan F, Robu A, Popescu L et al (2015) B subunit monomers of Cholerae toxin bind G1 ganglioside class as revealed by chip-nanoelectrospray multistage mass spectrometry. J Carbohydr Chem 34:388–408
11. Levery SB, Steentoft C, Halim A et al (2015) Advances in mass spectrometry driven *O*-glycoproteomics. Biochim Biophys Acta 1850:33–42
12. Lassman ME, McAvoy T, Chappell DL et al (2016) The clinical utility of mass spectrometry based protein assays. Clin Chim Acta 459:155–161

13. Zaia J (2010) Mass spectrometry and glycomics. OMICS 14:401–418
14. Hirabayashi J (2014) Lectin-based glycomics: how and when was the technology born? Methods Mol Biol 1200:225–242
15. Nakayama N, Bando Y, Fukuda T et al (2016) Developments of mass spectrometry-based technologies for effective drug development linked with clinical proteomes. Drug Metab Pharmacokinet 31:3–11
16. Longuespée R, Casadonte R, Kriegsmann M et al (2016) MALDI mass spectrometry imaging: a cutting-edge tool for fundamental and clinical histopathology. Proteomics Clin Appl 10:701–719
17. Morbioli GG, Mazzu-Nascimento T, Aquino A et al (2016) Recombinant drugs-on-a-chip: the usage of capillary electrophoresis and trends in miniaturized systems – a review. Anal Chim Acta 935:44–57
18. Bîndilă L, Peter-Katalinić J (2009) Chip-mass spectrometry for glycomic studies. Mass Spectrom Rev 28:223–253
19. Sarbu M, Robu A, Peter-Katalinić J, Zamfir AD (2015) Automated chip-nanoelectrospray mass spectrometry for glycourinomics in Schindler disease type I. Carbohydr Res 398:90–100
20. Zamfir AD, Flangea C, Serb A et al (2012) Brain chondroitin/dermatan sulfate, from cerebral tissue to fine structure: extraction, preparation, and fully automated chip-electrospray mass spectrometric analysis. Methods Mol Biol 836:145–159
21. Căpitan F, Robu AC, Schiopu C et al (2015) β-Lactoglobulin detected in human milk forms noncovalent complexes with maltooligosaccharides as revealed by chip-nanoelectrospray high-resolution tandem mass spectrometry. Amino Acids 47:2399–2407
22. Almeida R, Mosoarca C, Chirita M et al (2008) Coupling of fully automated chip-based electrospray ionization to high capacity ion trap mass spectrometer for ganglioside analysis. Anal Biochem 378:43–52
23. Bu G, Luo Y, Jing L, Zhang Y (2010) Reduced antigenicity of β-lactoglobulin by conjugation with glucose through controlled Maillard reaction conditions. Food Agric Immunol 21:143–156
24. Domon B, Costello CE (1988) A systematic nomenclature for carbohydrate fragmentation in FAB-MS/MS spectra of glycoconjugates. Glycoconj J 5:397–409
25. Lopez PH, Schnaar RL (2006) Determination of glycolipid-protein interaction specificity. Methods Enzymol 417:205–220
26. Aureli M, Mauri L, Ciampa MG et al (2016) GM1 ganglioside: past studies and future potential. Mol Neurobiol 53:1824–1842
27. Zamfir AD, Fabris D, Capitan F et al (2013) Profiling and sequence analysis of gangliosides in human astrocytoma by high-resolution mass spectrometry. Anal Bioanal Chem 405:7321–7335
28. Ariga T (2014) Pathogenic role of ganglioside metabolism in neurodegenerative diseases. J Neurosci Res 92:1227–1242
29. Suzuki M, Cheung NK (2015) Disialoganglioside GD2 as a therapeutic target for human diseases. Expert Opin Ther Targets 19:349–362
30. Horwacik I, Rokita H (2015) Targeting of tumor-associated gangliosides with antibodies affects signaling pathways and leads to cell death including apoptosis. Apoptosis 20:679–688
31. Todeschini AR, Dos Santos JN, Handa K, Hakomori SI (2008) Ganglioside GM2/GM3 complex affixed on silica nanospheres strongly inhibits cell motility through CD82/cMet-mediated pathway. Proc Natl Acad Sci U S A 105:1925–1930
32. Komura N, Suzuki KG, Ando H et al (2016) Raft-based interactions of gangliosides with a GPI-anchored receptor. Nat Chem Biol 12:402–410
33. Yanagisawa K (2015) GM1 ganglioside and Alzheimer's disease. Glycoconj J 32:87–91
34. Ohkawa Y, Momota H, Kato A et al (2015) Ganglioside GD3 enhances invasiveness of gliomas by forming a complex with platelet-derived growth factor receptor α and Yes kinase. J Biol Chem (26):16043–16058
35. Gustafsson T, Hua YJ, Dahlgren MW et al (2013) Direct interaction between cholera toxin and dendritic cells is required for oral adjuvant activity. Eur J Immunol 43:1779–1788
36. Kim CS, Seo JH, Cha HJ (2012) Functional interaction analysis of GM1-related carbohydrates and vibrio cholerae toxins using carbohydrate microarray. Anal Chem 84:6884–6890
37. Jayaraman N, Maiti K, Naresh K (2013) Multivalent glycoliposomes and micelles to study carbohydrate-protein and carbohydrate-carbohydrate interactions. Chem Soc Rev 42:4640–4656
38. Li J, Fan X, Kitova EN et al (2016) Screening glycolipids against proteins in vitro using picodiscs and catch-and-release electrospray ionization-mass spectrometry. Anal Chem 88:4742–4750

39. Lin H, Kitova EN, Klassen JS (2014) Measuring positive cooperativity using the direct ESI-MS assay. Cholera toxin B subunit homopentamer binding to GM1 pentasaccharide. J Am Soc Mass Spectrom 25:104–110
40. Tian R, Jin J, Taylor L et al (2013) Rapid and sensitive MRM-based mass spectrometry approach for systematically exploring ganglioside-protein interactions. Proteomics 13:1334–1338
41. Han L, Kitova EN, Li J et al (2015) Protein-glycolipid interactions studied in vitro using ESI-MS and nanodiscs: insights into the mechanisms and energetics of binding. Anal Chem 87:4888–4896
42. Syka JE, Coon JJ, Schroeder MJ et al (2004) Peptide and protein sequence analysis by electron transfer dissociation mass spectrometry. Proc Natl Acad Sci U S A 101:9528–9533
43. Hanisch FG (2012) O-glycoproteomics: site-specific O-glycoprotein analysis by CID/ETD electrospray ionization tandem mass spectrometry and top-down glycoprotein sequencing by in-source decay MALDI mass spectrometry. Methods Mol Biol 842:179–189
44. Sarbu M, Ghiulai RM, Zamfir AD (2014) Recent developments and applications of electron transfer dissociation mass spectrometry in proteomics. Amino Acids 46:1625–1634
45. Roepstorff P, Fohlman J (1984) Proposal for a common nomenclature for sequence ions in mass spectra of peptides. Biomed Mass Spectrom 11:601
46. Costello CE, Juhasz P, Perreault H (1994) New mass spectral approaches to ganglioside structure determination. Prog Brain Res 101:45–61
47. Gallegos KM, Conrady DG, Karve SS et al (2012) Shiga toxin binding to glycolipids and glycans. PLoS One 7:e30368

Chapter 8

Studying Protein–Protein Interactions by Biotin AP-Tagged Pulldown and LTQ-Orbitrap Mass Spectrometry

Zhongqiu Xie, Yuemeng Jia, and Hui Li

Abstract

The study of protein–protein interactions represents a key aspect of biological research. Identifying unknown protein binding partners using mass spectrometry (MS)-based proteomics has evolved into an indispensable strategy in drug discovery. The classic approach of immunoprecipitation with specific antibodies against the proteins of interest has limitations, such as the need for immunoprecipitation-qualified antibody. The biotin AP-tag pull-down system has the advantage of high specificity, ease of use, and no requirement for antibody. It is based on the high specificity, high affinity interaction between biotin and streptavidin. After pulldown, in-gel tryptic digestion and tandem mass spectrometry (MS/MS) analysis of sodium dodecyl sulfate polyacrylamide gel electrophoresis (SDS-PAGE) protein bands can be performed. In this work, we provide protocols that can be used for the identification of proteins that interact with FOXM1, a protein that has recently emerged as a potential biomarker and drug target in oncotherapy, as an example. We focus on the pull-down procedure and assess the efficacy of the pulldown with known FOXM1 interactors such as β-catenin. We use a high performance LTQ Orbitrap MSn system that combines rapid LTQ ion trap data acquisition with high mass accuracy Orbitrap analysis to identify the interacting proteins.

Key words Biotin labeled, AP-Tag, SDS-PAGE, LTQ Orbitrap

1 Introduction

Mass spectrometry-based proteomics has evolved into an indispensable tool in biological sample analysis [1, 2]. In typical proteomics experiments, proteins are proteolytically cleaved into peptides, separated based on specific physicochemical properties, and subsequently analyzed by a mass spectrometer [3, 4]. Tandem affinity purification is one of the methods commonly used for sample preparation in proteomics. However, tandem affinity purification often results in protein losses due to the need for multiple washing steps. Moreover, weak and transient interactions between proteins are easily missed [5]. To establish a more efficient and convenient protein purification and detection method for the study of protein interactions by mass spectrometry, an easier system with stronger

Iulia M. Lazar et al. (eds.), *Proteomics for Drug Discovery: Methods and Protocols*, Methods in Molecular Biology, vol. 1647, DOI 10.1007/978-1-4939-7201-2_8, © Springer Science+Business Media LLC 2017

affinity is needed. The biotin AP-tag system utilizes an AviTag or Acceptor Peptide (AP) of 15 amino acids that can be added to the N- or C-terminus of the protein of interest. The AP-tag can be recognized by a small biotin ligase (BirA) protein that specifically labels the lysine residues of the tag. Biotinylated proteins can then be pulled down by streptavidin beads. The avidin-biotin complex has the strongest known noncovalent interaction ($K_d = 10^{-15}$ M). The bond between avidin and biotin forms very rapidly, and, once formed, it is unaffected by extremes of pH, organic solvent, temperature, and denaturing agents. These features make the avidin-biotin system one of the most widely applied systems in the field of biotechnology [6–8]. The biotinylated proteins can be further purified and used to study protein function and protein interactions. Here, we use the FOXM1 AP-tag system to demonstrate the procedure. As FOXM1 isoforms act as both transcriptional activators and repressors with roles in cell proliferation, the potential in oncotherapy has been recently recognized [9–11]. Li et al. provided evidence that supports FOXM1 as a novel therapeutic target, FOXM1 inhibition being a new therapeutic strategy against glioma [9]. Briest et al. proposed FOXM1 as a novel therapeutic target in gastroenteropancreatic neuroendocrine tumors [10]. Gu et al. established FOXM1 as a high-risk myeloma gene and provided support for the design and testing of FOXM1-targeted therapies specifically for the FOXM1 high subset of myeloma patients [11].

2 Materials

Use analytical grade reagents and prepare all solutions using ultrapure water. Organic solvents should all be HPLC grade. Prepare and store all reagents and solvents at room temperature (unless indicated otherwise). Diligently follow all waste disposal regulations.

2.1 AviTag™ Related Biotinylation Reagents

1. Biotinylation strains: Efficient biotinylation *E. coli* B Strain AVB101.
2. Avidity vectors: PAN4, Biotin AviTag™ Vectors (N-terminal).
3. Biotin Ligase: BirA biotin–protein ligase.
4. Plasmid: pcDNA3.1-BirA and pcDNA3.1-AP-FOXM1.
5. Streptavidin Sepharose™: streptavidin-horse radish peroxidase.

2.2 Cell Transfection and Pull-Down Assay

1. Human embryonic kidney cells HEK293T cells.
2. Opti-MEM™ cell culture medium (serum free).
3. Phosphate buffered saline (PBS) solution.
4. Transfection reagent: polyethylenimine (PEI).

5. Immunoprecipitation (IP) lysis buffer: 50 mM Tris–HCl (pH 7.4), 150 mM NaCl, 1% NP-40, 0.25% sodium-deoxycholate, and protease and phosphatase inhibitors.
6. High Capacity Streptavidin Agarose.
7. Binding buffer: 20 mM sodium phosphate, 0.15 M NaCl, pH 7.5.
8. Dilution buffer: 50 mM Tris–HCl, pH 8.0.
9. Elution buffer: 10 mM glutathione, pH 8.0. Add 0.154 g reduced glutathione to 50 mL dilution buffer (*see* **Note 1**).
10. Protease and phosphatase inhibitors: 100 mM PMSF, 1% sodium pyrophosphate, 1% phosphatase inhibitor cocktail, 10 μg/mL aprotinin, leupeptin and pepstatin.
11. SDS loading buffer (2×): 100 mM Tris–HCl (pH 6.8), 4% (w/v) SDS, 0.2% (w/v) bromophenol blue, 20% (v/v) glycerol, 200 mM DTT.

2.3 SDS-PAGE: Immunoblotting and Coomassie Blue Assay

1. SDS-PAGE system: Mini-PROTEAN Electrophoresis System.
2. SDS-PAGE gel: 10% PAGE gel. Resolving gel mix: 4 mL H_2O, 3.3 mL 30% Acr-Bis (29:1), 2.5 mL1.5 M Tris–HCl pH 8.8, 100 μL 10% SDS, 100 μL 10% APS and 10 μL TEMED. Stacking gel mix: 1.4 mL H_2O, 0.5 mL 30% Acr-Bis (29:1), 0.38 mL 1.5 M Tris–HCl pH 6.8, 30 μL 10% SDS, 30 μL 10% APS and 5 μL TEMED.
3. Tris-buffered saline with Tween-20 (TBST) solution (1×): 50 mM Tris–HCl, pH 7.6, 150 mM NaCL, 0.1% Tween-20.
4. Milk blocking buffer: 5% nonfat dry milk (w/v) in TBST (1×). Prepare 150 mL solution by adding 7.5 g nonfat dry milk to 150 mL TBST (1×); mix well.
5. Trans-Blot Turbo Transfer System (*see* **Note 2**).
6. Antibody: anti-β-Catenin (rabbit) (1:1000).
7. Secondary antibody: Anti-Rabbit IgG for β-catenin.
8. Streptavidin-HRP.
9. Molecular weight markers (Bio-Rad).
10. Coomassie Blue solution: 0.1% Coomassie Blue R250 (w/w), 30% methanol, 5% acetic acid (*see* **Note 3**).
11. Fixing solution: 50% methanol, 10% acetic acid, 40% H_2O.
12. Destain solution: 5% methanol, 7.5% acetic acid, 87.5% H_2O.

2.4 In-Gel Digestion

1. Ammonium bicarbonate solution: 100 mM NH_4HCO_3, pH 8.9. Dissolve 0.79 g ammonium bicarbonate in 95 mL distilled water. Adjust the pH with concentrated ammonia to pH 8.9 (*see* **Note 4**).
2. DTT solution: 10 mM DTT. Dissolve 1.5 mg DTT in 1.0 mL of 100 mM ammonium bicarbonate solution.

3. Iodoacetoamide (IAA): 100 mM IAA. Dissolve 18 mg of iodoacetamide in 1.0 mL of 100 mM ammonium bicarbonate solution (*see* **Note 5**).
4. Trypsin (sequencing grade): 20 ng/μL. Add 1.0 mL of ice cold 50 mM ammonium bicarbonate to 20 μg of sequencing-grade trypsin. Dissolve the trypsin by pipetting up and down. Keep the trypsin solution on ice until use (*see* **Note 6**).
5. Wash solution: 50 mM ammonium bicarbonate, 50% acetonitrile.
6. Extraction buffer: 50% acetonitrile, 5% formic acid. Add 10 mL of acetonitrile to 5 mL of water. Add 1 mL of formic acid and adjust the total volume to 20 mL with water.

2.5 Liquid Chromatography-Mass Spectrometry

1. Nano-HPLC system. A typical nano LC/MS/MS operating at a flow rate of 200–350 nL/min.
2. Reversed phase nano-LC trap: Compatible with the nano-HPLC system and nano-LC column (check manufacturer's recommendations).
3. Reversed phase nano-LC column: 75 μm I.D. × 7.5 cm length, packed with 3 μm C18 particles.
4. Solvent A: 98% water, 2% acetonitrile, 0.1% formic acid.
5. Solvent B: 12% water, 88% acetonitrile, 0.1% formic acid (*see* **Note 7**).
6. Mass spectrometer: Orbitrap Velos Pro (Thermo Scientific) or other mass spectrometer that performs similar functions.
7. Data processing software: Proteome Discoverer (PD) version 1.4 (Thermo Scientific), or other software package that performs similar functions.

3 Methods

The eukaryotic expression vectors pcDNA3.1-AP-FOXM1 (AP-tagged FOXM1) and pcDNA3.1-BirA containing *E. coli* biotin ligase (BirA) were constructed respectively and co-transfected into HEK 293 T cells. This system produces biotinylated FOXM1 in cells. The expression of the biotin-labeled FOXM1 in cells was confirmed by silver nitrate staining and Western-blot with HRP-labeled Streptavidin. Biotinylated FOXM1 was purified with streptavidin agarose beads and the efficiency of the pulldown of known FOXM1-interacting proteins, such as β-catenin, was assessed. This expression system provided a technical foundation for the identification of FOXM1-interacting proteins and the study of FOXM1 functions [12–14].

3.1 AP-Tagged FOXM1 and BirA Plasmids Transfection

1. One day before transfection, plate human embryonic kidney cells HEK293T cells in 60 mm cell culture dish.
2. Let the cells attach by an overnight culture.
3. Dilute 20 μL PEI in 250 μL Opti-MEM™ medium (serum free) (*see* **Note 8**).
4. Take 2.5 μg of each plasmid DNA, and dilute them in 250 μL Opti-MEM™ medium (*see* **Note 9**).
5. Add diluted 5 μg DNA to the tube of diluted PEI.
6. Incubate for 10–15 min at room temperature.
7. Add DNA/PEI complex to cells.
8. Incubate the cells for 2–4 days at 37 °C and assay as required.

3.2 Biotinylated Protein Pull-Down Assay

1. At the time for harvesting cells, the supernatant of the cell culture is discarded.
2. Use precooled 4 °C PBS to wash the cells two times, then add 1 mL fresh PBS. Quickly scrape off the cells with a cell scraper and transfer the solution to a 1.5 mL microcentrifuge tube. Centrifuge at 15,000 × *g* for 20–30 s, discard the supernatant, and add 300 μL IP lysis buffer containing protease and phosphatase inhibitors to the pellet. Incubate for 30 min on ice (*see* **Note 10**).
3. Centrifuge at 15,000 × *g* for 10 min at 4 °C.
4. Transfer the supernatant to a new microcentrifuge tube (*see* **Note 11**).
5. Aspirate 200 μL lysate, and mix with 100 μL streptavidin agarose, and incubate for 1 h at 4 °C (*see* **Note 12**).
6. Centrifuge at 1000 × *g* for 2 min. Discard the supernatant.
7. Use 1 mL of ice-cold 4 °C PBS to wash four times. Discard the supernatant.
8. Add 100 μL IP lysis buffer and 100 μL 2× SDS loading buffer into the streptavidin–biotin bead mix.
9. Boil the bead mix at 95–100 °C, 20 min (*see* **Note 13**).
10. Centrifuge at 15,000 × *g* for 1 min. Collect the supernatant. Divide the supernatant into two aliquots, one for SDS-PAGE and staining with Coomassie blue, and the other for Western-blotting for the evaluation of a successful IP (*see* **Note 14**).

3.3 SDS-PAGE: Western Blotting to Evaluate the Expression of FOXM1 and Successful Pulldown of Known Interacting Protein β-Catenin

1. Load proper amount of samples into the wells of the SDS-PAGE gel, along with Bio-RAD molecular weight markers.
2. Run the gel at 85–95 V for about 10 min.
3. Increase the voltage to 110–120 V to finish in about 1.5 h.
4. Transfer the protein from the gel to the membrane.
5. Use TBST with 5% milk blocking buffer for 1 h at room temperature.

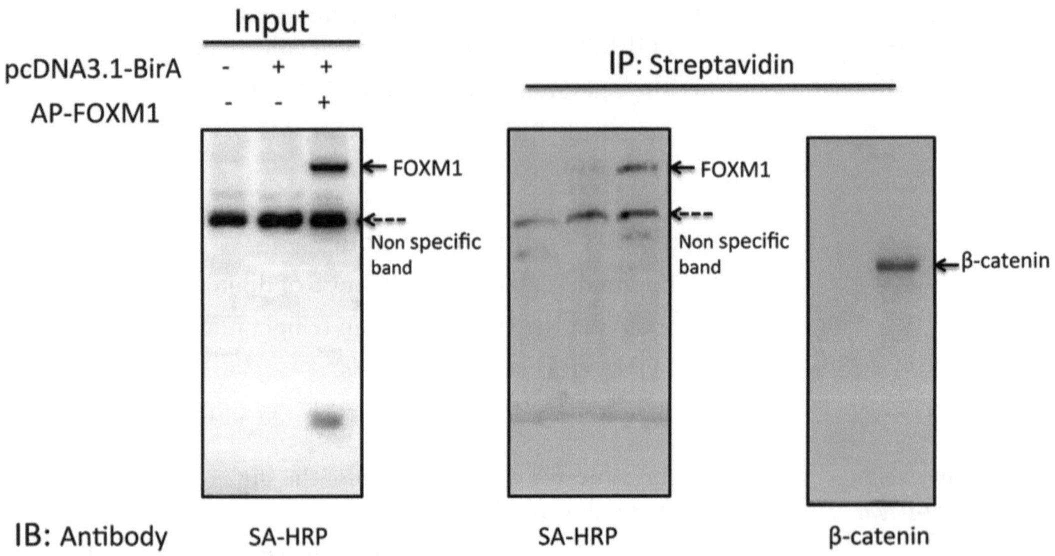

Fig. 1 Detection of β-catenin after pulldown. The eukaryotic expression vectors pcDNA3.1-AP-FOXM1 and pcDNA3.1-BirA containing *E. coli* biotin ligase (BirA) were co-transfected into HEK 293 T cells. IP-Western confirmed the detection of biotin-labeled FOXM1 with Streptavidin-HRP. Broken arrow points to a nonspecific band. β-catenin is a positive control as its interaction with FOXM1 has been reported [17]

6. Add primary antibody and incubate (anti-β-catenin for 1 h; Streptavidin-HRP for 30 min).
7. Use TBST to rinse two times, each time 5 min.
8. Add secondary antibody Anti-Rabbit IgG (for β-catenin) and incubate for 1 h. No need for Streptavidin-HRP.
9. Use TBST to rinse three times, each time 5 min.
10. Acquire image using normal methods for colorimetric detection (Fig. 1).

3.4 SDS-PAGE: Gel Staining with Coomassie Blue G-250

1. A separate SDS-PAGE gel is prepared for this purpose. The gel may be prefixed in fixing solution for 30 min overnight.
2. Stain the gel in the above solution, with 0.1% Coomassie Blue G-250 for 2–4 h, until the gel is a uniform blue color. Staining is complete when the gel is no longer visible in the dye solution. Prior to completing the staining, the gel will appear as a lighter area against the dark staining solution.
3. Destain the gel in destaining solution for 4–24 h.
4. Store the gels in 7% acetic acid at 4 °C (Fig. 2)

3.5 In-Gel Digestion

1. Rinse the SDS-PAGE gel with water for a few hours. Excise the bands of interest with a clean scalpel (*see* **Note 15**).
2. Cut the excised bands into ~1 mm × 1 mm cubes.

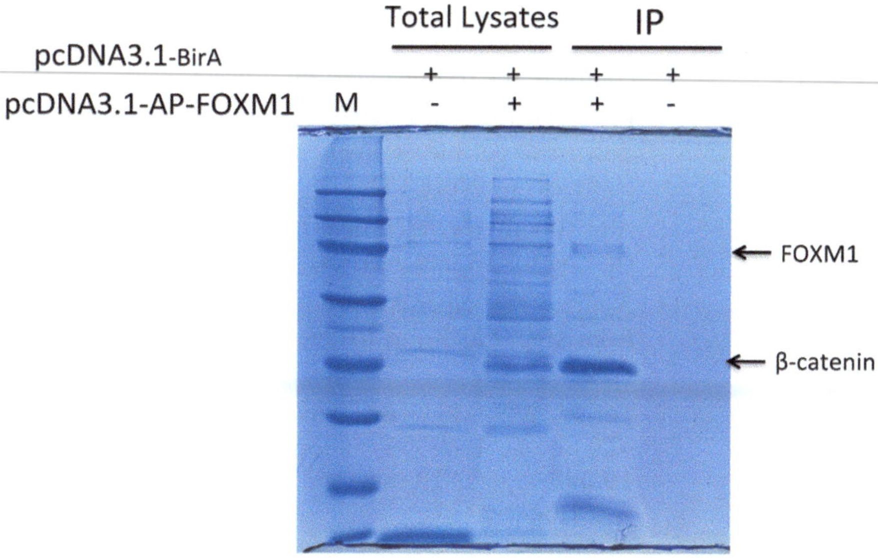

Fig. 2 Coomassie blue staining of FOXM1 and its interacting proteins. The expression of the biotin-labeled FOXM1 in the cells was confirmed by Coomassie blue staining. Other bands in the IP lane with both plasmids transfected but absent in the other IP lane will be subject to in-gel digestion and MS analysis

3. Transfer the gel pieces into a microcentrifuge tube and spin down.
4. Add 100 μL of 100 mM ammonium bicarbonate/acetonitrile, 1:1 (v/v), and incubate for 30 min. Vortex occasionally.
5. Add 500 μL acetonitrile and incubate at room temperature until the gel pieces become white and shrink. Vortex occasionally. Remove the acetonitrile by centrifugation. This step will remove the Coomassie Blue stain.
6. Add 50 μL of 10 mM DTT solution to reduce the cystine disulfide bonds, and incubate for 30 min at 56 °C. The DTT solution should fully cover the gel pieces.
7. Remove the DTT solution, add 100 μL of 100 mM IAA solution, and incubate for 30 min at room temperature in the dark. The IAA will alkylate the newly released cysteine sulfhydryl groups.
8. Remove the IAA solution, add 500 μL of wash solution, and incubate at room temperature with shaking for 15 min.
9. Dehydrate the gel pieces with 500 μL of acetonitrile for 10 min, and remove the supernatant by centrifugation.
10. Add trypsin solution to fully cover the dry gel pieces and preserve on ice or in the refrigerator. Allow the gel to incubate with trypsin for 20–30 min, and add more trypsin in case that the solution was absorbed by the gel. Incubate further the gel with trypsin for another 90 min to fully saturate the gel with trypsin (*see* **Note 16**).

11. Add 10–20 μL of 100 mM ammonium bicarbonate buffer to cover the gel and incubate the samples overnight at 37 °C to complete the tryptic digestion process.
12. Cool the tubes with the gel to room temperature and spin down the gel pieces. Aspirate the supernatant that contains peptides (*see* **Note 17**).
13. Add 100 μL of extraction buffer to each tube and incubate with shaking for 15 min at 37 °C.
14. Aspirate the supernatant and combine with the supernatant from **step 12**. Bring to dryness in a SpeedVac and store at −20 °C.
15. For LC-MS/MS analysis, redissolve the extracted peptides in ~20 μL of 0.1% (v/v) trifluoroacetic acid, vortex, centrifuge at 10,000 × *g* for 10 min, and load the sample in the nano-LC autosampler.

3.6 Nano-LC–MS/MS Analysis

The peptide mix from tryptic digestion can be analyzed by LC-MS/MS on a LTQ-Orbitrap Velos.

1. Load the peptide mixture onto the reversed-phase trap with 1% mobile phase B at a flowrate of 10 μL/min, and rinse the trap for 3 min with the same composition mobile phase. Sample enrichment on the trap may not be necessary for concentrated samples.
2. Transfer the sample from the trap to the nano-LC column for separation.
3. Heat the nano-LC column to 52 °C to improve the chromatographic resolution and reproducibility, and run the LC separation with a pre-optimized LC solvent flow and gradient.
4. Use a data-dependent acquisition method and set the MS^1 survey scan (m/z 300–2000) to a resolution of 60,000, followed by ten MS^2 scans using collision induced dissociation (CID) activation, to fragment the top ten most abundant precursors in the survey scan. The target values for the MS^1 by Orbitrap and MS^2 by ion trap scans were 6×10^6 and 1×10^4.
5. Enable dynamic exclusion with the following settings: repeat count 1, repeat duration 30 s, exclusion list size 500, and exclusion duration 40 s. Set the activation time at 30 ms, with an isolation width of 3 Da for the ion trap (IT) MS, the normalized activation energy at 35%, and the activation (q) at 0.25.

3.7 Database Search and Bioinformatics Analysis

Proteome Discoverer (PD) version 1.4, or other similar software package, can be used to perform the database search against respective Swiss-Prot protein databases for the raw data files. The search engines SEQUEST-HT, Mascot (version 2.4.0), and MS Amanda (version 1.4.4.2822) can be implemented in PD as previously described [15, 16].

3.8 Data Interpretation

A total of 253 proteins were found to be candidates that interact with FOXM1. Among them, β-Catenin and P53 have been reported before as known FOXM1 interactors [17]. Some tyrosine kinase and Ras-related proteins were found through this initial stage of investigation. They represent potential drug targets if proven by independent assays and experiments for their oncogenic effects.

4 Notes

1. Water and chemicals used for buffer preparation should be of high purity. It is recommended to filter the buffers by passing them through a 0.45 μm filter before use.
2. Using BIO-RAD Trans-Blot Turbo, transfer for 10–15 min.
3. Colloidal Coomassie G-250 falls between classical Coomassie and silver staining, with relatively high sensitivity, easiness to perform, as well as compatibility with mass spectrometry.
4. Prepare ammonium bicarbonate buffer daily in large (50–100 mL) volumes and discard after use.
5. Prepare fresh before use to prevent oxidation of the solutions.
6. Use $H_2{}^{18}O$ instead of regular water for peptide quantification or de novo sequencing.
7. Trained personnel should operate the system only. Become familiar with instrument-specific warnings and safety precautions!
8. PEI is a low toxicity reagent. There is no need to change fresh culture medium after transfection. If required, perform a medium change at least 4 h posttransfection.
9. pcDNA3.1-BirA and pcDNA3.1-AP-FOXM1 plasmids 1:1 mix, 2.5 μg each one.
10. Prepare fresh IP lysis buffer. Add protease inhibitors before use.
11. Do not vortex. Shear stress can damage the enzyme. It is better to mix gently with pipette.
12. Use Streptavidin Sepharose™, and rotating with a co-rotating mixer in 4 °C cabinet.
13. Ensure the mix has boiled fully, and avoid the tube cover pop open.
14. First ensure the success of the protein pull-down process by Western blot analysis. Use Coomasie blue to stain the other gel. Centrifuge all boiled protein to ensure spin down. Aliquot the protein samples for long-term storage at −20 °C. Repeated freeze and thaw cycles can cause protein degradation and should be avoided.

15. Wear gloves at all times during processing the gels to avoid keratin contamination that will heavily interfere with MS analysis.
16. Prolonged incubation of the gel with trypsin will enable a thorough diffusion of the enzyme into the polyacrylamide gel and will increase the yield in tryptic peptides.
17. Nonextracted gels can be stored at −20 °C for a few months.

Acknowledgment

We thank Drs. Guixiang Tan and Yongjun Tan in Hunan University for providing the experimental data.

References

1. Aebersold R, Mann M (2003) Mass spectrometry-based proteomics. Nature 422:198–207
2. Angel TE, Aryal UK, Hengel SM et al (2012) Mass spectrometry-based proteomics: existing capabilities and future directions. Chem Soc Rev 41:3912–3928
3. Cravatt BF, Simon GM, Yates JR 3rd (2007) The biological impact of mass-spectrometry-based proteomics. Nature 450:991–1000
4. Papin JA, Hunter T, Palsson BO, Subramaniam S (2005) Reconstruction of cellular signalling networks and analysis of their properties. Nat Rev Mol Cell Biol 6:99–111
5. He YF, Bao HM, Xiao XF et al (2009) Biotin tagging coupled with amino acid-coded mass tagging for efficient and precise screening of interaction proteome in mammalian cells. Proteomics 9:5414–5424
6. Robinson CV, Sali A, Baumeister W (2007) The molecular sociology of the cell. Nature 450:973–982
7. Cognet I, Guilhot F, Gabriac M et al (2005) Cardiotrophin-like cytokine labelling using Bir A biotin ligase: a sensitive tool to study receptor expression by immune and non-immune cells. J Immunol Methods 301:53–65
8. Kocher T, Superti-Furga G (2007) Mass spectrometry-based functional proteomics: from molecular machines to protein networks. Nat Methods 4:807–815
9. Li Y, Zhang S, Huang S (2012) FoxM1: a potential drug target for glioma. Future Oncol 8:223–226
10. Briest F, Berg E, Grass I et al (2015) FOXM1: a novel drug target in gastroenteropancreatic neuroendocrine tumors. Oncotarget 6:8185–8199
11. Gu C, Yang Y, Sompallae R et al (2016) FOXM1 is a therapeutic target for high-risk multiple myeloma. Leukemia 30:873–882
12. Shi X, Xie Z, Song Y et al (2012) Superlocalization spectral imaging microscopy of a multicolor quantum dot complex. Anal Chem 84:1504–1509
13. Tan Y, Xie Z, Ding M et al (2010) Increased levels of FoxA1 transcription factor in pluripotent P19 embryonal carcinoma cells stimulate neural differentiation. Stem Cells Dev 19:1365–1374
14. Xie Z, Tan G, Ding M et al (2010) Foxm1 transcription factor is required for maintenance of pluripotency of P19 embryonal carcinoma cells. Nucleic Acids Res 38:8027–8038
15. Babiceanu M, Qin F, Xie Z et al (2016) Recurrent chimeric fusion RNAs in non-cancer tissues and cells. Nucleic Acids Res 44:2859–2872
16. Yuan H, Qin F, Movassagh M et al (2013) A chimeric RNA characteristic of rhabdomyosarcoma in normal myogenesis process. Cancer Discov 3:1394–1403
17. Zhang N, Wei P, Gong A et al (2011) FoxM1 promotes beta-catenin nuclear localization and controls Wnt target-gene expression and glioma tumorigenesis. Cancer Cell 20:427–442

Chapter 9

Post-Translational Modification Profiling-Functional Proteomics for the Analysis of Immune Regulation

Avital Eisenberg-Lerner, Ifat Regev, and Yifat Merbl

Abstract

Posttranslational modifications (PTMs) of proteins are an integral part of major cellular regulatory mechanisms dictating protein function, localization, and stability. The capacity to screen PTMs using protein microarrays has advanced our ability to identify their targets and regulatory role. This chapter discusses a unique procedure that combines functional extract-based activity assay with large-scale screening utilities of protein microarrays. This "PTM-profiling" system offers advantages in quantitatively identifying modifications in an unbiased manner in the context of specific cellular conditions. While the possibilities of studying PTMs in different settings are enormous, the immune system presents an attractive model for studying the effects of perturbations in PTMs, and specifically the ubiquitin system, as these were already implicated in both immune function and dysfunction. This chapter discusses the significance of PTM profiling in addressing basic questions in immunology. We describe detailed protocols for the preparation of functional cell extracts from immune cell cultures, following differentiation or induced signals, and screening PTMs on protein arrays, as well as basic guidelines for data analysis and interpretation.

Key words Protein arrays, Proteomics, Posttranslational modifications, Ubiquitin, Immune system, Autoimmunity

1 Introduction

Posttranslational modifications (PTMs) exponentially increase the complexity of the genome, from approximately 25,000 genes to over a million protein species that comprise the proteome. While the amino acid sequence of a protein defines its basic structural properties, PTMs often control its function and specify stability, activity, or cellular localization. In fact, the highly dynamic nature of the proteome is largely dependent on the myriad of PTMs, which allow for rapid regulation of various protein structure and function features. Probing PTMs, as well as defining their regulatory code, on a proteome-wide scale still presents a challenge. While the advent of protein microarrays greatly improved our ability to identify targets, such analyses are often limited in providing

Iulia M. Lazar et al. (eds.), *Proteomics for Drug Discovery: Methods and Protocols*, Methods in Molecular Biology, vol. 1647, DOI 10.1007/978-1-4939-7201-2_9,

physiologically relevant information. In this chapter, we discuss a unique procedure for activity-based identification of PTMs on a large-scale protein array platform [1]. This system enables the detection of dynamic changes in PTM patterns in their biological context, in response to specific signals and across cell types. The PTM profiling system is amenable to biochemical manipulations, thereby offering a platform for drug discovery by assessing the PTM landscape in response to perturbations such as introduction of various compounds, while closely recapitulating the robustness of the physiological system. While we focus here on the significance of analyzing ubiquitination in the immune system, this technique may be applicable to various mammalian cells and conditions, for addressing diverse biological questions regarding health and disease.

1.1 The Ubiquitin-System

Among PTMs, the ubiquitin family was shown to regulate a variety of central cellular processes such as cell cycle, degradation, differentiation, stress responses, cellular trafficking, immune responses, and more [2]. Accordingly, alterations in this system have been associated with multiple human pathologies including cancer and immunological disorders [3–8]. Ubiquitination is carried out by the consequential activation of ubiquitin activating (E1), conjugating (E2), and ligating (E3) enzymes [2, 9]. In brief, E1 catalyzes the formation of a high-energy thiol ester intermediate with the C-terminal Gly residue of ubiquitin that is then transferred to one of several ubiquitin conjugating E2 enzymes. In the final step of the cascade, an E3 ligase mediates the attachment of ubiquitin by an isopeptide bond between the C-terminal Gly of ubiquitin and mostly a ε-NH2 group of an internal Lys of the substrate. Several hundreds of ubiquitin E3 ligases in the proteome confer the substrate specificity of the system. Additional ubiquitin molecules may attach sequentially to one another via an isopeptide bond involving one of the seven internal Lys residues in the ubiquitin moiety, generating a linkage-specific polyubiquitin chain. K48-linked polyubiquitin chains were shown to target proteins for proteasomal degradation, while other chain linkages, as well as conjugation by ubiquitin-like molecules (Ubls) [10], have different cellular functions [10]. Proper function of the ubiquitin system depends on the regulated and concerted activation of multiple factors, including removal of ubiquitin moieties via de-ubiquitinating (DUBs) enzymes that remove ubiquitin from substrates [11, 12]. Thus, with over 600 E3 ligases and 100 DUBs in the proteome, deregulation of the ubiquitin system may produce detrimental cellular effects.

1.2 Ubiquitin in Immunity and Autoimmunity

Among the different cellular processes that are mediated by ubiquitination, the control of immune responses is an intricate example as it involves balancing activation with tolerance [5, 13–15]. Proper execution of ubiquitination events is essential for amending functions of the immune system and not surprisingly, aberrations in the ubiquitin system have been implicated in various immune pathologies. For example, disruption of ubiquitin-mediated control of NFkB activity [16] is associated with both dysfunction and hyper reactivity of the immune system. Essential steps in NFkB activation include K63-linked ubiquitination of NEMO, which mediates IkK activation, and K48-linked ubiquitination of the NFkB inhibitor IkB, which induces its degradation and the release of active NFkB molecules. A mutation causing disruption of ubiquitin binding to NEMO is known to cause immunodeficiency by improper NFkB activation [17–19]. Aberrations in ubiquitin-dependent termination of NFkB signaling, such as mutations or deletions in the DUBs CYLD or A20 [20, 21], were also shown to induce immune irregularities such as promotion of inflammation and tumorigenesis, and were found in various lymphomas [22–25].

Interestingly, several E3 ubiquitin ligases were shown to mediate T-cell anergy, a peripheral tolerance mechanism, characterized by block of proliferation and effector functions in the lack of co-stimulation [26]. Uncoupling the break of tolerance from innate immune signals is linked to the development of autoimmunity. Indeed, aberrations in the function of anergy-mediating E3 ligases have been implicated in autoimmune diseases. For example, mutations or loss of expression of the Cbl-b E3 ligase uncoupled proliferation and IL-2 production from CD28 co-stimulation, and induced spontaneous autoimmunity with rheumatoid arthritis-like characteristics that include auto-antibody production, infiltration of activated T and B lymphocytes into multiple organs, and parenchymal damage [27, 28]. Further, the loss of tolerance in Cbl-b-null mice resulted in massive lethality following re-challenging with the tolerizing antigen [29]. This was suggested to be mediated via ubiquitination-dependent downregulation of PLCγ1 activity and stability, leading to reduced calcium mobilization in response to restimulation [29]. Finally, both reduced Cbl-b gene expression [30] and sequence polymorphisms in the Cbl-b gene [31] were found in T-lymphocytes of patients with systemic lupus erythematosus and multiple sclerosis [30–32]. In addition to Cbl-b, the E3 ligases ITCH and GRAIL have also been implemented in anergy induction and autoimmune disorders [29, 33–36]. For example, deficiency in ITCH, which by mediating the ubiquitination and degradation of the T-cell receptor and JUN-B affects T-cell function and proliferation, causes autoimmune disease characterized by inflammatory cell infiltration of the lungs, liver, and gut [37]. Thus, dysfunction of the ubiquitin machinery in the immune system can lead to unselective T-cell activation and loss of tolerance to self-antigens.

1.3 Understanding the Complexity of the System

These few examples are merely the tip of the iceberg of the potential numerous ubiquitin-dependent factors that may be involved in regulating proper induction and termination of immune responses. Most often, E3 ligases have more than a single substrate and at times, a specific target may be regulated by more than one E3 ligase. With the additional regulation of ubiquitin removal by DUBs, the ubiquitin system presents a highly complex network of events. Gaining true comprehension of the system's regulators, plethora of substrates, and their specificities requires ample proteomic approaches that facilitate systematic analyses on a proteome-wide scale. In addition, the combinatorial nature of PTMs, which involves intra-molecular cross-regulation of different modifications, poses the need to generate global assessments of the PTM landscape. However, while genomic and transcriptional changes have been widely studied, fundamental questions regarding the protein modification landscape, in general and specifically in immunity, remained relatively untouched. Partly this is because despite the importance of PTMs, there are few analytical tools to analyze them. In the next sections, we will discuss our recently developed PTM profiling approach, a functional assay for the detection and evaluation of PTMs on a large-scale platform [1], which offers novel opportunities in probing the activity of the ubiquitin system in immune responses.

1.4 PTM Profiling Using Mammalian Cell Extracts and Protein Microarrays

PTM profiling is an activity-based assay for the identification and comparative evaluation of posttranslational modifications under different cellular states. The assay is based on the preparation of functional cell extracts, made under detergent-free conditions that preserve biochemical activity and many aspects of the biological context. PTM profiling utilizes protein microarrays that contain thousands of proteins that are absorbed to a nitrocellulose surface. Different conformations for every bound protein permit high probability of reactivity. The active extracts are incubated on pre-blocked protein arrays, allowing the enzymes in the extracts to modify the proteins attached to the surface of the microarray. A great advantage of PTM profiling is the fact that in contrast to conventional in-vitro assays that depend on exogenously added enzymes, in PTM profiling protein modifications result from the intrinsic enzymatic activities of the extract. Upon completion of the reaction, the microarrays are washed extensively to remove nonspecific interactions, and the modified proteins are detected by incubating the arrays with modification-specific antibodies (e.g., poly-ubiquitin antibody), followed by fluorescently labeled secondary antibodies, and then scanning the arrays on a microarray scanner to acquire fluorescent images. Using a quantification software, signal intensity values, signal-to-noise ratios (SNR), and local background, are determined per individual spots in the array. A reactivity level, which reflects the extent of modification that was detected on each protein, is calculated. Thus, profiling modifications from functional extracts prepared under specific cellular conditions may reveal differential PTM patterns that are associated with different physiological states.

1.5 Advantages (+) and Caveats (−) of PTM Profiling in Studying Immune Regulation

(+) *Efficient platform for large-scale studies:* PTM profiling offers the possibility to probe the response to specific signals in vitro, such as differentiation or activation of immune cell subtypes. The ability to quantitatively evaluate at least two modifications on thousands of proteins in parallel on a single protein array provides an efficient large scale platform for studying diverse processes.

(+) *Non-genetic manipulation possibilities:* Deciphering the underlying mechanisms of cellular processes largely depends on the ability to manipulate molecular networks. However, such a line of research is often challenging in immune cells due to difficulties on genetic manipulations of such cells. PTM profiling offers a non-genetic manipulation approach for studying signaling dynamics in immune cells. The system's amenability to biochemical manipulations allows for supplementation of the functional extract with various substances such as mutant proteins including enzymes and substrates, chemicals, antibodies and more, and evaluation of their effect on downstream modifications. For example, identification of an E3 ubiquitin ligase's specific targets may be achieved by competitively inhibiting the endogenous enzyme by introducing a dominant negative mutant (e.g., catalytically dead HECT E3 mutant that can still bind the substrate) into the extract during the reaction. In this case, targets whose ubiquitination will be affected will be presumed as putative substrates of the perturbed E3 ligase, compared to those that remain unchanged. In a similar fashion, PTM profiling offers a platform for screening drugs or inhibitors in a large-scale manner.

(−) Currently, protein microarrays are produced only for human recombinant proteins. However, previous studies using cell-free systems in other organisms (e.g., Xenopus) have shown that functional conservation of regulatory events may occur, for example in mitotic regulation. Probing mice extracts on human microarrays may therefore provide novel insight into the conservation of regulation across species that should be further validated experimentally. This would be particularly relevant to studies of the immune system, which often rely on mice as model systems.

(−) PTM profiling has not been validated for use with minute amount of protein. Thus, studies of primary cells whose quantity is often limited may require specific calibration. Alternatively, several mice may be pooled for a single reaction.

(−) Indirect signals may be detected in cases of modification of an interacting partner of a target on the protein array. However, this difficulty may be overcome by increasing the stringency of the washes.

2 Materials

2.1 Cell Culture

When working with human cells, all appropriate bio-safety practices must be followed.

All solutions and equipment coming into contact with living cells must be sterile, and aseptic techniques should be used. Monocyte activation and differentiation into macrophages is described as an example.

1. U937 cells (ATCC) grown in suspension.
2. RPMI-1640 medium.
3. Fetal bovine serum (FBS).
4. Antibiotic/antimycotic solution: Penicillin-Streptomycin 100 U/mL, or other products of similar effectiveness.
5. APC/Cy7 anti-human CD11c antibody (Biolegend): Concentration is lot-specific, the recommended dilution is indicated in the Subheading 3.
6. PE anti-human CD14 antibody (Biolegend): 0.2 mg/mL.
7. Energy mix stock (50×): 375 mM creatine phosphate, 50 mM ATP at pH 8.0, 50 mM $MgCl_2$. Store small aliquots at −80 °C. Handle the mixture on ice.
8. Swelling buffer: Add 24.3 mL ice-cold Milli-Q water or equivalent into 50 mL canonical tube, 500 μL 1 M HEPES at pH 7.5, 37.5 μL 1 M $MgCl_2$, 125 μL 1 M KCl. Vortex to mix and store at −20 °C. Prior to use, freshly add 25 μL of 1 M dithiothreitol (DTT) and 1 tablet EDTA-free protease-inhibitor (Roche).
9. Tris buffered saline (TBS): 50 mM Tris–HCl (pH 7.6), 150 mM NaCl. Store at 4 °C.
10. TBS-T: 0.05% Tween-20 in TBS (v/v).
11. Phosphate buffered saline (PBS): 137 mM NaCl, 2.7 mM KCl, 8 mM Na_2HP0_4, pH 7.4. Store at 4 °C.
12. Phorbol 12-myristate 13-acetate (PMA) solution: PMA in DMSO. Stock concentration may vary, final working concentration is indicated in the Subheading 3.
13. Interferon-γ (IFN-γ): Stock concentration may vary, final working concentration is indicated in the Subheading 3.
14. Rubber policeman.
15. 175 cm^2 tissue culture flasks.
16. Sterile 20-1/2G syringe needles.
17. LSRII flow cytometer or equivalent.

2.2 Protein Microarray

1. ProtoArray (Life Technologies): while protein microarrays are available from other suppliers, the protocol described herein is based on the use of this particular product.
2. ArrayIt blocking buffer, cold (ArrayIt Corporation).
3. Recombinant proteins to supplement extracts: May be purified or purchased from a commercial supplier.
4. Primary antibodies to detect PTMs of interest (e.g., FK2 polyubiquitin antibody): 1 mg/mL.
5. Fluorescently labeled secondary antibodies.
6. Forceps.
7. PBS solution: solution prepared as described in Subheading 2.1, **item 11**.
8. Cover slips, 24 mm × 60 mm.
9. Eppendorf centrifuge 5810 and rotor A-4-81 for 50-mL conical tubes (or equivalent).
10. GenePix 4000B microarray scanner (Molecular Devices).
11. Gene Pix Pro software (Molecular Devices).
12. Optional: ProtoArray Prospector v.5 (available at *http://www.lifetechnologies.com/us/en/home/life-science/protein-expression and-analysis/biomarker-discovery/protoarray/resources/data-analysis.html*).

3 Methods

3.1 Cell Culture and In-Vitro Differentiation

1. Culture four 175 cm^2 tissue culture flasks of U937 cells in suspension in RPMI-1640 medium with 10% (v/v) FBS and 1% (v/v) antibiotics (*see* **Note 1**).
2. Induce cell differentiation by supplementing the medium with 5 ng/mL PMA for 24–48 h.
3. Differentiated cells will change cell morphology and adhere to the surface of the flask. Examine full differentiation state of the cells by examining cell surface markers for activated macrophages, CD14 and CD11c, using flow cytometry (LSR-II). Recommended working concentrations are 1–5 μL per million cells in 100 μL staining volume. Since antibody concentrations are lot-specific, calibration may be required.
4. Stimulate cells as desired to induce specific signals.
5. To harvest cells for extraction, gently scrape cells with a rubber policeman (cells in this state are mostly insensitive to trypsinization). Centrifuge cells for 5 min at 125 × *g*, discard the supernatant.
6. Wash cells two times to remove residual medium. In each wash, suspend the cells in PBS and centrifuge cells for 5 min at 125 × *g*. Remove the supernatant carefully and place cells on ice.

3.2 Functional Cell Extract Preparation

1. Resuspend cells with ice-cold PBS at a volume equivalent to the cell pellet and centrifuge for 5 min at 125 × *g*, 4 °C. Decant the supernatant carefully.
2. Lyse the cells by adding prechilled swelling buffer at a volume equivalent to about 80% of the recovered pellet. Supplement the cells with 50× energy mix at 1/50 of suspension volume. Incubate on ice for 30 min, mix every few minutes by inverting the tubes.
3. Break cell membranes with repeated freeze-thaw cycles. Freeze cells in liquid nitrogen, followed by thawing in a 30 °C water bath. Repeat freeze-thaw cycles three times. After the final thaw, place the tube on ice (*see* **Note 2**).
4. Shear the DNA by passing the cells through a 20-1/2G needle, seven to eight times. It is important to avoid forming bubbles.
5. Centrifuge extracts for 5 min at 2300 × *g*, 4 °C. Collect the supernatant and centrifuge again for 30 min at 16,000 × *g*, 4 °C.
6. Divide the extracts into Eppendorf tubes and freeze aliquots in liquid nitrogen. Extracts can be stored at this step at −80 °C (*see* **Note 3**).
7. It is recommended to test enzymatic activities of the extracts before incubation on the protein array (for experimental procedures for testing enzymatic activity in cell extracts *see* [38]).

3.3 Block Arrays and Incubate with Functional Extracts

1. Human ProtoArrays are stored at −20 °C. Prior to use transfer arrays to 4 °C for 15 min, according to the manufacturer's instructions (*see* **Note 4**).
2. Block the arrays by gently shaking them in ArrayIt blocking buffer for 1 h at 4 °C. The surface area of the array should be well covered. Other blocking buffers such as 1% BSA may be used for this step (*see* **Notes 5** and **6**).
3. Wash three times in TBS at 4 °C, with gentle shaking, for 3 min each wash (*see* **Note** 7).
4. If the functional cell extracts were frozen, thaw on ice while washing the arrays.
5. Centrifuge extracts for 10 min at 16,000 × *g* at 4 °C to remove debris (*see* **Note 8**).
6. Apply a volume of 100 μL of cell extract (10 mg/mL) onto each microarray and allow it to cover the whole membrane surface by gently laying a cover slip on top of it (*see* **Notes 9** and **10**).
7. Incubate the covered arrays for 1 h at room temperature in a humidified chamber (*see* **Note 11**).

8. Stop the reaction by adding a large volume of TBS into the array containers. This will dissociate the cover slips from the arrays. Remove the floating coverslips carefully without touching the microarray surface (*see* **Notes 12** and **13**).
9. Wash the microarrays twice in TBS-T and a third time in TBS (*see* **Note 14**).

3.4 Probe Arrays with Antibodies

1. Relevant primary antibodies against the modification of interest should be diluted in blocking buffer. For example, for the detection of poly and mono-ubiquitin, dilute 1 μL of FK2 anti-ubiquitin antibody in 250 μL blocking buffer. A 1:250 dilution ratio is typically recommended, but may require calibration to specific antibodies.
2. Add 120 μL diluted primary antibodies onto the array and cover it with a coverslip. Incubate covered arrays for 1 h at room temperature in a humidified chamber. This step may be carried out overnight at 4 °C (*see* **Note 15**).
3. Wash twice in TBS-T and a third time in TBS with gentle shaking for 5 min each wash. If the primary antibody was fluorescently labeled, proceed to scanning the array as described in subheading 3.5.
4. Dilute fluorescently labeled secondary antibodies in blocking buffer. Dilutions of 1:250-1:1000 are recommended (*see* **Note 16**).
5. Add 120 μL diluted secondary antibodies onto each array and cover with a coverslip. Incubate covered arrays for 1 h at room temperature in a humidified chamber.
6. Remove excess secondary antibodies by washing three times in TBS-T with gentle shaking for 5 min each wash. Wash the slides again in PBS for 1 min.

3.5 Scan Arrays and Calculate Signal Intensities

1. Dry each array by vertically positioning it in a 50 mL conical tube and centrifuging for 5 min at low speed, 200 × *g;* allow excess buffer to collect at the bottom of the tube (*see* **Note 17**).
2. Immediately proceed to scanning the microarrays using a suitable microarray scanner (e.g., GenePix 4000B) (*see* **Note 18**). Gain and laser intensity should be adjusted to obtain maximal signals while avoiding saturation. Recommended starting parameters for scanning are: PMT gain value of 600, laser power of 100%, and focus point of 0 μm (*see* **Note 19**).
3. Assign a reactivity level value to each spot on the array based on the measured signals, using quantification software (e.g., GenePix Pro) (*see* **Note 20**).
4. Signal intensity data should be filtered to reduce unreactive proteins, normalize between arrays, and analyze using designated software (e.g., ProteinArray Prospector v.5) (*see* **Notes 21** and **22**).

4 Notes

1. The PTM profiling protocol is suitable for different human cells. Calibration may be needed to determine the number of cells that yield at least 1 mg protein during functional extract preparation. Recommended protein concentration: 10 mg/mL.
2. When performing cell rupture by repeated freezing and thawing cycles, rapidly thaw the cells in 30 °C and return to liquid nitrogen immediately after thawing.
3. Repeated freezing and thawing of the extract should be avoided. Extracts can usually be stored up to 6 months.
4. Gentle handling of the microarray is recommended; use forceps to lift the array. Place the forceps on the barcode array area, avoid touching the nitrocellulose membrane surface throughout the procedures.
5. Blocking buffer should be prepared fresh before use, saved in 1 mL aliquots at 4 °C for antibody dilutions in later steps. Blocking of the membrane will prevent nonspecific binding and removes unadsorbed proteins from the nitrocellulose surface. Effective blocking conditions may be calibrated based on the specific extracts or probes to be used. Blocking time may also vary depending on the assay.
6. Blocking and washing buffers should be filtered to remove small particles that might adhere to the nitrocellulose surface.
7. Washing should be carried out with gentle shaking and using buffer volumes large enough to completely cover the surface area.
8. Thaw the extracts on ice prior to use, this is critical to retain full activity of the extract. The desired amount of extract is 100 μL (at 10 mg/mL) per array. Take into consideration the material loss in the prior centrifugation step. Avoid transferring any debris from the pellet of the extract, as it might interfere with protein-protein interactions on the surface of the array.
9. To promote or inhibit specific activities, the extract may be supplemented with specific enzymes, inhibitors, and modifiers, immediately prior to incubation on the array. For example, to promote ubiquitination of spotted proteins, the extract could be supplemented with tagged-ubiquitin (1–2 μg) and with E1-activating enzymes (100 nM) or E2-conjugating enzymes (500 nM).
10. Place the thawed extract on one side of the nitrocellulose membrane. Use forceps to gently place a cover slip onto the

extract while lowering slowly the cover slip, allowing the extract to spread on the protein array surface by capillary forces. Avoid introducing bubbles in this step. Once the cover slip is placed, do not remove it as it may scratch the nitrocellulose membrane.

11. Long incubations may dry the extract on the microarray and interfere with the reaction. To maintain a humid environment, microarrays should be kept during this step in a closed container with a wet Whatman paper.
12. When comparing multiple microarrays it is important to stop the reaction by washing the extracts off the microarrays following equal incubation periods.
13. If the coverslips fail to detach from the array, gently tilt the container or remove the coverslips using forceps while the arrays are still inside the TBS filled containers.
14. To detect only covalent modifications add 0.5-1% SDS to the first TBS-T wash.
15. Overnight incubation with primary antibodies may increase the signal.
16. Secondary probes should be chosen according to the detected wavelengths in the microarray scanner. Note that most microarray scanners will have only two excitation wavelengths, at 532 nm and at 635 nm.
17. Avoid ununiformed drying of residual washing buffer on the slide surface by quickly proceeding to the centrifugation step.
18. Scanning at 10 μm resolution is recommended as it produces sufficient quality images in most cases, with smaller file sizes than the scan at 5 μm resolution.
19. The dynamic range of a standard experiment may vary from 500 to 65,000 signal intensity units. However, the dynamic range and signal sensitivity may differ, depending on the modifications being analyzed and antibodies used to detect them, for both biological and technical reasons, respectively.
20. During the scan, an image file is generated. This file holds the signal intensity values for each spot in the array as well as their local background level. By calculating the signal-to-noise ratio (SNR) for each spot, usually calculated by the quantitation analysis software, we can determine the reliability of the signal and filter out data points below a certain SNR value. Low signal intensity could have a high SNR value, suggesting that even spots of low reactivity may yield meaningful information.
21. For analysis of large-scale microarray experiments, online tools and software packages such as "R" Bioconductor or MATLAB

(MathWorks) are available. Data analysis should be conducted according to standard statistical tests.

22. Protein abundance on the protein array should be considered when comparing the reactivity levels of different proteins. Thus, the amount of protein that is deposited on each spot in the array should be used to normalize the detected signal intensity of the posttranslational modification of interest (e.g., poly-ubiquitin). Proteins on the array are GST-tagged and a Relative Fluorescence Unit (RFU) value, which represents the protein's abundance in each spot, is given for every ProtoArray batch production, based on the spot's reactivity towards anti-GST antibodies.

Acknowledgments

Yifat Merbl is supported by the I-CORE Program of the Planning and Budgeting Committee and The Israel Science Foundation (grant No. 1775/12), and by the European Research Council (ERC) under the European Union's Horizon 2020 research and innovation programme.

References

1. Merbl Y, Refour P, Patel H et al (2013) Profiling of ubiquitin-like modifications reveals features of mitotic control. Cell 152:1160–1172. doi:10.1016/j.cell.2013.02.007
2. Hershko A, Ciechanover A (1998) The ubiquitin system. Annu Rev Biochem 67:425–479. doi:10.1146/annurev.biochem.67.1.425
3. Park Y, Jin H, Aki D et al (2014) The ubiquitin system in immune regulation. Adv Immunol 124:17–66. doi:10.1016/B978-0-12-800147-9.00002-9
4. Wang J, Maldonado MA (2006) The ubiquitin-proteasome system and its role in inflammatory and autoimmune diseases. Cell Mol Immunol 3:255–261
5. Liu Y-C (2004) Ubiquitin ligases and the immune response. Annu Rev Immunol 22:81–127. doi:10.1146/annurev.immunol.22.012703.104813
6. Davis ME, Gack MU (2015) Ubiquitination in the antiviral immune response. Virology 479-480:52–65. doi:10.1016/j.virol.2015.02.033
7. Gómez-Martín D, Díaz-Zamudio M, Alcocer-Varela J (2008) Ubiquitination system and autoimmunity: the bridge towards the modulation of the immune response. Autoimmun Rev 7:284–290. doi:10.1016/j.autrev.2007.11.026
8. Ciechanover A, Schwartz AL (2004) The ubiquitin system: pathogenesis of human diseases and drug targeting. Biochim Biophys Acta 1695:3–17. doi:10.1016/j.bbamcr.2004.09.018
9. Komander D, Rape M (2012) The ubiquitin code. Annu Rev Biochem 81:203–229. doi:10.1146/annurev-biochem-060310-170328
10. Kirkin V, Dikic I (2007) Role of ubiquitin- and Ubl-binding proteins in cell signaling. Curr Opin Cell Biol 19:199–205. doi:10.1016/j.ceb.2007.02.002
11. Yau R, Rape M (2016) The increasing complexity of the ubiquitin code. Nat Cell Biol 18:579–586. doi:10.1038/ncb3358
12. Komander D, Clague MJ, Urbé S (2009) Breaking the chains: structure and function of the deubiquitinases. Nat Rev Mol Cell Biol 10:550–563. doi:10.1038/nrm2731
13. Popovic D, Vucic D, Dikic I (2014) Ubiquitination in disease pathogenesis and treatment. Nat Med 20:1242–1253. doi:10.1038/nm.3739

14. Schartner JM, Fathman CG, Seroogy CM (2007) Preservation of self: an overview of E3 ubiquitin ligases and T cell tolerance. Semin Immunol 19:188–196. doi:10.1016/j.smim.2007.02.010
15. Liu Y-C, Penninger J, Karin M (2005) Immunity by ubiquitylation: a reversible process of modification. Nat Rev Immunol 5:941–952. doi:10.1038/nri1731
16. Chen ZJ (2005) Ubiquitin signalling in the NF-κB pathway. Nat Cell Biol 7:758–765. doi:10.1038/ncb0805-758
17. Smahi A, Courtois G, Vabres P et al (2000) Genomic rearrangement in NEMO impairs NF-kappaB activation and is a cause of incontinentia pigmenti. The international Incontinentia Pigmenti (IP) consortium. Nature 405:466–472. doi:10.1038/35013114
18. Döffinger R, Smahi A, Bessia C et al (2001) X-linked anhidrotic ectodermal dysplasia with immunodeficiency is caused by impaired NF-kappaB signaling. Nat Genet 27:277–285. doi:10.1038/85837
19. Orange JS, Brodeur SR, Jain A et al (2002) Deficient natural killer cell cytotoxicity in patients with IKK-gamma/NEMO mutations. J Clin Invest 109:1501–1509. doi:10.1172/JCI14858
20. Majumdar I, Paul J (2014) The deubiquitinase A20 in immunopathology of autoimmune diseases. Autoimmunity 47(5):307–319. doi:10.3109/08916934.2014.900756
21. Ma A, Malynn BA (2012) A20: linking a complex regulator of ubiquitylation to immunity and human disease. Nat Rev Immunol 12:774–785. doi:10.1038/nri3313
22. Compagno M, Lim WK, Grunn A et al (2009) Mutations of multiple genes cause deregulation of NF-kappaB in diffuse large B-cell lymphoma. Nature 459:717–721. doi:10.1038/nature07968
23. Kato M, Sanada M, Kato I et al (2009) Frequent inactivation of A20 in B-cell lymphomas. Nature 459:712–716. doi:10.1038/nature07969
24. Novak U, Rinaldi A, Kwee I et al (2009) The NF-{kappa}B negative regulator TNFAIP3 (A20) is inactivated by somatic mutations and genomic deletions in marginal zone lymphomas. Blood 113:4918–4921. doi:10.1182/blood-2008-08-174110
25. Schmitz R, Stanelle J, Hansmann M-L, Küppers R (2009) Pathogenesis of classical and lymphocyte-predominant Hodgkin lymphoma. Annu Rev Pathol 4:151–174. doi:10.1146/annurev.pathol.4.110807.092209
26. Schwartz RH (2003) T cell anergy. Annu Rev Immunol 21:305–334. doi:10.1146/annurev.immunol.21.120601.141110
27. Bachmaier K, Krawczyk C, Kozieradzki I et al (2000) Negative regulation of lymphocyte activation and autoimmunity by the molecular adaptor Cbl-b. Nature 403:211–216. doi:10.1038/35003228
28. Paolino M, Thien CBF, Gruber T et al (2011) Essential role of E3 ubiquitin ligase activity in Cbl-b-regulated T cell functions. J Immunol 186:2138–2147. doi:10.4049/jimmunol.1003390
29. Jeon M-S, Atfield A, Venuprasad K et al (2004) Essential role of the E3 ubiquitin ligase Cbl-b in T cell Anergy induction. Immunity 21:167–177. doi:10.1016/j.immuni.2004.07.013
30. Gómez-Martín D, Ibarra-Sánchez M, Romo-Tena J et al (2013) Casitas B lineage lymphoma b is a key regulator of peripheral tolerance in systemic lupus erythematosus. Arthritis Rheum 65:1032–1042. doi:10.1002/art.37833
31. Doníz-Padilla L, Martínez-Jiménez V, Niño-Moreno P et al (2011) Expression and function of Cbl-b in T cells from patients with systemic lupus erythematosus, and detection of the 2126 a/G Cblb gene polymorphism in the Mexican mestizo population. Lupus 20:628–635. doi:10.1177/0961203310394896
32. Stürner KH, Borgmeyer U, Schulze C et al (2014) A multiple sclerosis-associated variant of CBLB links genetic risk with type I IFN function. J Immunol 193:4439–4447. doi:10.4049/jimmunol.1303077
33. Anandasabapathy N, Ford GS, Bloom D et al (2003) GRAIL: an E3 ubiquitin ligase that inhibits cytokine Gene transcription is expressed in Anergic CD4+ T cells. Immunity 18:535–547. doi:10.1016/S1074-7613(03)00084-0
34. Heissmeyer V, Macián F, Im S-H et al (2004) Calcineurin imposes T cell unresponsiveness through targeted proteolysis of signaling proteins. Nat Immunol 5:255–265. doi:10.1038/ni1047
35. Sande OJ, Karim AF, Li Q et al (2016) Mannose-capped Lipoarabinomannan from mycobacterium tuberculosis induces CD4+ T cell Anergy via GRAIL. J Immunol 196:691–702. doi:10.4049/jimmunol.1500710
36. Gu H, Chiang YJ, Kole HK et al (2000) Cbl-b regulates the CD28 dependence of T-cell activation. Nature 403:216–220. doi:10.1038/35003235

37. Lohr NJ, Molleston JP, Strauss KA et al (2010) Human ITCH E3 ubiquitin ligase deficiency causes syndromic multisystem autoimmune disease. Am J Hum Genet 86:447–453. doi:10.1016/j.ajhg.2010.01.028

38. Williamson A, Jin L, Rape M (2009) Preparation of synchronized human cell extracts to study ubiquitination and degradation. Methods Mol Biol 545:301–312. doi:10.1007/978-1-60327-993-2_19

Chapter 10

Reverse Phase Protein Arrays and Drug Discovery

Kenneth G. Macleod, Bryan Serrels, and Neil O. Carragher

Abstract

Reverse Phase Protein Arrays (RPPA) represent a sensitive antibody-based proteomic approach, which enables simultaneous quantification of the abundance of multiple proteins and posttranslational modifications across multiple samples. Here, we provide protocols for RPPA performed on two distinct protein-binding substrates associated with two most commonly used RPPA platform technologies. We compare and contrast the respective advantages and limitations of each platform within the context of drug discovery applications.

Key words Protein, Array, Antibody, Multiplex, Proteomics

1 Introduction

The Reverse Phase Protein Array (RPPA) technology is based upon the principles of immunoassays; however, the term "reverse-phase" describes the distinction between RPPA and other sandwich-style methods such as enzyme-linked immunoassays (ELISA) and bead-based immunoassay methodologies which employ a capture antibody in tandem with a detection antibody for each analyte (antigenic protein) under study [1, 2]. In the RPPA method, analytes are immobilized on a solid substrate and subsequently probed with a single detection antibody toward a specific protein target (Fig. 1). Each sample is deposited across multiple arrays, which are physically separated from each other and each array is exclusively dedicated to a specific detection antibody ensuring no cross-reactivity between antibodies (Fig. 1). Recent advances in RPPA methods, combined with more sophisticated sample handling, optical detection and better quality antibody reagents have facilitated robust and routine application of RPPA to conduct large-scale multiplex analysis of posttranslational markers across in vitro, preclinical, or clinical samples [3–9]. RPPA is a complementary proteomic method to mass spectrometry. In contrast to *de-novo* identification of protein markers, RPPA provides in depth posttranslational analysis of

Iulia M. Lazar et al. (eds.), *Proteomics for Drug Discovery: Methods and Protocols*, Methods in Molecular Biology, vol. 1647, DOI 10.1007/978-1-4939-7201-2_10, © Springer Science+Business Media LLC 2017

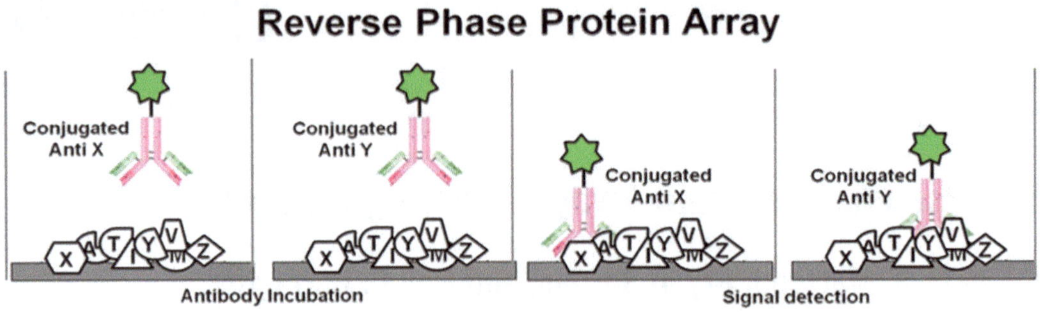

Fig. 1 Schematic of the RPPA principle. A complex mixture of protein lysates extracted from cells or tissues is deposited onto a solid substrate such as a Nitrocellulose coated slide or functionalized protein binding optical planar waveguide chip. Small volumes of each lysate sample are deposited across multiple arrays, which are physically separated by automated printing. Each array is incubated with a separate antibody to enable paralleled multiplex analysis. Antibody binding is detected by conjugated antibody labels or addition of secondary antibody detection reagents followed by fluorescent signal detection using an appropriate slide scanner

known pathways. Tangible benefits of using RPPA include: precise quantification of multiple pathway responses at a posttranslational level including ratiometric analysis of the activation state of druggable pathways; sample numbers are not limited by reagent costs or instrument throughput enabling application across multiple dose-response and time-series studies; and, ultrasensitive detection of low-abundant proteins and phospho-epitope markers from small samples.

Key drug discovery applications of RPPA include:

1. Lead/candidate drug mechanism-of-action profiling: Quantifying on- and off-target activities at the posttranslational level.
2. Compound library screening across selected pathway nodes.
3. Biomarker discovery: In clinical and preclinical studies linked to drug sensitivity, pharmacodynamic dose-response and phenotypic outcome.
4. Predictive pharmacodynamics: Monitoring organ- and tissue-specific pathway effects of a compound in vivo.
5. Drug combinations: Elucidate feed-back and compensatory signaling mechanisms to identify rational drug combination hypothesis. Investigate the mechanistic basis of additive and synergistic activity through pharmacodynamic analysis of pathway response.
6. Exploring new target biology: Profiling downstream signaling events and pathway signaling cross-talk following si/shRNA/CRISPR-Cas9 or other genetic knockout studies.

Key factors for the generation of optimal RPPA results include: careful consideration of experimental design and procedures to ensure high quality sample preparation; reliable printing technology with precise sample positioning and consistent sample distribution; defined and tested print surface chemistry; robustly tested and validated detection antibodies; and, a sensitive detection system coupled with appropriate image analysis software and quality control measures to ensure robustness of the pathway data. Outlined here are protocols for RPPA performed on two distinct RPPA platforms (Zeptosens and Nitrocellulose-based RPPA), which employ different protein-binding substrates, reagents, and procedures. Both platforms allow high-throughput analysis of multiple samples, unlimited multiplexing of detection antibodies, and flexibility of target selection between studies. Although the platforms perform broadly similar functions, each system has unique strengths (and limitations) summarized here, which should be considered and harnessed to the best advantage for each RPPA study.

2 Materials

2.1 Nitrocellulose Arrays

1. Complete cell lysis buffer (1×): 1% Triton X-100, 50 mM HEPES (pH 7.4), 150 mM NaCl, 1.5 mM $MgCl_2$, 1 mM EGTA, 100 mM NaF, 10 mM $Na_4P_2O_7$, 1 mM Na_3VO_4, 10% glycerol, freshly added protease and phosphatase inhibitors (Complete Mini and PhosStop, Roche Applied Science) (*see* **Note 1**).
2. Sample buffer (4×): 40% Glycerol, 8% SDS, 0.25 M Tris–HCL, pH 6.8. Prepare 100 mL stock solution by adding 40 mL glycerol to 8 g SDS dissolved in 25 mL of 1 M Tris buffer stock (pH 6.8) and 25 mL deionized water. Filter the 4× sample buffer by using a 0.2 μm filter and store the stock solution at room temperature. Prior to use, add 100 μL beta mercaptoethanol to 900 μL of 4× sample buffer.
3. Hot block at 95 °C.
4. Sample dilution buffer: 90% PBS, 10% glycerol (v/v).
5. Tissue disruption: FastPrep-24 tissue and cell homogenizer (MP Biomedicals).
6. Lysing matrix-D, 2 mL vials (MP Biomedicals) (*see* **Note 2**).
7. Scalpel, forceps and cutting tile.
8. Slides, Oncyte Avid single pad (Grace Bio-labs) (*see* **Note 3**).
9. 96-well v-bottomed polypropylene plates.
10. 384-well polystyrene plates (Genetix).
11. Array printing: Aushon 2470 micro arrayer (Aushon Biosystems), fitted with 185 μm pins.

12. Slide processing trays (ideally, with compartments just larger than the slides).
13. Antigen retrieval reagent (ReBlot Plus Strong Solution (10×), Millipore Corp.)
14. Validated antibody panel (*see* **Note 4**).
15. Dylight-800-labeled anti-species antibody (i.e., anti-rabbit IgG and anti-mouse IgG).
16. Slide blocking: SuperBlock T20 (TBS) (Thermo Scientific).
17. Phosphate buffered saline (PBS) solution: PBS, pH 7.3. Dissolve 10 PBS tablets (e.g., Oxoid™) in 1 L of deionized water and autoclaved to sterilize.
18. PBS-Tween 20 solution (PBST, 1×): PBS supplemented with Tween-20 at 0.1% v/v.
19. Tris buffered saline-Tween 20 solution (TBST, 1×): TBS 10× stock is prepared as follows: 24 g Tris–HCl, 5.6 g Tris-Base, and 88 g of NaCl are dissolved in 900 mL of deionized water. The pH is adjusted to 7.6 with concentrated HCl and the volume is made up to 1 L with deionized water. To make 1× TBST dilute 10× TBS with deionized water and add Tween-20 at 0.1% v/v).
20. Antibody hybridization cassettes: 4 × 16 pad array format, 4 × 8 pad format (ArrayIt).
21. Array scanning: Innoscan 710 slide scanner (Innopsys).
22. Array analysis: Mapix software (Innopsys).

2.2 Zeptosens Arrays

1. Cell lysis buffer: CLB1 buffer (Bayer Technology Services) (*see* **Note 5**).
2. Sample spotting buffer: CSBL1 buffer (Bayer Technology Services).
3. Tissue disruption: FastPrep-24 tissue and cell homogenizer (MP Biomedicals).
4. Lysing matrix D, 2 mL vials (MP Biomedicals) (*see* **Note 2**).
5. Scalpel, forceps, and cutting tile.
6. Micro-centrifuge tubes, 1.5 mL.
7. 96-well v-bottomed polypropylene plates.
8. 384-well polypropylene plates (Greiner Bio One).
9. Slides: ZeptoMARK chips (Bayer Technology Services).
10. Array printing: Gesim Nano-plotter 2.1E.
11. Slide blocking: Buffer BB1 (Bayer Technology Services).
12. CAB1 buffer (9030 Bayer Technology Services).
13. CAB2 buffer (9035 Bayer Technology Service).

Table 1
Test solutions of BSA-fluorofore conjugate in RDB1 and CSBR1

Solution B (μL)	RDB1 buffer (μL)	CSBR1 buffer (μL)
125	875	9000
250	750	9000
375	625	9000
500	500	9000
625	375	9000

14. ZeptoFOG blocking station (Bayer Technology Services).
15. Validated antibody panel (*see* **Note 4**).
16. Fluorescent BSA standards: Prepare a stock solution of BSA-fluorophore conjugate (AlexaFluor 647 BSA conjugate, Invitrogen) by dissolving 5 mg/mL BSA-fluorophore in PBS supplemented with 0.1% sodium azide and freeze at −20 °C in 10 μL aliquots. Perform a 2-step dilution of the BSA-fluorophore stock solution by adding 2 μL of the BSA stock solution to 198 μL of buffer RDB1 dilution buffer (Bayer Technology Services) (solution A) and further dilute (1:100) by adding 20 μL of this solution to 1980 μL of RDB1 buffer (solution B). The optimal concentration of the labeled BSA standard in RDB1 and CSBR1 reference spotting buffer (Bayer Technology Services) is determined by performing a dilution series of standards as shown in Table 1. The dilution that gives the best resolution and exposure times within the dynamic range of the ZeptoReader in test printing is adopted as the reference BSA standard and aliquots are frozen at −20 °C for use (they may be stored for up to 6 months).
17. Array processing: ZeptoCARRIER hybridization cassette.
18. Array scanning: ZeptoREADER, planar wave guide technology.
19. Array analysis: ZeptoVIEW 3.1 image analysis software.

2.3 Bradford Assay

1. Coomassie Plus Protein Assay (Thermo Scientific).
2. BSA protein standard stock: 2 mg/mL in distilled water.
3. BSA protein standard serial dilution set: Prepare a set of 8 BSA standards at a concentration of 0, 0.05, 0.1, 0.15, 0.2, 0.3, 0.4, and 0.6 mg/mL by adding 0, 2.5, 5, 10, 15, 20 or 30 μL of the 2 mg/mL BSA stock solution to PBS to give a final volume of 95 μL. Add to each standard solution 5 μL of the lysis buffer used for sample preparation (1:20 by volume).
4. Flat-bottomed 96-well plate.
5. Microplate reader fitted with 595 nm filter.

2.4 Fast Green Staining

1. Destaining solution: 30% Methanol, 7% glacial acetic acid, 63% distilled water (v/v). Mix 300 mL methanol, 630 mL distilled water and 70 mL glacial acidic acid.
2. Fast Green stock solution (400×): Add 0.1 g Fast Green to 9.9 mL destaining solution.
3. Fast Green staining solution (1×): Add 2.5 mL Fast Green stock solution (400×) to 997.5 mL destaining solution.
4. Sodium hydroxide solution: 1% (i.e., 0.25 M). Add 1.0 g NaOH pellets per 100 mL of deionized water.

3 Methods

Prior to performing any experiment it is advisable to perform a test lysis of a 6-well plate in which cell(s) have been seeded at a range of densities (e.g., 2.5×10^5, 5×10^5, and 1×10^6 cells/well). This will establish the protein yield per well and will inform the design of the experiment. The cell number per well is dependent on the cell size, cell growth rate, and experimental design (e.g., duration of treatment). Generally, $0.3–0.5 \times 10^6$ cells/3 mL growth medium is recommended in each well of 6-well plates. Once the preliminary test has been successfully performed, and protein concentrations within the required range are achieved, cell plates should be prepared at the predetermined optimal seeding density required to provide adequate protein concentrations between 1 and 2 mg/mL and treated according to the experimental design. Where possible, positive and negative control samples should be included in each sample set. The power of RPPA to process (relatively) high sample numbers lends itself well to the analysis of both time and concentration-dependent effects of the agent(s) under investigation.

3.1 Nitrocellulose RPPA Sample Preparation from Cultured Cells

1. Remove the culture medium from plates and wash the cells twice with PBS. Remove all PBS by aspiration (*see* **Note 6**). For cells grown in suspension centrifuge the cell suspension for 2 min at 300 × *g*. Discard the supernatant and resuspend pellets in cold PBS (using about 1/10 of starting volume). Recentrifuge the cell suspension for 2 min at 300 × *g* at 4 °C. For cells grown in high serum content this wash should be repeated once more.
2. Add lysis buffer to the cells (suggested volumes are 75–150 μL for each well of a 6-well plate and 250–400 μL for a 10 cm dish). Keep plates on ice at all times.
3. Ensure the lysis buffer is spread over the entire plate surface (*see* **Note 7**). Scrape cells off the plates and collect the cell lysate into micro-centrifuge tubes. Incubate the samples on ice for 20 min with occasional shaking every 5 min.

4. Centrifuge the cell lysate at maximum speed (16,000 × *g*) for 10 min at 4 °C.
5. Carefully collect the supernatant, discarding the pellet.
6. Determine the cellular protein concentration by the Bradford assay (*see* Subheading 3.5), and adjust all samples to the same concentration (e.g., 2 mg/mL).
7. Mix the cell lysate with 4× sample buffer (3 parts of cell lysate plus one part of 4× sample buffer). Heat the samples to 95 °C for 5 min. The samples are ready for RPPA processing and may be stored at −80 °C.

3.2 Nitrocellulose RPPA Sample Preparation from Tissue Material

1. Where possible, a fragment of frozen tumor tissue is cut on dry-ice using a scalpel, forceps and a precooled cutting tile (*see* **Note 8**). The cut sample is kept frozen in a labeled micro-centrifuge tube on dry-ice.
2. Precool the lysing matrix-D tubes on ice (*see* **Note 2**). Add the frozen tissue material immediately prior to disruption using the FastPrep-24 homogenizer. Use the recommended speed and duration from the Fastprep-24 manual for specific tissues.
3. Place the vials immediately back on ice after homogenization. The process may need to be repeated if disruption is incomplete (*see* **Note 9**).
4. Add the lysis buffer to lysis matrix-D tubes and incubate on ice for 30 min with occasional vortex mixing.
5. Centrifuge the samples at 4 °C for 10 min using a benchtop micro-centrifuge at maximum speed (16,000 × *g*).
6. Collect the supernatant (tumor lysates) and transfer to a fresh set of pre-labeled micro-centrifuge tubes (*see* **Note 10**).
7. Determine the cellular protein concentration by the Bradford assay, and adjust all samples to the same concentration (e.g., 2 mg/mL).
8. Mix the cell lysate with 4× sample buffer (3 parts of cell lysate plus one part of 4× sample buffer). Heat the samples to 95 °C for 5 min. The samples are ready for RPPA processing and may be stored at −80 °C.

3.3 Zeptosens RPPA Sample Preparation from Cultured Cells

1. Remove the culture medium from plates and wash the cells twice with PBS. Remove all PBS by aspiration (*see* **Note 6**). For cells grown in suspension, centrifuge the cell suspension for 2 min at 300 × *g*. Discard the supernatant and re-suspend pellets in cold PBS (using about 1/10 of starting volume). Re-centrifuge the cell suspension for 2 min at 300 × *g* at 4 °C. For cells grown in high serum content this wash should be repeated once more.

2. Add lysis buffer to the cells (suggested volumes are 75–150 µL for each well of a 6-well plate and 250–400 µL for a 10 cm dish) (*see* **Note 5**).
3. Ensure the lysis buffer is spread over the entire plate surface (*see* **Note 7**). Scrape cells off the plates and collect the cell lysate into micro-centrifuge tubes. Incubate the samples at room temperature for 30 min with occasional shaking every 5 min.
4. Centrifuge the cell lysate at 16,000 × *g* for 10 min at room temperature.
5. Carefully collect the supernatant, discarding the pellet.
6. Determine the cellular protein concentration by the Bradford assay, and adjust all samples to the same concentration (e.g., 2 mg/mL).
7. The samples are ready for RPPA processing and may be stored at −80 °C.

3.4 Zeptosens RPPA Sample Preparation from Tissue Material

1. Where possible, a fragment of frozen tumor tissue is cut on dry-ice using a scalpel, forceps and a pre-cooled cutting tile (*see* **Note 8**). The cut sample is kept frozen in a labeled micro-centrifuge tube on dry-ice.
2. Pre-cool the lysing matrix-D tubes on ice (*see* **Note 2**). Add the frozen tissue material immediately prior to disruption using the FastPrep-24 homogenizer. Use the recommended speed and duration from the Fastprep-24 manual for specific tissues.
3. Place vials immediately back on ice after homogenisation. The process may need to be repeated if disruption is incomplete (*see* **Note 9**).
4. Once homogenisation is achieved, remove the vials from ice and allow to warm up briefly (2–5 min should be sufficient), and add CLB1 Buffer to the lysis matrix-D tubes and incubate at room temperature for 30 min with occasional vortex mixing.
5. Centrifuge the samples at room temperature for 10 min using a benchtop micro-centrifuge at maximum speed (16,000 × *g*).
6. Collect the supernatant (tumor lysates) and transfer to a fresh set of pre-labeled micro-centrifuge tubes (*see* **Note 10**).
7. Determine cellular protein concentration by the Bradford assay, and adjust all samples to the same concentration (e.g., 2 mg/mL).
8. The samples are ready for RPPA processing and may be stored at −80 °C.

3.5 Bradford Assay

1. Prepare 50 µL of each sample (*see* Subheadings 3.1–3.4) by diluting the sample 1:20 (v/v) in PBS (i.e., add 2.5 µL sample to 47.5 µL of PBS).

2. Add 10 μL of the BSA standards and the diluted samples into duplicate wells of the 96-well plate.
3. Add 240 μL of Coomassie reagent (pre-warmed to room temperature) to each well, avoiding air bubbles. Incubate for 10 min at room temperature.
4. Read the absorbance in a plate reader at 595 nm.
5. Import the data into Excel for plotting of standard curve and calculation of sample concentrations.
6. Check that the standard curve falls within the defined criteria and ensure that all diluted samples fall within the range between the lowest and highest concentration BSA standards.
7. Adjust the protein concentrations of samples by adding cell lysis buffer to each sample as determined from the protein assay.

3.6 Nitrocellulose RPPA

1. Transfer the samples to be assayed to a 96-well plate and use a multi-channel pipette to prepare a dilution series of 100, 50, 25, and 12.5% for each sample using PBS supplemented with 10% glycerol as the diluent. A minimum volume of 10 μL is required for each dilution step per printing pin in the Aushon arrayer print head (e.g., if printing with 8 pins a minimum volume of 80 μL is required for all sample dilutions). Dilution steps are performed in a 96-well plate with an 8 tip multichannel pipette as shown below. A sample dilution series of 100, 50, 25, and 12.5% is thus generated for all samples in the study.
   Row 1: add 160 μL of undiluted samples 1 to 8.
   Row 2: transfer 80 μL from row 1 and add to 80 μL of PBS/10% glycerol, mix well.
   Row 3: transfer 80 μL from row 2 and add to 80 μL of PBS/10% glycerol, mix well.
   Row 4: transfer 80 μL from row 3 and add to 80 μL of PBS/10% glycerol, mix well.
2. Transfer the diluted samples using a multi-channel pipette to a 384-well plate in a layout to match the pin configuration and slide/array layout.
3. Print the diluted samples onto slides using a customized printing program; all samples are printed in triplicate on all arrays using the Aushon 2470 miroarray printing platform.
4. Printing is performed at 70% humidity and samples are distributed in source plates such that each plate is open for printing for a maximum of 1 h (*see* **Note 11**).
5. Incubate the slides after sample printing for 1 h to ensure sample capture on the nitrocellulose (*see* **Note 12**).
6. Wash the RPPA slides (four times, 15 min each) in dH_2O water on a rocking platform.

7. Incubate the RPPA slides with antigen retrieval reagent (Reblot Strong diluted 1:10 with dH_2O) for 10 min; the reagent is removed and the slides are washed with dH_2O (two times, 5 min each) ensuring that the slides are covered (*see* **Note 13**).
8. Remove slides from the dH_2O wash and place in the ArrayIt cassette, filling wells with fresh deionized water to prevent the slides from drying out.
9. Wash slides with 1× PBST (two times, 5 min each).
10. Incubate slides for 10 min with SuperBlock T20 (TBS).
11. Wash slides with 1× TBST (two times, 5 min each).
12. Incubate individual RPPA arrays with primary antibody diluted 1:250 in SuperBlock T20 (60 min on a rocking platform) (*see* **Note 14**).
13. Wash RPPA arrays with 1× TBST (two times, 5 min each).
14. Incubate slides for 10 min with SuperBlock T20 (TBS).
15. Wash slides with 1× TBST (three times, 5 min each).
16. Incubate RPPA arrays with secondary antibody (Dylight-800-labeled anti-species antibody diluted 1:2500 in SuperBlock T20) for 30 min on a shaking platform, in the dark.
17. Wash slides with 1× TBST (two times, 5 min each).
18. Wash slides with dH_2O and centrifuge to remove excess water.
19. Incubate slides at room temperature, in the dark for 10 min to dry prior to data acquisition.
20. RPPA arrays are imaged using the slide scanner with settings adjusted for maximal signal but avoiding saturation.

3.7 Zeptosens RPPA

1. For each Zeptosens assay parameter files are created with details of all samples and antibodies to be tested (*see* **Note 15**).
2. Transfer 20 μL of all samples to be assayed to a 96-well plate and dilute 1:10 with spotting buffer (CSBL1). A 4-step dilution series is performed using a multi-channel pipette to give a dilution series of 100, 75, 50, and 25% for each sample using CSBL1 supplemented with 10% CLB1 (lysis buffer) as the diluent. Dilution steps are performed in a 96-well plate using an 8-tip multichannel pipette as shown below. A sample dilution series of 100, 75, 50, and 25% is thus generated for all samples.
   Row 1: add 20 μL of undiluted samples (in order 1 to 8) to 180 μL of CSBL1 buffer, mix well.
   Row 2: transfer 60 μL from row 1 and add to 20 μL of CSBL1/CLB1 (90:10), mix well.
   Row 3: transfer 40 μL from row 1 and add to 40 μL of CSBL1/CLB1 (90:10), mix well.

Row 4: transfer 20 μL from row 1 and add to 60 μL of CSBL1/CLB1 (90:10), mix well.

3. Transfer 40 μL of the diluted samples, using a multi-channel pipette, to a 384-well plate in a layout to match the pin configuration and ZeptoChip/array layout.
4. Print samples onto ZeptoChips using a standard printing program; all slides are printed with fluorescently labeled BSA standards using the GeSim Nanoplotter 2.1E.
5. Printing is performed at 70% humidity and the slide platform is maintained at 14 °C to minimize sample evaporation during the printing process.
6. Air-dry ZeptoChips following sample printing for 2 h.
7. Place ZeptoChips in blocking racks with the spots facing upward and block for 1.5 h using the ZeptoFOG blocking station and 100 mL of blocking buffer BB1.
8. Wash ZeptoChips (three times, 2 min each) in distilled water and subsequently centrifuge at 200 × *g* for 5 min at room temperature to remove excess water.
9. Place ZeptoChips into custom-designed Zeptocarriers for subsequent addition of buffers and antibodies according to handling instructions in the Zeptosens manual.
10. Add 90 μL of CAB1 buffer to each array and remove by aspiration; repeat this step three times (*see* **Note 16**).
11. Add 90 μL of appropriately diluted primary antibody to each array (in a predefined pattern as determined by the study parameter file) and incubate arrays at room temperature overnight.
12. Remove primary antibody by aspiration from Zeptocarriers and wash arrays three times with CAB1 as described previously.
13. Add 90 μL of the appropriately diluted secondary antibody and incubate arrays for 2.5 h in the dark.
14. Remove unbound secondary antibody by aspiration and wash arrays three times with CAB1 as described above. The last CAB1 wash is left on the slide.
15. Slides are scanned using the ZeptoReader.

3.8 Data Analysis

1. Zeptosens data analysis. Fluorescence readout of antibody binding to samples deposited on the Zeptosens arrays is conducted using the planar waveguide ZeptoREADER system (Bayer Technology Services), typically in the red-end of the visible spectrum at an excitation wavelength of 635 nm and an emission wavelength of 670 nm. Multiple fluorescence signal exposures are acquired for each array and integrated over a predefined time period depending on the signal intensity.

Array images are stored as 16-bit TIFF files and analyzed with the ZeptoVIEW3.1 software package (Bayer Technology Services). Each sample 4-point concentration series is spotted onto the microarray chip between Alexa Fluor-conjugated BSA standards. Fluorescence intensity signals of each sample are calculated by optimized image analysis algorithms and normalized to intensity values of BSA standards through a local 2D quadratic function. The ZeptoView software selects the brightest image exposure for each array, which is not saturating the detection camera to ensure optimal sensitivity and linearity across each dilution series for each antibody. A single relative fluorescence intensity (RFI) value is obtained by a weighted linear fit through sample dilutions representing the relative protein abundance across the sample series [2]. Quality control parameters for each sample are obtained by Sahpiro-Wilk statistical test of intensity distributions across each dilution range.

2. Aushon data analysis. Analysis of the nitrocellulose arrays can be performed across multiple colorimetric and fluorescent slide scanning detection platforms. Irrespective of which slide scanner is applied, each nitrocellulose array requires an array mask file (e.g., a GAL file) which includes all the details of the array layout, sample identification, sample positions as well as relevant dilution information for all samples. In the case of a fluorescent detection slide scanner, measurement of fluorescence of each spot/feature is adjusted automatically by subtracting the local background value (the median background value of the area within 2× spot diameter for each spot) from the average fluorescence value within the spot. The linearity of signal for each antibody for all the samples is determined by plotting the fluorescence values for at least four sample dilutions against the known protein concentrations and checking the R^2 value is acceptable. For a perfect line, an R^2 value of 1 is expected and a cut-off value of 0.9 can be used to flag up problems with either samples or antibodies. Nonlinear values for samples are indicative of a problem either with the samples or antibody and should be investigated further. Various other linear, log-linear, logistic, and nonparametric models for calculating a single value from the sample dilutions series to represent protein abundance have been proposed [10–14]. Care should be taken to ensure the method of analysis selected is most appropriately tailored to each specific RPPA dataset and method of data generation [8].
3. Further comparison between Zeptosens and Nitrocellulose-based RPPA platforms. Zeptomark chips feature planar waveguide technology which dramatically reduces the depth of signal measurement to 100 nm (as opposed to 200 μm field depth with conventional slide scanners), resulting in virtually

no background fluorescence and very high sensitivity. Nitrocellulose slides are widely reported to have background fluorescence around 600 nm but this can be minimized by using detection antibodies labeled with either red or far-red or non-fluorescent-based chromophores. Zeptomark slides are printed with samples diluted in the range of 0.2–0.05 mg/mL with a limit of detection of around 2000 protein molecules per array spot. This makes this technology extremely useful when very high sensitivity is required (e.g., for small expression changes or for detection of relatively low abundance targets). Nitrocellulose slides by comparison have increased background fluorescence, but this is offset somewhat by the increased binding capacity (and therefore sample load) per array spot. Oncyte Avid 3D nitrocellulose slides (Grace Biolabs) are capable of binding 40 μg of protein per cm^2 (approximately 2 ng per 270 μm diameter spot). These considerations promote the use of Zeptosens for low yield samples and when great sensitivity of detection is required (e.g., limited clinical material small 96-well/384-well), but allow for the use of nitrocellulose arrays when sample material is not limited (e.g., cell-based assays). The layout of Zeptosens arrays is largely fixed and constrained by the system analysis software which places limits both on experimental design and sample number per RPPA study. Nitrocellulose arrays are somewhat more flexible and are better adapted to deal with very small or very large sample sets. Compared with Zeptosens arrays, Nitrocellulose RPPA assays are cheaper to perform, mainly because of markedly reduced slide costs.

Zeptosens RPPA is a proprietary system that provides for an all-in-one solution of optimized reagents and consumables to conduct a RPPA study from start-to-finish and thus has very stringent steps covering sample generation, processing, printing, assaying, and analysis. Assay integrity is ensured by using a barcode tracking system linked to locked parameter files which are generated for each study. It is possible to perform nitrocellulose arrays to the same standards but considerable workup is required by users in developing a comparable start-to-finish system. With standard operating procedures in place for all the assay steps, both platforms are capable of delivering a service that is compatible with GLP (Good Laboratory Practice), GCLP (Good Clinical Laboratory Practice), CAP (College of American Pathologists), and CLIA (Clinical Laboratory Improvement Act) certifications, and, as such, are potentially invaluable tools in both research laboratory and clinical laboratory settings.

4 Notes

1. For ease of handling, 2× stock buffer may be prepared, filtered to sterilize, and stored at 4 °C. Concentrated stocks of NaF, $Na_4P_2O_7$, and Na_3VO_4 may be prepared and frozen at −20 °C. Just prior to use the inhibitors, cocktail tablets and glycerol are added to the 2× stock buffer and the volume is adjusted with dH_2O. The buffer is subsequently kept ice cold. The buffer stock solutions are prepared as follows:
   (a) HEPES (2-[4-(2-hydroxyethyl)piperazin-1-yl]ethanesulfonic acid) solution: 0.5 M HEPES, pH 7.4. Dissolve 23.83 g HEPES in 150 mL deionized water and adjust the pH to 7.4 with 3 M NaOH. Adjust the volume to 200 mL with deionized water.
   (b) EGTA (Ethylene glycol-bis(2-aminoethylether)-*N*,*N*,*N*′,*N*′-tetraacetic acid) solution: 0.5 M EGTA. Add 19 g EGTA to 90 mL deionized water. To facilitate the EGTA dissolution in water, add 3.5–4 g of solid NaOH pellets. Check for the pH to be in the range of 7.5–8. Adjust further the pH, if necessary, by adding more NaOH. Adjust the volume to 100 mL with deionized water.
   (c) Sodium chloride solution: 1.0 M NaCl. Add 14.61 g NaCl per 250 mL deionized water.
   (d) Magnesium chloride solution: 1.0 M $MgCl_2$. Add 20.33 g $MgCl_2$ per 100 mL deionized water.
   (e) Lysis buffer (2×): Prepare 250 mL lysis buffer by measuring 118.25 mL deionized water, and adding 5 mL of Triton X-100, 50 mL of 0.5 M HEPES, 75 mL of 1.0 M NaCl, 1 mL of 0.5 M EGTA, and 750 μL of 1.0 M $MgCl_2$. The 2× stock lysis buffer is filtered through a 0.2 μm filter to sterilize and is stored at 4 °C.
   (f) Sodium orthovanadate solution (100×): 100 mM Na_3VO_4. Add 0.9195 g to 40 mL deionized water and stir to dissolve. Adjust the pH to 10 by dropwise addition of concentrated HCl. The solution will now be yellow. Boil the solution until it becomes clear and allow it to cool. Check the pH and adjust again to pH 10 by the addition of 1 M HCl and re-boil until the solution becomes colorless. This step may need to be repeated several times until the solution is colorless when cooled and at pH 10. The inhibitor can be aliquoted and stored at −20 °C.

(g) Sodium pyrophosphate tetrabasic (10×): 100 mM $Na_4P_2O_7$. Add 2.65 g $Na_4P_2O_7$ to 100 mL deionized water, aliquot and store at −20 °C.

(h) Sodium fluoride (10×): 1 M NaF. Add 4.2 g NaF per 100 mL deionized water, aliquot and store at −20 °C.

(i) Complete lysis buffer (1×): Measure 5.0 mL of 2× stock buffer, add 1 mL of glycerol, 1.0 mL of $Na_4P_2O_7$ (10×), 1.0 mL of NaF (10×), 0.1 mL of Na_3VO_4 (100×), 1.9 mL of deionized water, 1 complete mini protease inhibitor tablet and 1 PhosStop phosphatase inhibitor tablet. Stir until dissolved and keep on ice.

2. The lysis matrix tubes are cooled on ice (not on dry ice as this makes them brittle and liable to crack) and the tissue fragment is added immediately prior to disruption. Samples are pulverized without lysis buffer and immediately placed on ice (for nitrocellulose RPPA) whereupon 300 μL of lysis buffer is added according to the protocol.

3. Different nitrocellulose slide formats (e.g., 16 pad slides, 8 pad slides) are available, but using a single pad format allows flexibility of slide layout and we have not seen any evidence of leakage using this format using the ArrayIt hybridization cassettes.

4. Antibody validation is a fundamental and key component of successful RPPA technology. Lists of validated antibodies may be found online (MD Anderson RPPA core facility: https://www.mdanderson.org/education-and-research/resources-for-professionals/scientific-resources/core-facilities-and-services/functional-proteomics-rppa-core/antibody-lists-protocols/functional-proteomics-reverse-phase-protein-array-core-facility-antibody-lists-and-protocols.html). The more exhaustive and extensive the validation process the better in terms of confidence in the data produced. In reality, however, testing is always limited by resources and time, but a minimum test set involving several cell lines across two or more species, and at least two different tissue samples, should be tested by western blot before a new antibody is tentatively added to a validated list. It is also advised that any key or significant expression changes observed in an RPPA data-set should be corroborated by testing some of the samples by subsequent western blotting. The accepted minimum criteria for antibody specificity are explainable bands in a Western blot and good correlation between Western blot and RPPA results [8].

5. For Zeptosens analyses, all extractions involving the use of CLB1 buffer must be performed at room temperature and not on ice to avoid buffer precipitation.

6. Removal of all PBS is vital to ensure that the lysis buffer is not diluted and also to maximize protein concentration. Placing the plate at a slight angle on ice for 1–2 min will allow the PBS to drain off to the edge of the well where it can be removed with a pipette or a pastette.
7. Because the volume of lysis buffer is typically small, it is advised to scrape the cells immediately after the addition of lysis buffer. In this way, all cells are lysed and degradation of the sample proteins is kept to a minimum.
8. Small tissue fragments (approximately the size of a grain of rice) may be cut from the tissue to be assayed using a scalpel and forceps on a cooled cutting tray (or tile on a bed of dry ice). To prevent loss of material when cutting frozen tissue samples, the cutting tile may be placed on the top of a layer of tissue on dry-ice inside a polystyrene box to contain fragments of cut tissue.
9. If three rounds of homogenization have not completely disrupted the tissue samples, the lysis buffer may be added and the process repeated. This will result in some frothing of the lysis buffer and a reduction in recovered volume but is normally sufficient for the assay to be performed.
10. If debris from the bead mill is visible, or the sample appears to be cloudy, repeat the centrifugation step to clear the sample before proceeding with the assay.
11. Sample evaporation during printing is reduced by ensuring that source plates are open for less than 1 h each during the printing process. In addition, all the wells that do not contain sample material are filled with distilled water to increase humidity at the plate surface.
12. Printed slides are now ready for processing or may be stored under desiccating conditions at 4 °C until required.
13. Do not perform the antigen retrieval step in the ArrayIt cassette as the reagent can react with the cassette coating and the arrays are compromised.
14. Nonspecific signals are determined for each slide by omitting the incubation step with primary antibody on one array per slide (i.e., Superblock T20 only is added for the 1 h incubation). Fluorescent tracer secondary antibody is added as normal for the second antibody incubation step.
15. Parameter files are linked to printed barcodes that are attached to all Zeptocarriers, ensuring sample tracking throughout the assay and subsequent data analysis.
16. Do not remove the last CAB1 wash until ready to add antibodies to the ZeptoChips.

References

1. Paweletz CP, Charboneau L, Bichsel VE et al (2001) Reverse phase protein microarrays which capture disease progression show activation of pro-survival pathways at the cancer invasion front. Oncogene 20(16):1981–1989
2. Pawlak M, Schick E, Bopp MA, Schneider MJ, Oroszlan P, Ehrat M (2002) Zeptosens' protein microarrays: a novel high performance microarray platform for low abundance protein analysis. Proteomics 2(4):383–393
3. Weissenstein U, Schneider MJ, Pawlak M et al (2006) Protein chip based miniaturized assay for the simultaneous quantitative monitoring of cancer biomarkers in tissue extracts. Proteomics 6(5):1427–1436
4. Boyd ZS, Wu QJ, O'brien C et al (2008) Proteomic analysis of breast cancer molecular subtypes and biomarkers of response to targeted kinase inhibitors using reverse-phase protein microarrays. Mol Cancer Ther 7(12):3695–3706
5. Voshol H, Ehrat M, Traenkle J, Bertrand E, Van Oostrum J (2009) Antibody-based proteomics: analysis of signaling networks using reverse protein arrays. FEBS J 276(23):6871–6879
6. Cardnell RJ, Feng Y, Diao L et al (2013) Proteomic markers of DNA repair and PI3K pathway activation predict response to the PARP inhibitor BMN 673 in small cell lung cancer. Clin Cancer Res 19(22):6322–6328
7. Girotti MR, Lopes F, Preece N et al (2015) Paradox-breaking RAF inhibitors that also target SRC are effective in drug-resistant BRAF mutant melanoma. Cancer Cell 27(1):85–96
8. Akbani R, Becker KF, Carragher N et al (2014) Realizing the promise of reverse phase protein arrays for clinical, translational, and basic research: a workshop report: the RPPA (reverse phase protein array) society. Mol Cell Proteomics 13(7):1625–1643
9. Cancer Genome Atlas Network (2012) Comprehensive molecular portraits of human breast tumours. Nature 490(7418):61–70
10. Mircean C, Shmulevich I, Cogdell D et al (2005) Robust estimation of protein expression ratios with lysate microarray technology. Bioinformatics 21(9):1935–1942
11. Nishizuka S, Charboneau L, Young L et al (2003) Proteomic profiling of the NCI-60 cancer cell lines using new high-density reverse-phase lysate microarrays. Proc Natl Acad Sci U S A 100(24):14229–14234
12. Hu J, He X, Baggerly KA, Coombes KR, Hennessy BT, Mills GB (2007) Non-parametric quantification of protein lysate arrays. Bioinformatics 23(15):1986–1994
13. Troncale S, Barbet A, Coulibaly L et al (2012) NormaCurve: a SuperCurve-based method that simultaneously quantifies and normalizes reverse phase protein array data. PLoS One 7(6):e38686
14. Sun M, Lai D, Zhang L, Huang X (2015) Modified SuperCurve method for analysis of reverse-phase protein Array data. J Comput Biol 22(8):765–769

Check for updates

Chapter 11

Probing Protein Kinase-ATP Interactions Using a Fluorescent ATP Analog

Leslie E.W. LaConte, Sarika Srivastava, and Konark Mukherjee

Abstract

Eukaryotic protein kinases are an intensely investigated class of enzymes which have garnered attention due to their usefulness as drug targets. Determining the regulation of ATP binding to a protein kinase is not only critical for understanding function in a cellular context but also for designing kinase-specific molecular inhibitors. Here, we provide a general procedure for characterizing ATP binding to eukaryotic protein kinases. The protocol can be adapted to identify the conditions under which a particular kinase is activated. The approach is simple, requiring only a fluorescent ATP analog such as TNP-ATP or MANT-ATP and an instrument to monitor changes in fluorescence. Although the interaction kinetics between a kinase and a given ATP analog may differ from that of native ATP, this disadvantage is offset by the ease of performing and interpreting this assay. Importantly, it can be optimized to probe a large variety of conditions under which the kinase-nucleotide binding might be affected.

Key words Kinase, Nucleotide, ATP binding, CASK, Pseudokinase, TNP-ATP, Fluorescence

1 Introduction

Eukaryotic protein kinases are one of the largest gene families and constitute 2% of the proteome [1]. Protein kinases function as versatile molecular switches which play a critical role in the fundamental biology of the cell. Dysregulation of kinase activity is the cause of various diseases including certain forms of cancer [2, 3]. The enzymatic functions of these phosphotransferases are tightly regulated by various molecular strategies, a prominent one being obstruction of nucleotide binding [4]. The regulation of nucleotide binding to protein kinases therefore is a field of intense investigation both in basic science and in drug development [5, 6]. In fact, kinases represent the most druggable component of the human proteome [7]. A number of screening technologies for kinase

The original version of this chapter was revised. An erratum to this chapter can be found at https://doi.org/10.1007/978-1-4939-7201-2_20

Iulia M. Lazar et al. (eds.), *Proteomics for Drug Discovery: Methods and Protocols*, Methods in Molecular Biology, vol. 1647, DOI 10.1007/978-1-4939-7201-2_11, © Springer Science+Business Media LLC 2017

profiling have been reported. Many of the kinase assays use radioactive reagents such as γ^{32}P-ATP [8] or are based on luminescence from the luciferase system [9]. Both the systems require some knowledge of the kinase's substrate, which may not always be available. An alternative approach to testing kinase activity is to examine the affinity of nucleotide binding to its hydrophobic pocket. Indeed, most of the kinase inhibitors discovered to date inhibit ATP binding to kinases either by directly binding to the ATP binding pocket (type I inhibitors) or to adjacent sites (type II inhibitors) [10]. A high throughput and rapid method for testing nucleotide binding to kinases may therefore be a very useful and versatile tool for the study of kinases and for drug discovery. Here, we describe the use of fluorescent ATP analogs like 2′-(or-3′)-O-(trinitrophenyl) adenosine 5′-triphosphate (TNP-ATP) or (2′-(or-3′)-O-(N-Methylanthraniloyl) adenosine 5′-triphosphate (MANT-ATP) to measure the ATP interaction with kinases. This method has been effectively used for a number of different purposes: to test specific binding conditions of ATP to proteins initially described as pseudokinases such as CASK and JAK2 [11, 12], to test how specific mutations affect ATP binding properties [13], to test nucleotide binding in known eukaryotic protein kinases [14], and as a high-throughput assay to screen for inhibitors for bacterial kinases [15]. The advantages of the method include ease of experimental design, no requirement for radioactive reagents, and ease of interpretation. Additionally, there is no need for large quantities of protein, and detection is not dependent upon the protein undergoing a large conformational change, as is required by some assays but does not happen with all kinases.

In the described assay, we use TNP-ATP as a fluorescent probe to detect the ATP interaction with eukaryotic protein kinases. Upon the addition of a protein which can bind TNP-ATP, there is a three to fivefold increase in fluorescence, and a clear blue shift in emission maxima is observed (Fig. 1a) [16]. In the example data provided from the unusual kinase, CASK, a distinct leftward shift and increase in peak fluorescence intensity is observed between the spectrum of TNP-ATP alone shown in blue and the spectrum of TNP-ATP bound by CASK shown in yellow. The increase in fluorescence is instantaneous and does not require incubation time. To ensure specificity, ATP can be used to compete with TNP-ATP binding to the protein of interest, resulting in a measurable reversal of the change in fluorescence intensity and emission maximum (Fig. 1a; note difference between green and yellow curves). This assay is also useful for identifying and investigating kinase inhibitors; for example, the surprising finding that the high-affinity interaction of the CaMK domain of CASK with TNP-ATP is inhibited by magnesium (Fig. 1a, d) laid the groundwork for establishing that the kinase activity of the intact CASK can be also inhibited by magnesium [11, 13, 17].

Although the binding kinetics and binding affinity of TNP-ATP are different from that of ATP, competition assays and K_i

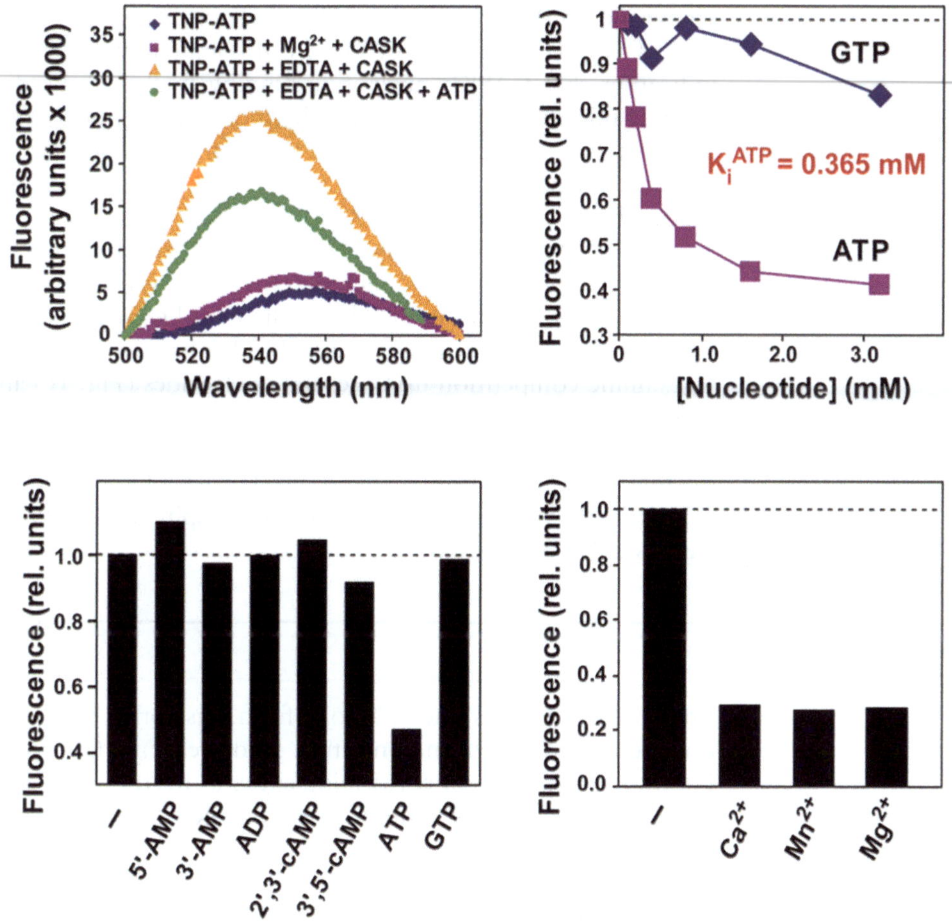

Fig. 1 The TNP-ATP kinase binding assay can be used to compare affinities of different nucleotides to the ATP binding pocket as well as testing for inhibitors. (**a**) Fluorescence emission spectra (excitation = 410 nM) of TNP-ATP (1 μM) in Tris–HCl buffer containing 2 mM $MgCl_2$ in the absence or presence of the CASK CaM-kinase domain (1 μM), after subsequent addition of EDTA (4 mM, to remove the Mg^{2+}) and of Na^+-ATP (500 μM) to the same cuvette, as indicated. (**b**) Titration of the TNP-ATP (1 μM) fluorescence as a function of the GTP and ATP concentration (as indicated) in two parallel cuvettes containing CASK CaM-kinase domain (1 μM) in Tris–HCl buffer supplemented with 4 mM EDTA (excitation = 410 nM; emission = 540 nM). Data shown represents fluorescence units after the subtraction of the TNP-ATP background fluorescence followed by normalization to the initial reading (relative units). (**c**) Inhibition of the TNP-ATP interaction with the CASK CaM-kinase domain by adenine nucleotides. The fluorescence (excitation = 410 nM; emission = 540 nM) of the CASK CaM-kinase domain/TNP-ATP complex (1 μM each) was measured in EDTA (4 mM) before and after the addition of the indicated nucleotides (500 μM each). Fluorescence units were normalized to the respective initial readings following background subtraction (relative units). (**d**) Inhibition of TNP-ATP binding to the CASK CaM-kinase domain by divalent ions. The fluorescence (excitation = 410 nM; emission = 540 nM) of the CASK CaM-kinase domain/TNP-ATP complex (1 μM each) was measured in Tris–HCl buffer supplemented with EDTA (4 mM) or 2 mM of the indicated divalent cations. The data represent normalized fluorescent units following background subtraction (relative units) [11]. Reprinted with permission from ref. 11

determination can be of tremendous value when comparing other nucleotides to establish binding-site specificity of kinases (Fig. 1b, c). This is especially important because some kinases (e.g., casein kinase and CaMKII) can utilize GTP as a substrate [18, 19]. In the case of CASK, ATP, but not GTP, can compete with TNP-ATP, indicating that CASK's nucleotide binding pocket is specific for adenine nucleotides (Fig. 1b). Furthermore, ATP is more effective in competing with TNP-ATP for binding to CASK than other adenine nucleotides, suggesting that CASK has a higher affinity for and binds preferentially to ATP.

The TNP-ATP binding assay can be easily converted to a screen for identifying conditions under which a kinase can bind to TNP-ATP [15]. These assays can be done reliably either in a cuvette or in a plate if a suitable instrument is available. We have used these assays to examine competition using several nucleotides (Fig. 1c) and to determine the effect of divalent ions (Fig. 1d). We have also used this assay to effectively screen for mutant CASK CaMK domains capable of coordinating magnesium (Fig. 2). Our results were subsequently validated by X-ray crystallography and enzymology studies [11, 13].

2 Materials

Prepare all solutions and dilutions fresh, just prior to the assay, using ultrapure water and analytical grade reagents (*see* **Note 1**). Follow steps specific for performing either a cuvette-based binding assay, microplate-based binding assay, or microplate-based saturation assay (*see* **Note 2**).

2.1 Solutions and Sample Preparation for All Assays

1. Tris–HCl buffer: 50 mM Tris–HCl, 50 mM KCl, pH 7.2 (*see* **Note 3**). Dissolve appropriate amounts of Tris base and KCl in water nearly to desired volume with stirring. Use a pH meter to monitor pH during gradual dropwise addition with stirring of a solution of hydrochloric acid until pH is 7.2. Bring Tris–HCl buffer to final volume with water. The buffer can be stored at room temperature for several months.
2. Lysozyme control stock solution: 420 μM (*see* **Note 4**). Prepare the protein control sample by dissolving 6 mg of lyophilized lysozyme in 1 mL of Tris–HCl buffer (molecular weight of lysozyme is 14.3 kDa).
3. Lysozyme control assay solution: 4 μM. Add 10 μL of the 420 μM lysozyme control stock solution to 990 μL of Tris–HCl buffer.
4. Protein sample: 1–4 μM protein sample (*see* **Note 5**). Prepare the kinase sample of interest by determining the concentration of the purified (*see* **Notes 6–8**) protein (using the Bradford assay or similar method of protein concentration determination). For best results, initial concentration should be at least 50–100 μM. The protein sample should be diluted from the

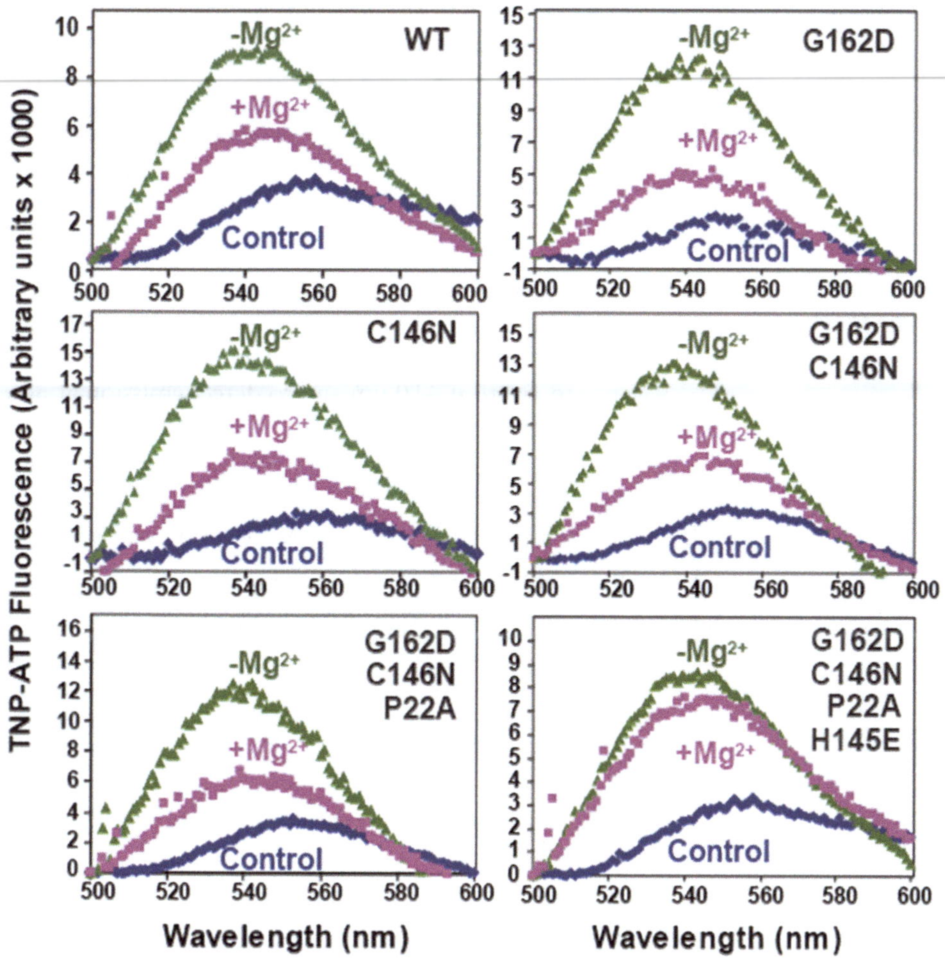

Fig. 2 The TNP-ATP binding assay provides a robust method to test the effects of point mutations in a kinase domain. Fluorescence emission spectra of TNP-ATP binding in the presence of wild-type (WT) and mutant CASK. The protein (WT or mutant) is indicated in the upper right corner. Amino acid mutations at specific locations are shown. Control spectrum of TNP-ATP (1 μM) in Tris–HCl buffer (pH 7.0) with EDTA (4 mM) (control). Spectra of samples containing 1 μM of the indicated recombinant CASK CaM-kinase domain, TNP-ATP (1 μM), and EDTA (4 mM) in Tris–HCl buffer (pH 7.0) ($-Mg^{2+}$). Spectra of samples containing 1 μM of the indicated recombinant CASK CaM-kinase domain, TNP-ATP (1 μM) and 100 μM $MgCl_2$ in Tris–HCl buffer (pH 7.0) ($+Mg^{2+}$). Samples were excited at 410 nm and spectra were recorded between 500 nm and 600 nm. The spectra are representatives of experiments repeated three times with essentially identical results [13]

stock solution in Tris–HCl buffer to achieve a final concentration of 1–4 μM. A final volume of 2 mL is needed for the cuvette assay. A total volume of 600 μL (for triplicates) is needed for the microplate assay.

5. TNP-ATP stock solution: 6.4 mM of 2′,3′-*O*-(2,4,6-trinitrophenyl) adenosine 5′-triphosphate, trisodium salt (TNP-ATP; Invitrogen) (*see* **Note 9**).

2.2 Microplate Binding Assay

1. TNP-ATP solution, dilution 1: 0.64 mM. Add 10 μL TNP-ATP stock solution (6.4 mM) to 90 μL Tris–HCl buffer.

2.3 Microplate Saturation Assay

1. TNP-ATP solution, dilution 2: 2 mM. Add 62.5 μL stock TNP-ATP solution to 137.5 μL Tris–HCl buffer.
2. TNP-ATP solution, dilution 3: 0.2 mM. Add 20 μL of 2 mM TNP-ATP solution to 180 μL Tris–HCl buffer.

2.4 Equipment for Cuvette and Microplate Assay

1. pH meter and electrode compatible with Tris–HCl buffers.
2. Spectrofluorometer equipped with magnetic stirrer (for the cuvette assay). The instrument to be used should be set up in advance to excite at 410 nm and to detect emission at 540 nm, as well as for wavelength scan from 500 nm to 600 nm. The excitation and emission slits should be set at 3 nm and 5 nm respectively.
3. 10 mm × 10 mm cuvette suitable for fluorescence.
4. Magnetic stir bar for cuvette.
5. Spectrofluorometer with plate-reading capability (for the microplate assay). The instrument to be used should be set up in advance to excite at 410 nm and to detect emission at 540 nm, as well as for wavelength scan from 500 nm to 600 nm. The excitation and emission slits should be set at 3 nm and 5 nm respectively. The instrument should be set up to read a 96-well plate and the readings should be taken from the top.
6. 96-well microplate with a black bottom to minimize background fluorescence.
7. Scientific graphing program (e.g., GraphPad Prism® for the microplate assay).

3 Methods (*See* Notes 10 and 11)

3.1 Standard Cuvette-Based Binding Assay

1. Place magnetic stir bar in cuvette.
2. Add 2 mL of Tris–HCl buffer to cuvette.
3. Add 1.6 μL of 6.4 mM stock TNP-ATP solution to cuvette to make final TNP-ATP concentration equal to 5 μM.
4. Place cuvette in spectrofluorometer with magnetic stirring turned on.
5. Excite sample at 410 nm.
6. Perform scan from 500 nm to 600 nm, and confirm peak fluorescence at 561 nm (*see* **Note 12**).
7. Remove and rinse cuvette.

8. Add protein sample solution of interest (1–4 μM) to cuvette. Total volume should equal 2 mL.
9. Add 1.6 μL of 6.4 mM stock TNP-ATP solution to the cuvette to make final TNP-ATP concentration equal to 5 μM.
10. Place cuvette in spectrofluorometer with magnetic stirring turned on.
11. Excite sample at 410 nm.
12. Read and record emission maximum at 540 nm or perform scan from 500 nm to 600 nm. Examine the emission scan for peak fluorescence at 540 nm (blue-shifted from peak observed in **step 6**) (*see* **Note 13**).
13. Remove and rinse cuvette.
14. Add 2 mL of lysozyme control assay solution (4 μM) to cuvette.
15. Add 1.6 μL of 6.4 mM stock TNP-ATP solution to the cuvette to make final TNP-ATP concentration equal to 5 μM.
16. Place cuvette in spectrofluorometer with magnetic stirring turned on.
17. Excite sample at 410 nm.
18. Read and record emission maximum at 540 nm or perform scan from 500 nm to 600 nm.
19. Export data for analysis.
20. Correct experimental values for kinase of interest by subtracting emission values obtained with lysozyme control readings (representing background fluorescence).

3.2 Microplate-Based Binding Assay

1. Prepare buffer control wells by pipetting 200 μL of Tris–HCl buffer into three wells (triplicates).
2. Add 1.6 μL of 0.64 mM TNP-ATP solution into these three wells to make final TNP-ATP concentration equal to 5 μM.
3. Place the plate in microplate reader.
4. Set microplate reader to shake plate prior to acquiring data.
5. Set up reader for excitation at 410 nm and emission scan from 500 nm to 600 nm.
6. Prepare protein control wells by pipetting 200 μL of lysozyme control assay solution (4 μM) into three wells.
7. Add 1.6 μL of 0.64 mM TNP-ATP solution into these three wells to make final TNP-ATP concentration equal to 5 μM.
8. Add 200 μL of protein sample solution (1–4 μM) to three wells.
9. Add 1.6 μL of 0.64 mM TNP-ATP solution into these three wells to make final TNP-ATP concentration equal to 5 μM.

10. Place the plate in a microplate reader.
11. Set the microplate reader to shake the plate prior to acquiring data.
12. Set up the reader for excitation at 410 nm and either single wavelength emission reading at 540 nm or emission wavelength scans from 500 nm to 600 nm (*see* **Notes 12** and **13**).
13. Export data for analysis.
14. Correct experimental values for kinase of interest by subtracting emission values obtained with lysozyme control readings (representing background fluorescence).

3.3 Microplate-Based TNP-ATP Saturation Assay (20 Min) (See Note 14)

1. Prepare buffer control wells by pipetting 190 μL of Tris–HCl buffer into three wells (triplicates).
2. Add 10 μL of 2 mM TNP-ATP solution into these three wells to make final TNP-ATP concentration equal to 100 μM.
3. Place the plate in a microplate reader.
4. Set the microplate reader to shake plate prior to acquiring data.
5. Set up the reader for excitation at 410 nm and emission scan from 500 nm to 600 nm (*see* **Notes 12** and **13**).
6. Prepare protein control wells by pipetting 190 μL of lysozyme solution (4 μM) into three wells.
7. Add 10 μL of 2 mM TNP-ATP solution into these three wells to make final TNP-ATP concentration equal to 100 μM.
8. Prepare triplicates of protein sample wells for each concentration of TNP-ATP to be tested (1 μM, 5 μM, 10 μM, 25 μM, 50 μM, and 100 μM) for a total of 18 wells. If target concentration is, for example, 2 μM, add 100 μL of a 4 μM solution to each well. Final well volume will equal 200 μL, yielding a final protein concentration of 2 μM.
9. Add TNP-ATP and Tris–HCl buffer into the 18 protein sample wells according to Table 1.
10. Place the plate in a microplate reader.
11. Set the microplate reader to shake plate prior to acquiring data.
12. Set up the reader for excitation at 410 nm and obtain single wavelength emission reading at 540 nm (*see* **Notes 12** and **13**).

3.4 Data Analysis

1. Correct experimental values for kinase of interest by subtracting emission values obtained with lysozyme control readings (representing background fluorescence).
2. For the high-throughput assays or single-wavelength emission competition assays, a simple bar graph analysis can be used to display differences between controls and experimental values.

Table 1
Sample well preparation for microplate-based TNP-ATP saturation assay

[TNP-ATP] in well (μM)	TNP-ATP solution (mM)	Volume of TNP-ATP solution (μL)	Volume of Tris–HCl buffer (μL)
1	0.2	2.5	97.5
5	0.2	5	95
10	0.2	10	90
25	2	2.5	97.5
50	2	5	95
100	2	10	90

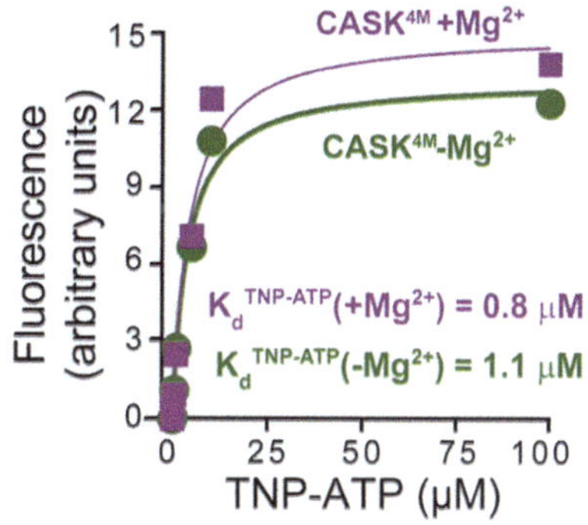

Fig. 3 The TNP-ATP binding assay can be used to compare nucleotide binding affinities under different conditions. Increasing amounts of TNP-ATP were added to cuvettes containing 10 mM Tris–HCl pH 7.0, 1 μM quadruple-mutant CASK ($CASK^{4M}$, a mutant engineered to bind nucleotide in the presence of magnesium), and either 4 mM EDTA (closed circle; $-Mg^{2+}$) or 200 μM Mg^{2+} (closed square; ($+Mg^{2+}$). The TNP-ATP fluorescence of the samples (excitation: 410 nm; emission: 540 nm) is plotted after subtracting background TNP-ATP fluorescence obtained with control samples, which contained the same TNP-ATP, Tris–HCl, EDTA, or Mg^{2+} concentrations but lysozyme instead of CASK. The plot is representative of three independent experiments [13]

3. For the 500–600 nm scan measurements, ASCII files are opened in Microsoft Excel as data files. The reading at 500 nm is considered to be zero and blanked from the data set. The scan is plotted as XY scatter plot between the wavelength (X-axis) and the reading (Y-axis).

3.5 Data Analysis for TNP-ATP Saturation Assay in GraphPad Prism®

1. Nonlinear curve fitting can be performed on data from the saturation assay using a program such as GraphPad Prism® to determine $K_d^{TNP\text{-}ATP}$ (Fig. 3) (*see* **Note 15**). GraphPad Prism® contains a built-in model for saturation binding using the equation: $Y = Bmax*X/(K_d + X) + NS*X + background$, in which

Bmax is the maximum specific binding, Kd is the equilibrium binding constant, NS is the slope of nonspecific binding, and background is the amount of nonspecific binding [20].

2. In Prism, create an XY data table, with TNP-ATP concentration in the X column and fluorescence units in the Y column.
3. Choose nonlinear regression from the Analyze menu.
4. Choose saturation binding equations.
5. Choose *One site—Total.*
6. Constrain the "background" parameter to zero since the data has already been corrected for background fluorescence.
7. Perform curve fitting to determine $K_d^{TNP\text{-}ATP}$.

4 Notes

1. TNP fluoresces brightly in a hydrophobic environment; therefore, it is critical to avoid hydrophobic contaminants in the assay buffer.
2. Use of a microplate-based assay allows the investigator to set up relatively high-throughput assays to test a wide array of conditions (nucleotide variants, cofactors, protein binding partners) under which a protein binds the nucleotide. The microplate also reduces the amount of protein sample needed for an assay, even when done in triplicate.
3. We use 50 mM KCl in our experiments but different proteins may have different salt preferences; buffer composition should therefore be optimized for the protein of interest. We do not recommend the use of 150 mM NaCl because complete dissociation of chloride (not an intracellular anion) produces higher electroconductivity, which may interfere with the interactions under study. If higher salt concentrations are used, we recommend the use of potassium glutamate or potassium acetate at 150 mM, which more closely resembles the intracellular ionic composition.
4. The control protein has to be carefully selected. Proteins such as albumin bind to ATP, confronting investigators with what they presume to be high background. In our hands, lysozyme works well.
5. It may be prudent to express and purify the full-length protein since other domains may affect ATP binding. In the case where only the kinase domain is expressed, careful selection of domain demarcation has to be performed, and may include several trials since unstructured regions of a protein fragment may affect ATP binding.

6. Although in an enzymatic assay, such as a phosphotransferase assay, the purity of protein preparation is of highest priority due to turnover rates, in ATP binding assays such as the one described, which have a finite and easily saturable endpoint, it is more important to obtain a well-folded protein. Multiple steps of protein purification which may affect the conformation of the protein should therefore be avoided. We typically use single-step affinity purification of the sample protein under physiological concentrations of electrolytes and pH.
7. Proteins that are purified especially from eukaryotic sources like insect cells may be posttranslationally modified, which can affect the nucleotide binding activity.
8. Activated kinases rapidly toggle between a closed and an open conformation [21]. Kinases operate in a cellular milieu replete with nucleotides including ATP. Purification of kinase domains without nucleotide may trap a conformation with low affinity for nucleotides, so it may be useful to include non-hydrolyzable analogs of ATP such as AMPPNP in the protein purification buffer. Using ATP itself should be avoided since this may cause autophosphorylation in vitro.
9. Freeze TNP-ATP immediately upon arrival and protect from long-term light exposure.
10. Rapidly purified proteins with high nucleotide affinity may have nucleotide bound to its pocket, which will interfere with ATP binding assays. For example, we found that the CASK CaM-kinase domain binds to nucleotides such as 3′AMP present in bacterial lysates. Occupation of the nucleotide binding pocket after purification can be ruled out by acquiring an absorbance spectra prior to beginning the assay; the presence of a peak at 260 nm indicates a bound nucleotide.
11. TNP-ATP may display much higher affinity than ATP itself, so when screening for inhibitors, false negatives are more likely than false positives using this assay.
12. If peak fluorescence is not at 561 nm, contaminants may be present and must be eliminated before continuing with experiment.
13. We have observed that a mere increase in fluorescence may be an artifact; it is essential to determine that in the presence of the protein sample of interest, there is a blue shift and the emission maximum is at 540 nm.
14. High affinity for the adenine nucleotide may not always translate into a low Michaelis constant for ATP in an enzymatic reaction, since rapid phosphotransfer may be limited by the displacement of ADP by the incoming ATP molecule. We observe a very high Km (~1 mM) for ATP in the case of CASK.

15. A similar approach can be used to determine values such as K_i in a competition assay.

Acknowledgments

The work was supported by startup funds to KM from VTCRI. Experiments were designed and performed by KM. The manuscript was conceived and written by LL, SS, and KM. KM and his laboratory are supported by R01EY024712 from NEI.

References

1. Manning G (2005) Genomic overview of protein kinases. WormBook 1–19. doi:10.1895/wormbook.1.60.1
2. Casado P, Rodriguez-Prados JC, Cosulich SC et al (2013) Kinase-substrate enrichment analysis provides insights into the heterogeneity of signaling pathway activation in leukemia cells. Sci Signal 6(268):rs6. doi:10.1126/scisignal.2003573
3. Dancey J, Sausville EA (2003) Issues and progress with protein kinase inhibitors for cancer treatment. Nat Rev Drug Discov 2 (4):296–313. doi:10.1038/nrd1066
4. Huang D, Zhou T, Lafleur K et al (2010) Kinase selectivity potential for inhibitors targeting the ATP binding site: a network analysis. Bioinformatics 26(2):198–204. doi:10.1093/bioinformatics/btp650
5. Cohen P (2002) Protein kinases—the major drug targets of the twenty-first century? Nat Rev Drug Discov 1(4):309–315. doi:10.1038/nrd773
6. Johnson L (2007) Protein kinases and their therapeutic exploitation. Biochem Soc Trans 35(Pt 1):7–11. doi:10.1042/BST0350007
7. Defert O, Boland S (2015) Kinase profiling in early stage drug discovery: sorting things out. Drug Discov Today Technol 18:52–61. doi:10.1016/j.ddtec.2015.10.002
8. Hastie CJ, McLauchlan HJ, Cohen P (2006) Assay of protein kinases using radiolabeled ATP: a protocol. Nat Protoc 1(2):968–971. doi:10.1038/nprot.2006.149
9. Zegzouti H, Zdanovskaia M, Hsiao K, Goueli SA (2009) ADP-Glo: a bioluminescent and homogeneous ADP monitoring assay for kinases. Assay Drug Dev Technol 7 (6):560–572. doi:10.1089/adt.2009.0222
10. Liu Y, Gray NS (2006) Rational design of inhibitors that bind to inactive kinase conformations. Nat Chem Biol 2(7):358–364. doi:10.1038/nchembio799
11. Mukherjee K, Sharma M, Urlaub H et al (2008) CASK functions as a Mg2+–independent neurexin kinase. Cell 133 (2):328–339. doi:10.1016/j.cell.2008.02.036
12. Ungureanu D, Wu J, Pekkala T et al (2011) The pseudokinase domain of JAK2 is a dual-specificity protein kinase that negatively regulates cytokine signaling. Nat Struct Mol Biol 18 (9):971–976. doi:10.1038/nsmb.2099
13. Mukherjee K, Sharma M, Jahn R et al (2010) Evolution of CASK into a Mg2+–sensitive kinase. Sci Signal 3(119):ra33. doi:10.1126/scisignal.2000800
14. Cheng K, Koland JG (1996) Nucleotide binding by the epidermal growth factor receptor protein-tyrosine kinase. Trinitrophenyl-ATP as a spectroscopic probe. J Biol Chem 271 (1):311–318
15. Guarnieri MT, Blagg BS, Zhao R (2011) A high-throughput TNP-ATP displacement assay for screening inhibitors of ATP-binding in bacterial histidine kinases. Assay Drug Dev Technol 9(2):174–183. doi:10.1089/adt.2010.0289
16. Stewart RC, VanBruggen R, Ellefson DD, Wolfe AJ (1998) TNP-ATP and TNP-ADP as probes of the nucleotide binding site of CheA, the histidine protein kinase in the chemotaxis signal transduction pathway of Escherichia coli. Biochemistry 37(35):12269–12279. doi:10.1021/bi980970n
17. Piluso G, D'Amico F, Saccone V et al (2009) A missense mutation in CASK causes FG syndrome in an Italian family. Am J Hum Genet 84(2):162–177. doi:10.1016/j.ajhg.2008.12.018
18. Jakobi R, Traugh JA (1995) Site-directed mutagenesis and structure/function studies of casein kinase II correlate stimulation of activity by the beta subunit with changes in conformation and ATP/GTP utilization. Eur J Biochem 230(3):1111–1117

19. Bostrom SL, Dore J, Griffith LC (2009) CaMKII uses GTP as a phosphate donor for both substrate and autophosphorylation. Biochem Biophys Res Commun 390(4):1154–1159. doi:10.1016/j.bbrc.2009.10.107
20. Motulsky HJ. Equation: One site—Fit total and nonspecific binding. http://www.graphpad.com/guides/prism/7/curve-fitting/index.htm?reg_one_site_fit_total_and_ns.htm. Accessed 4 Aug 2016
21. Meharena HS, Chang P, Keshwani MM et al (2013) Deciphering the structural basis of eukaryotic protein kinase regulation. PLoS Biol 11(10):e1001680. doi:10.1371/journal.pbio.1001680

Chapter 12

Preparation of Disease-Related Protein Assemblies for Single Particle Electron Microscopy

A. Cameron Varano, Naoe Harafuji, William Dearnaley, Lisa Guay-Woodford, and Deborah F. Kelly

Abstract

Electron microscopy (EM) is a rapidly growing area of structural biology that permits us to decode biological assemblies at the nanoscale. To examine biological materials for single particle EM analysis, purified assemblies must be obtained using biochemical separation techniques. Here, we describe effective methodologies for isolating histidine (his)-tagged protein assemblies from the nucleus of disease-relevant cell lines. We further demonstrate how isolated assemblies are visualized using single particle EM techniques and provide representative results for each step in the process.

Key words Recombinant proteins, Histidine tag, Fibrocystin, Electron microscopy, Single particle analysis

1 Introduction

Electron microscopy (EM) allows us to examine and characterize biological entities ranging from relatively large, uniform virus structures to smaller, non-symmetrical proteins [1, 2]. Determining the intricate details of disease-related proteins using high-resolution EM imaging can reveal new targets for rational drug design. The presented protocols here are adapted from our recent EM structural studies on native BRCA1 protein assemblies formed in human cancer cells [3, 4]. A new aspect of the presented work is the application of these methods for disease-related protein assemblies beyond human cancer. Here, we demonstrate the utility of our established protein separation techniques for instances in which the target protein is expressed in low abundance. Specifically, we provide detailed methods to produce and characterize protein assemblies incorporating the C-terminal domain (CTD) of the fibrocystin/polyductin protein (FPC) [5, 6].

Iulia M. Lazar et al. (eds.), *Proteomics for Drug Discovery: Methods and Protocols*, Methods in Molecular Biology, vol. 1647, DOI 10.1007/978-1-4939-7201-2_12,

There are many known mutations in the FPC protein that are implicated in autosomal-recessive polycystic kidney disease. At the cellular level, the longest mRNA product of the FPC gene encodes a membrane-bound protein that undergoes Notch-like proteolytic cleavage to generate a functional carboxy-terminal domain (CTD-FPC). The CTD-FPC translocates to the cell's nucleus to carry out a variety of gene-related functions. As the FPC protein is naturally expressed in low abundance in the kidney epithelia, this presents a technical barrier to understand the molecular basis of disease-related mechanisms [7].

To address this issue and facilitate structural analysis on FPC assemblies, we employed a protein enrichment strategy by over-expressing a his-tagged (6×–His) version of the CTD-FPC in mouse kidney cells (mIMCD-3 line; ATCC; [8]). Nuclear assemblies that incorporated the His-tagged recombinant protein were isolated and imaged using previously established single particle EM techniques [3]. The overall steps of the process that we describe include: (1) a nuclear extraction step, (2) a nickel chromatography isolation procedure, (3) EM specimen preparation and data collection, and (4) processing the image data. Overall, this work demonstrates a means to assess disease-related protein assemblies for structural analysis while providing new avenues to explore protein interfaces for drug discovery purposes.

2 Materials

All the reagent solutions described in our procedures work optimally when prepared with ultrapure water. In general, working reagents should be prepared and stored on ice or at 4 °C. When working to obtain structural information of protein complexes, it is important to avoid excessive procedures that can create "bubbles" in solution or at the air-water interface, as these disturbances can affect protein integrity.

2.1 Cytoplasmic and Nuclear Extraction

Cellular nuclear material is obtained by using a commercially available kit, NE-PER Nuclear and Cytoplasmic Extraction Reagents (Thermo Scientific). For this procedure, we use a recommended packed cell volume of 100 μL. For different cell volumes, please consult the manufacturer's recommendations. Reagents are prepared according to the following steps, and we recommend following the procedures described by the manufacturer to obtain the nuclear fraction.

1. Protease inhibitor: 1 tablet of cOmplete mini protease inhibitor 100× (Roche Diagnostics). Using forceps, place 1 tablet of cOmplete mini protease inhibitor 100× on a piece of weigh paper, then crush the tablet. Carefully transfer the crushed

tablet to the 600 μL polypropylene tube. Add 100 μL of water to the crushed tablet. Pipette up and down gently to dissolve into solution.

2. Cytoplasmic extraction reagent one (CER 1) solution: Add 1000 μL of CER 1 reagent to a labeled, prechilled 1.5 mL polypropylene tube, followed by 10 μL of the protease inhibitor solution and 10 μL of Halt phosphatase inhibitor 100× (Thermo Scientific). Pipette up and down gently to mix the reagents.
3. Cytoplasmic extraction reagent two (CER 2) solution: Add 55 μL of CER 2 reagent to a labeled, prechilled 600 μL polypropylene tube.
4. Nuclear extraction reagent (NER) solution: Add 500 μL of NER solution into a prechilled 1.5 mL polypropylene tube, followed by 5 μL of protease inhibitor solution (*see* **Note 1**).

2.2 Nickel Chromatography Purification

Nickel chromatography utilizes the affinity of a recombinant protein with a polyhistidine (His) tag to associate with nickel cations. The protein of interest is then eluted off using buffers containing imidazole (*see* **Note 2**). To perform these steps, prepare the following buffers in 50 mL conical tubes using ultrapure water and chilled to 4 °C:

1. HEPES buffer: 20 mM, pH 7.4. Weigh 0.2383 g HEPES and transfer to a 50 mL cylinder. Add ultrapure water to a volume of 50 mL. Transfer to a beaker, mix and adjust the pH to 7.4 with 5 N NaOH solution. Store on ice in a 50 mL conical tube (*see* **Note 3**).
2. HEPES buffer with salts: 20 mM HEPES, 140 mM NaCl, 2 mM $MgCl_2$, and 2 mM $CaCl_2$, pH 7.4. Weigh 0.2383 g HEPES, 0.4091 g NaCl, 11.098 mg $CaCl_2$ and transfer to a 50 mL cylinder. Add 100 μL of 1 M $MgCl_2$. Add ultrapure water to a volume of 50 mL. Transfer to a beaker, mix and adjust the pH to 7.4 with 5 N NaOH solution.
3. Imidazole solution: 1 M. Weigh 0.34 g imidazole and transfer to a 5 mL cylinder. Add ultrapure water to a volume of 5 mL.
4. Wash buffer: 5 mM imidazole in HEPES buffer with salts. Add 50 μL of 1 M imidazole to 10 mL of 20 mM HEPES buffer with salts (pH 7.4).
5. Elution buffer: 200 mM imidazole in HEPES buffer with salts. Add 1 mL of 1 M imidazole to 4 mL of 20 mM HEPES buffer with salts (pH 7.4).
6. Ni-NTA agarose slurry (QIAGEN): Pipette 400 μL of the Ni-NTA agarose storage buffer solution into a 1.5 mL polypropylene centrifuge tube (*see* **Note 4**).

2.3 Transmission Electron Microscopy (TEM)

1. TEM for examining negatively stained CTD-FPC complexes: FEI Spirit Bio-Twin TEM equipped with a LaB_6 filament and operating at 120 kV.
2. CCD camera for recording images: Eagle 2 k HS CCD camera employing low-dose conditions (~1–5 electrons per Å^2).
3. PELCO easiGlow™ Glow Discharge Cleaning System (Ted Pella, Inc.)
4. DV502-A vacuum evaporator (Denton Vacuum, Inc.).
5. Commercial carbon-coated grids (Ted Pella, Inc.).
6. Copper glider grids 200 mesh (Ted Pella, Inc.).
7. Collodion 2% in amyl acetate (Electron Microscopy Sciences).
8. Whatman filter paper 1 (GE Healthcare).
9. PVDF filter, 0.2 μm.
10. Aluminum foil.
11. Uranyl formate, 99.9–100% purity (Electron Microscopy Sciences).

2.4 Software

1. SPIDER: Software package [2] which allows for 2D classification of protein complexes through a multivariate data analysis approach. Individual protein complexes (particles) are selected from micrographs. After selection, all the particles are subjected to a low-pass filter. Particles are then aligned through several iterations of vector alignment. The final step before classification is masking the particles, to remove the background noise. The parameters in each step are modulated by the user; however, the routines are standard.
2. RELION: Software package [9] which performs reconstruction and refinement calculations using an empirical Bayesian methodology.

3 Methods

3.1 Preparing TEM Grids

3.1.1 Coating TEM Grids

The following method describes how to prepare EM grids for sample application using conventional methods. Alternatively, carbon-coated grids may be purchased from commercial suppliers (e.g., Ted Pella, Inc.). All filter paper used in the preparation of EM specimens is Whatman Filter Paper 1.

1. Fill a clean PYREX dish (190 mm × 100 mm) with ultrapure water to the brim. Water should be free of any dust or particles.
2. Using a pasture pipette, place 1 drop of collodion 2% in amyl acetate into water.

3. Using forceps, place copper glider grids (200 mesh) dark side down on the liquid (*see* **Note 5**).
4. Cut a piece of linen paper matching the size of the layout of the grids.
5. Place paper onto of grids. Allow the liquid to fully soak.
6. In a petri dish place a piece of filter paper.
7. Use forceps to cut the polymer film around paper.
8. Smoothly and quickly lift paper with grids off of the water. Use forceps to grab the corners furthest from self then sweeping the paper up so the grids face away from oneself. Place the grids facing up on the petri dish. Leave petri dish lid ajar to allow for drying.
9. Dry grids for 1–2 days for optimal results.
10. Following the polymer coating of the copper grids, a fine layer of carbon is applied. Place the sheet of paper holding the grids under the dome of a DV502-A vacuum evaporator.
11. Pump the vacuum down to a pressure of 5×10^{-6} torr.
12. Once a stable vacuum is established, pass 40 amps through a conical graphite tip for about 30 s. This method creates a thin and even layer of atomized carbon on the grids.

3.1.2 Glow Discharge Procedure for Continuous Carbon Grids

1. Place continuous carbon-coated copper grids on glass slide wrapped in parafilm to keep the grids in place.
2. Open dome of Pelco easiGlow™.
3. Insert slide with attached carbon-coated grid into Pelco easiGlow™ assembly.
4. Wipe rubber rim of the dome and machine to remove any dust.
5. Replace glass dome.
6. Use stylist to tap auto run (*see* **Note 6**).
7. After the cycle is completed, open the dome and remove glow-discharged grids.

3.1.3 Preparing Uranyl Formate Heavy Metal Stain

1. Boil 3 mL of ultrapure water in a 10 mL beaker on a hot plate.
2. Using tongs, transfer the beaker to a stir plate and add 22.5 mg of uranyl formate and a stir bar. Stir for 5 min. This results in a 0.75% uranyl formate solution.
3. Add 4.2 μL of 5 N NaOH. Stir another 5 min.
4. Draw up in a 5 mL syringe.
5. Filter through a 0.2 μm PVDF filter into a 15 mL conical tube to remove any undissolved uranyl formate, and then cover tube with aluminum foil (*see* **Note 7**).

3.2 Cytoplasmic and Nuclear Extraction

The following information is based on the guidelines provided by Thermo Scientific for optimal use of their NE-PER Nuclear and Cytoplasmic Extraction Reagents.

1. Collect cells by scraping and transfer cell suspension to 50 mL conical tubes.
2. Centrifuge the tubes at 500 × *g* for 5 min at 4 °C.
3. Wash cells by resuspending the cell pellet with 1.5 mL PBS with phosphatase inhibitor.
4. Transfer cells to prechilled 2 mL polypropylene tube and pellet by centrifugation at 500 × *g* for 2 min.
5. Add ice-cold CER 1 to the resulting cell pellet.
6. Vortex the sample for 15 s followed by a 10 min incubation on ice.
7. Add 55 μL ice-cold CER 2 buffer to the sample solution.
8. Vortex the tube for 5 s and then incubate on ice for 1 min.
9. Vortex the tube for 5 s on the highest setting. Centrifuge the tube for 5 min at 16,000 × *g* in a microcentrifuge at 4 °C (*see* **Note 8**).
10. Carefully decant the supernatant containing the cytoplasmic components to a waste beaker (*see* **Note 9**).
11. Resuspend the insoluble pellet containing nuclear components in ice-cold NER solution.
12. Vortex for 15 s and then place on ice. Repeat the vortexing-incubation cycle in 10 min intervals for a total of 40 min (*see* **Note 10**).
13. Centrifuge at 16,000 × *g* for 10 min at 4 °C.
14. Carefully transfer the supernatant containing the nuclear extract (NE) to a prechilled 1.5 mL polypropylene tube and place on ice (*see* **Note 11**).

3.3 Nickel Chromatography Purification

Immobilized-metal affinity chromatography (IMAC) exploits the principles of specific protein interactions with chelated metal groups, held in place by immobilized surfaces or beads. One popular association is formed by the interaction of polyhistidine residues (His tag) with functionalized Nickel-Nitrilotriacetic acid (Ni-NTA) resin. Below we describe procedures for isolating His-tagged CTD-FPC protein assemblies from nuclear extracts prepared according to the protocol described in Subheading 3.2.

1. Dilute the NE in 20 mM HEPES buffer without salts and store on ice (*see* **Note 12**).
2. Centrifuge the mixture Ni-NTA agarose slurry for 2 min at 700 × *g*.

3. Carefully remove the supernatant and discard.
4. Add 1 mL of the wash buffer containing 20 mM HEPES buffer, and 5 mM imidazole to the resin and invert to mix.
5. Centrifuge for 2 min at 700 × *g*.
6. Remove the supernatant and discard.
7. Resuspend the Ni-NTA resin into 1 mL of wash buffer and centrifuge at 700 × *g* for 2 min. Remove the supernatant.
8. Add the diluted NE to Ni-NTA resin.
9. Gently mix the material on a clinical rotator at 4 °C for 60 min (*see* **Note 13**).
10. Wash the Ni-NTA resin using the wash buffer to remove background proteins (*see* **Note 14**). Collect the wash buffer that flows though the resin in a 15 mL conical tube.
11. Elute the His-tagged proteins of interest using the elution buffer. The high concentration of imidazole in this buffer will compete off the His-tagged FPC complexes. Typically, the eluted material is collected in multiple fractions. Each fraction is roughly equal to 1 bed volume, which is ~200 μL in our experiments.
    (a) Add 1 mL of the elution buffer, carefully pipetting it along the column wall.
    (b) Collect increments of 200 μL in prechilled 1.5 mL polypropylene tubes. A total of five fractions are usually collected. The His-tagged CTD-FPC protein often elutes in fractions 2 and 3 (*see* Fig. 1).
12. All the samples collected should be stored on ice and analyzed for total protein concentration using a standard Bradford assay.

3.4 Preparation of Negatively Stained EM Specimens

1. Using parafilm, place 3–200 μL drops of ultrapure water and 2–200 μL drops of 0.75% uranyl formate in a row. These drops will be used for specimen washing and staining steps (*see* **Note 15**).
2. Carefully, pick up a fresh glow-discharged grid with forceps. The tips of the forceps should only touch the edge of the grid (*see* **Note 16**).
3. Place 3 μL of sample on the grid and incubate at room temperature for 1 min (Fig. 2a).
4. Blot off the excess sample onto filter paper (Fig. 2b).
5. Wash the face of the grid three times with ultrapure water. Gently touch the face of the grid on each water droplet and then again blot off the ultrapure water using filter paper. Use a new drop for each wash step. Do not fully immerse the grid into the water droplet or wet the back of the grid.

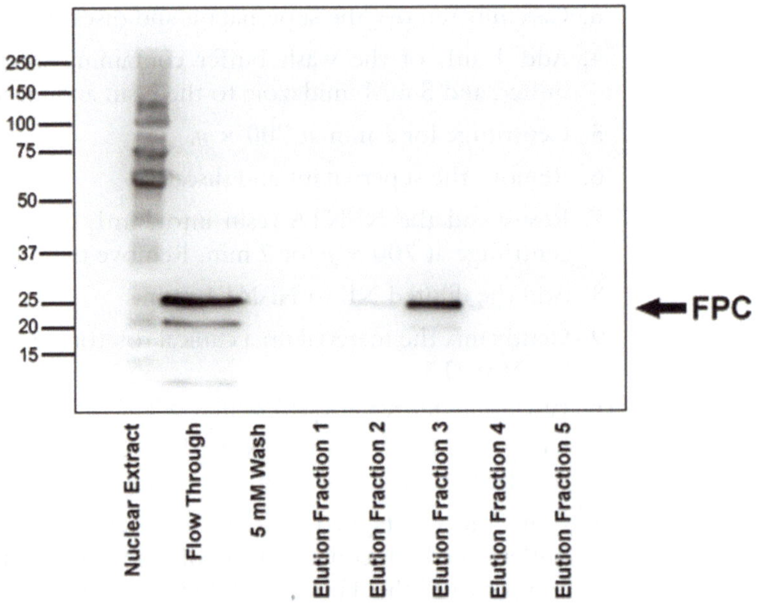

Fig. 1 Western blot probed with anti-His antibodies (Abcam) shows samples obtained from the Ni-NTA purification procedure. Lanes include the nuclear extract, column flow-through, wash and eluted fractions (1–5). His-tagged CTD-FPC elutes in fractions 2 and 3

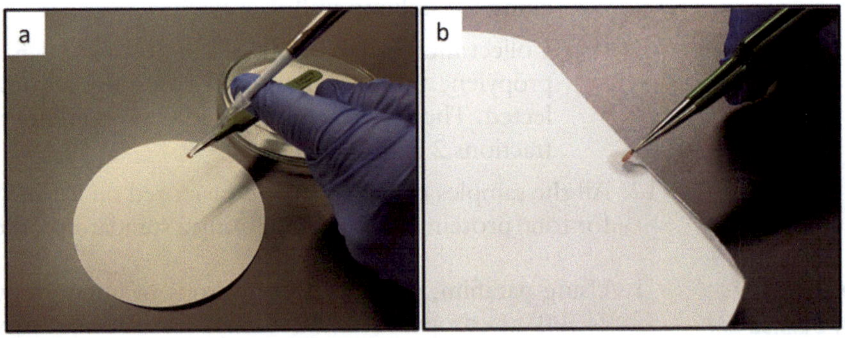

Fig. 2 Biological material was carefully added to glow-discharged continuous carbon grids. (**a**) Aliquots (3 μL each) of sample were applied to glow-discharged EM grids with the grid secured in forceps. (**b**) When washing each sample, turn the grid to the side to blot off excess solution onto Whatman #1 filter paper

6. Wash the face of the grid one time with 0.75% uranyl formate, utilizing the same method as the water wash.
7. Stain with 0.75% uranyl formate for 30 s (Fig. 3a). Keeping the sample on the stain for 30 s is what differs this from the previous wash step.
8. Blot away excess solution using a vacuum hose. Be cautious not to touch the vacuum hose directly to the grid as it may damage the sample (Fig. 3b). Store the samples in a labeled petri dish until ready to image.

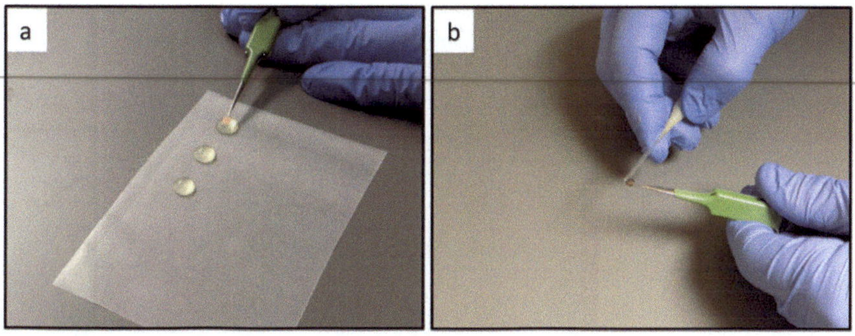

Fig. 3 (**a**) To negatively stain the samples with 0.75% uranyl formate, touch the face of the grid onto uranyl formate droplet without submersing the grid into the stain. **(b)** Dry the stained specimens with a vacuum hose from the backside of the grid. The vacuum hose should not come in contact with the grid as it may damage the sample

3.5 TEM Image Collection

1. Allow the scope to fully evacuate the column and cool.
2. Insert the sample via a single-tilt EM specimen holder at room temperature.
3. With the beam engaged the sample must be aligned along the axis of the beamline, in order for proper defocus to be attained. This is done at the center point of the sample.
4. The sample is now ready for imaging. A typical technique for this is to obtain an unbroken, well-stained grid square with wide ridges formed by the coating process. These appear as dark lines throughout the square.
5. The edge of one of these ridges is then magnified and the defocus value is changed until a clear focus is obtained (*see* **Note 17**). Scanning along the edge of a ridge generally gives a good contrast region to visualize the biological particles.
6. Once an area of good particle occupancy has been identified digital images are acquired (Fig. 4a). Save images in a 16-bit.tiff format for downstream image processing procedures.

3.6 Data Analysis and Representative CTD-FPC Results

1. Prior to particle selection, the original images are normalized using the standard routines in the SPIDER software package [2].
2. Individual complexes from the images are manually selected using the WEB interface of the SPIDER software package [2], and employing a box size that is approximately twice the diameter of the particles of interest.
3. Multi-reference alignment routines are implemented outputting 2D classes (Fig. 4b) as previously described [3].
4. Image stacks containing selected particles are imported into the RELION software package [9].

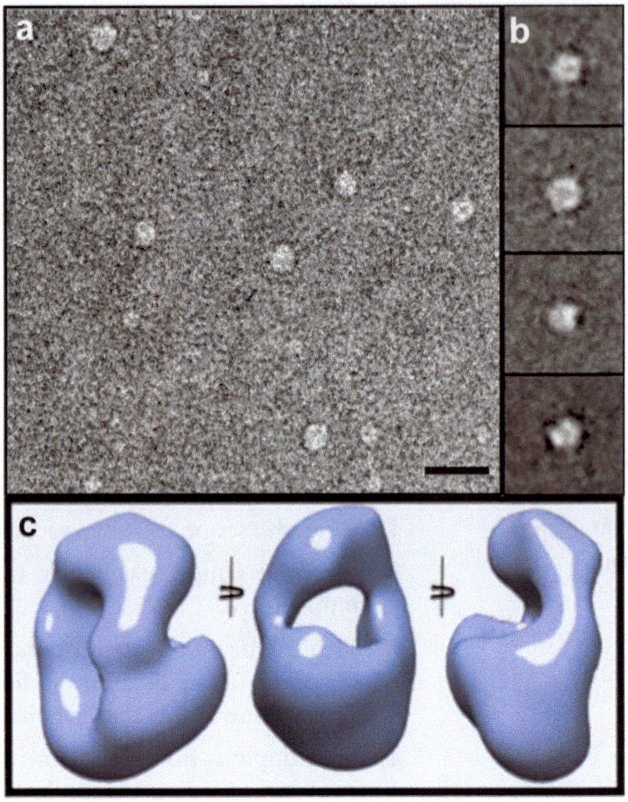

Fig. 4 (**a**) A representative micrograph of purified CTD-FPC assemblies taken at 68,000× magnification. Scale bar is 20 nm. **(b)** Class averages obtained from multi-reference alignment procedures implement in the SPIDER software package [2]. (**c**) 3D reconstruction of the CTD-FPC complex calculated using RELION software package [9] with each view rotated 90°

5. A spherical structure can be used as a reference map to reconstruct CTD-FPC complexes through 25 refinement iterations using an angular sampling interval of 7.5°. Other parameters input into RELION [9] include a magnification of 68,000×, a pixel size of 4.4 Å, and a regularization parameter of $T = 4$. A representative 3D reconstruction of the His-tagged CTD-FPC complex is shown in Fig. 4c.

4 Notes

1. Halt phosphatase inhibitor is not used when doing a nickel column purification, as it may interfere with the nickel resin's binding capacity.

2. Imidazole is a salt, which competes with the polyhistidine (His) tag for the Nickel-NTA resin. All the buffers and samples must be kept at 4 °C or on ice to maintain protein integrity.
3. This buffer is used to dilute the sample. It should be fully chilled to 4 °C before using it.
4. The slurry is a 50% resin and buffer mixture. Pipetting a volume of 400 μL results in a 200 μL bed volume of Ni-NTA resin. Before we begin the purification, the Ni-NTA slurry is separated from the storage buffer and equilibrated with the wash buffer.
5. The rim of the edge of the copper grid is darker on one side. Grids can be placed close together, but should not overlap.
6. Auto run should be set at 15 mA, glow 1 min and hold 10 s.
7. Uranyl formate solution is light sensitive. Covering the tube in aluminum foil minimizes the exposure to light. If stored in this manner, the uranyl formate solution can be kept up to 48 h.
8. Place the tube in a specific orientation, for example, so that the lid hinge is located outside of the rotator. This helps to locate the pellet in the tube since the pellet created is not always easily distinguishable.
9. Our protein of interest is a nuclear protein; therefore, we do not need to keep the cytoplasmic fraction.
10. The intermittent vortexing allows for a more complete release of the soluble nuclear components.
11. Extracts can be stored at −80 °C for 30 days or used immediately.
12. This dilution step lowers the salt concentration of the sample, bringing it closer to a physiological level.
13. The amount of volume remaining is a small quantity. The meniscus of the buffer solution should be just above the resin.
14. This volume is dependent upon the bed volume. The wash volume is ~15 times the bed volume.
15. Negative stain refers to the fact that proteins are embedded by the uranyl salt solution. This step produces amplitude contrast that indicates the presence of white particles on a dark background.
16. The grids are delicate and easy to crack with forceps. Keeping the forceps along the edge of the grid maximizes the potential imaging regions of the grid free of damage.
17. For FPC a nominal magnification of ~68,000 x was used, giving a final sampling at the specimen level of 4.4 Å per pixel. FPC images were acquired using a defocus value of −1.5 μm.

Acknowledgment

This work was supported by NIH/NCI grant R01CA193578 to D.F.K.

References

1. Taylor KA, Glaeser RM (2008) Retrospective on the early development of cryoelectron microscopy of macromolecules and a prospective on opportunities for the future. J Struct Biol 163 (3):214–223
2. Frank J, Radermacher M, Penczek P et al (1996) SPIDER and WEB: processing and visualization of images in 3D electron microscopy and related fields. J Struct Biol 116(1):190–199
3. Gilmore BL, Winton CE, Demmert AC et al (2015) A molecular toolkit to visualize native protein assemblies in the context of human disease. Sci Rep 5:14440
4. Winton CE (2016) A microchip platform for structural oncology applications. NPJ Breast Cancer 2:16016
5. Kim I, Li C, Liang D et al (2008) Polycystin-2 expression is regulated by a PC2-binding domain in the intracellular portion of fibrocystin. J Biol Chem 283(46):31559–31566
6. Boddu R, Yang C, O'Connor AK et al (2014) Intragenic motifs regulate the transcriptional complexity of Pkhdl/PKHD1. J Mol Med (Berl) 92(10):1045–1056
7. Guay-Woodford LM (2014) Autosomal recessive polycystic kidney disease: the prototype of the hepato-renal fibrocystic diseases. J Pediatr Genet 2:89–101
8. Rauchman MI, Nigam SK, Delpire E, Gullans SR (1993) An osmotically tolerant inner medullary collecting duct cell line from an SV40 transgenic mouse. Am J Phys 3(Pt 2):F416–F424
9. Scheres SH (2012) A Bayesian view on cryo-EM structure determination. J Mol Biol 415 (2):406–418

Chapter 13

Identification of Lipid Binding Modulators Using the Protein-Lipid Overlay Assay

Tuo-Xian Tang, Wen Xiong, Carla V. Finkielstein, and Daniel G.S. Capelluto

Abstract

The protein-lipid overlay assay is an inexpensive, easy-to-implement, and high-throughput methodology that employs nitrocellulose membranes to immobilize lipids in order to rapid screen and identify protein-lipid interactions. In this chapter, we show how this methodology can identify potential modulators of protein-lipid interactions by screening water-soluble lipid competitors or even the introduction of pH changes during the binding assay to identify pH-dependent lipid binding events.

Key words Lipid-protein overlay assay, Phospholipids, Phosphoinositides, Inositol 1,3-bisphosphate, EEA1 FYVE, Dishevelled-2 DEP, Phafin2, Tollip, pH effect

1 Introduction

Lipids not only function in membrane organization but are also engaged in the spatiotemporal regulation of protein-mediated signaling transduction pathways and intracellular membrane trafficking. The function of a lipid is highlighted by the structure and net charge of its head group as well as the length and saturation of its fatty acid chains. Abnormal changes in lipid metabolism are associated with genetic defects that trigger human diseases such as cancer, diabetes, and neurodegenerative diseases [1]. Thus, targeting lipid-protein interactions is an emerging area of biomedical research for drug development. Indeed, there is accumulated literature that emphasizes lipid binding proteins as prospective drug targets [2–4]. Protein-lipid binding can be simply assessed by various biochemical techniques, including liposome-based assays, such as the liposome flotation and the liposome microarray-based assays [5, 6]. Although these robust methodologies provide the ability for screening protein-lipid binding by using liposomes of different compositions, their preparation and stability, and the elevated cost

Iulia M. Lazar et al. (eds.), *Proteomics for Drug Discovery: Methods and Protocols*, Methods in Molecular Biology, vol. 1647, DOI 10.1007/978-1-4939-7201-2_13, © Springer Science+Business Media LLC 2017

of the instrumentation required for screening lipid binding limits their use. Biophysical techniques, including surface plasmon resonance, isothermal calorimetry, solution NMR titrations, and microscale thermophoresis, allow the quantification of protein-lipid interactions [7], but they also require specialized instrumentation. One powerful and relatively quick methodology to identify potential protein lipid ligands is the protein-lipid overlay assay (PLOA). This assay, described by Alessi and colleagues in 2002, consists of immobilizing serial dilutions of a lipid of interest onto a nitrocellulose membrane. This membrane is then incubated with the protein of interest (typically a fusion protein), washed, and protein binding to the lipid is detected by the sequential addition of an antibody against the epitope tag and a chemiluminescent antibody [8]. This 2-day method requires 1–10 μg of protein and microgram amounts of lipid. PLOA can be used as a first high-throughput screening to identify novel protein-lipid interactions. Indeed, PLOA has been used to investigate the specificity of 33 *Saccharomyces cerevisiae* PH domains for phosphoinositide binding [9]. Also, the assay has been modified by designing miniaturized nitrocellulose membranes arrays, which contained duplicated sets of 51 lipids to survey *S. cerevisiae* novel protein-lipid interactions [10]. Echelon Inc. markets a series of lipid strips and arrays, in which lipids are prespotted on nitrocellulose membranes at amounts as high as 100 pmol of lipid per spot. Alternatively, and as described in detail below, lipid-spotted nitrocellulose membranes can be made in the laboratory by spotting the lipid of interest onto the membrane. These lipid-containing membranes can last for at least 6 months if they are stored at 4 °C and protected from the light.

By using PLOA, however, one should consider that the lipid is not organized in a lipid bilayer as occurs under physiological condition and that other lipids from the membrane structure always accompany the lipid under investigation. Thus, despite its simplicity and sensitivity, the PLOA method exhibits some limitations including nonspecific binding, or, lack of it. Thus, the use of an alternative method to support the lipid binding results is recommended [7, 11]. However, if the lipid ligand is consistently identified using PLOA and other orthogonal assays, then additional features of protein-lipid interactions can be investigated using PLOA. For example, lipid binding-deficient mutants of a protein can be investigated to identify potential lipid binding sites [12, 13]. The presence of cofactors for protein-lipid interactions can also be studied using PLOA. For example, the requirement of the lectin-like oxidized low-density lipoprotein receptor-1 for Ca^{2+}, but not Mg^{2+}, in order to associate with phosphatidylserine could be identified using this methodology [14]. Also, PLOA can be used to identify cooperativity in lipid binding. Syntenin binds phosphatidylinositol 4,5-bisphosphate (PtdIns(4,5)P_2) through its PDZ domains, but enhancement of the binding in the presence of a syndecan-2 peptide can be observed using PLOA [15]. Using the PLOA method, the C-terminal DDHD domain of the COPII component protein, p125A, shows low specificity for lipids, but becomes more specific when it is expressed together with the central SAM domain of the same protein [16]. PLOA can also be used to identify specificity for fatty acid chains. Thus, it was demonstrated that during the light period, *Arabidopsis* florigen FT preferentially binds to phosphatidylcholine fatty acids that are less saturated [17]. Protein binding to phospholipids can be coupled to proton pumps as suggested for the peripheral protein EEA1, whose FYVE domain binds PtdIns(3)P in a pH-dependent manner [18]. Thus, PLOA was proven to be efficient at identifying the pH dependency for the binding of Dishevelled-2 DEP domain to phosphatidic acid [19]. Several reports also used PLOA to identify negative modulators or competitors. For example, the adaptor protein Disabled-2 (Dab2) binds phosphatidylinositol 4,5-bisphosphate (PtdIns(4,5)P_2) and sulfatides at overlapping sites within the N-terminal region of the protein [20]. Thus, sulfatide binding of Dab2 can be inhibited by an excess of soluble PtdIns (4,5)P_2 [21]. Likewise, binding of the adaptor protein Tom1 to Tollip specifically inhibits Tollip's PtdIns(3)P binding [22]. In the followings, we describe a protocol to use PLOA as a first step for screening molecules that can modulate protein-lipid interactions.

2 Materials

1. Nitrocellulose membrane (GE Healthcare Portrand Supported 0.45 μm NC).
2. Solution I: Prepare a solution of chloroform/methanol/water in a proportion of 65:35:8 (v/v) in a 1 mL glass vial (*see* **Note 1**).
3. Scissors, tweezers, and a soft pencil.
4. Microcentrifuge tubes, 0.5 mL.
5. Benchtop cooler rack.
6. Phospholipids (Echelon, Matreya, Cayman Chemicals, and Avanti Lipids products immobilize well on nitrocellulose membranes).
7. Refrigerated microcentrifuge and clinical centrifuge.
8. Water bath sonicator.
9. Bovine serum albumin (BSA) that is fatty acid-free (*see* **Note 2**).
10. Fusion proteins (*see* **Note 3**).
11. Washing buffer: 10 mM Tris–HCl (pH 8), 150 mM NaCl, 0.1% Tween-20 (*see* **Note 4**).
12. Blocking buffer: 3% fatty acid-free BSA in washing buffer (*see* **Note 5**).

13. Pre-developing buffer: 10 mM Tris–HCl (pH 8), 150 mM NaCl.
14. Aluminum foil.
15. Petri dishes of 100 and 40 mm diameters.
16. Primary rabbit anti-glutathione-*S*-transferase (GST) antibody: A commercial product can be obtained as a 0.51 μg/μL solution from Proteintech, and is recommended for usage in a dilution of 1:2000.
17. Donkey anti-rabbit Horseradish peroxidase (HRP)-coupled secondary antibody: A commercial product can be obtained from GE Healthcare, and is recommended for usage in a dilution of 1:10,000.
18. Modulator: Small drug or any macromolecule that alters the binding affinity of the lipid to the protein.
19. Chemiluminescent reagent kit: In this work, the Thermo Fisher SuperSignal West Pico Chemiluminescent Substrate product, which contains the stable peroxide solution and the luminol/enhancer solution, was used.
20. Platform shaker.
21. Cold room or cold box.
22. ChemiDoc XRS+ System and Image Lab software (Bio-Rad).

3 Methods

3.1 Spotting Lipids on Nitrocellulose Membranes

1. Cut pieces of nitrocellulose membrane that will fit five individual lipid spots (~4 cm × ~2.5 cm). Make sure to cut a corner to ascertain the orientation of the membrane. Keep the protective backing in place during the preparation of the lipid membranes. Take the pieces of nitrocellulose membrane and mark them with a soft pencil at the sites where the lipids will be placed.
2. Prepare microgram aliquot stocks of the phospholipid of interest. Centrifuge the original lipid vial inside a 50 mL centrifuge tube. Place tissue paper at the bottom of the centrifuge tube to protect the vial during centrifugation. Spin at 500 rpm for 3 min at 4 °C. This step prevents the loss of lipid during the vial opening. Resuspend the lipid (1 mg/mL) in Solution I, sonicate in a room temperature water bath for a few seconds, and aliquot micrograms of lipid solutions into 1 mL brown glass vials. Dry the solutions under N_2 gas in a fume hood. Store the dried lipids in glass vials sealed with Parafilm at −20 °C until use.
3. Resuspend the lipid stock at the desired concentration in Solution I. Mix gently using finger tapping. Prepare a serial dilution

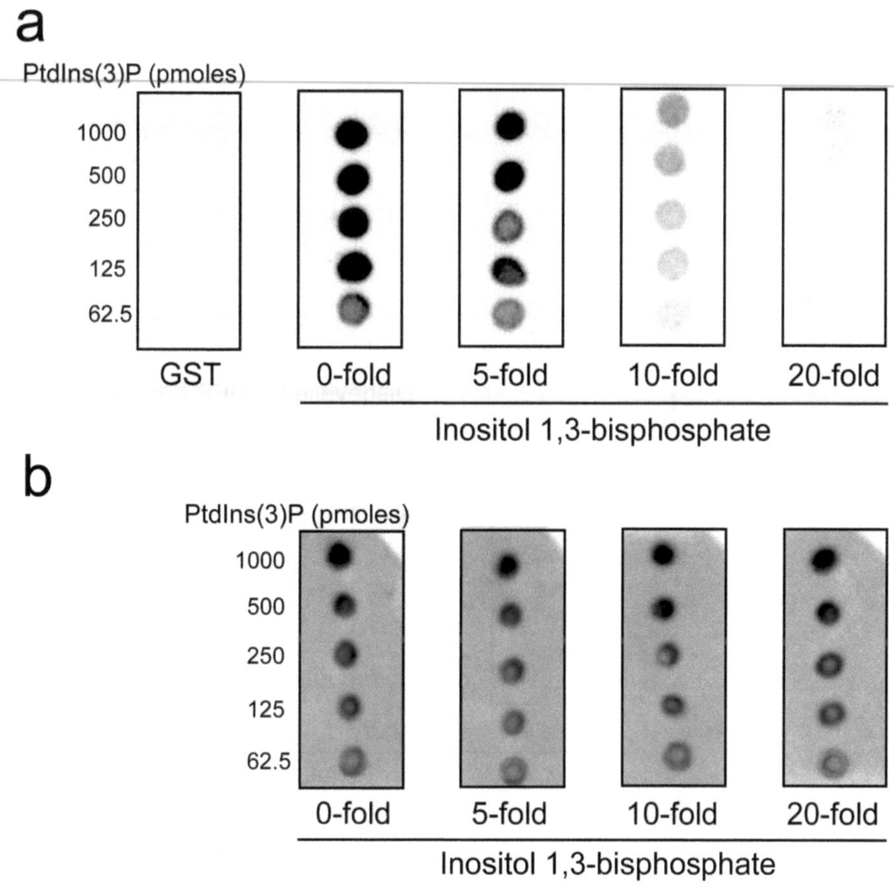

Fig. 1 Identification of modulators of protein-lipid binding using PLOA. In this experiment, GST-Tollip (**a**) and GST-Phafin2 (**b**) were probed for inhibition of PtdIns(3)P binding using the head group of the lipid, inositol 1,3-bisphosphate. As opposed to that observed for Phafin2, the results suggest that the head group of PtdIns (3)P competes with the lipid for Tollip's binding

of the lipid in the nano-picomolar range in solution I. Use 0.5 mL microcentrifuge tubes prechilled in the cooled benchtop rack. Keep all the solutions in the cooled benchtop rack.

4. Pipette 1 μL of each lipid onto the pre-labeled nitrocellulose membranes with the highest concentration of the lipid on top (*see* Figs. 1 and 2; *see* **Note 6**).
5. Dry the spotted lipid membranes for 1 h at room temperature, protecting them from light using aluminum foil. Membranes can be stored in petri dishes, wrapped with aluminum foil, at 4 °C.

3.2 Screening of Modulators of Lipid Binding Using LPOA

1. Remove the protecting backing from each piece of nitrocellulose membrane and, using tweezers, individually transfer the membranes (with the lipid spots facing down) into a 10 mm petri dish.

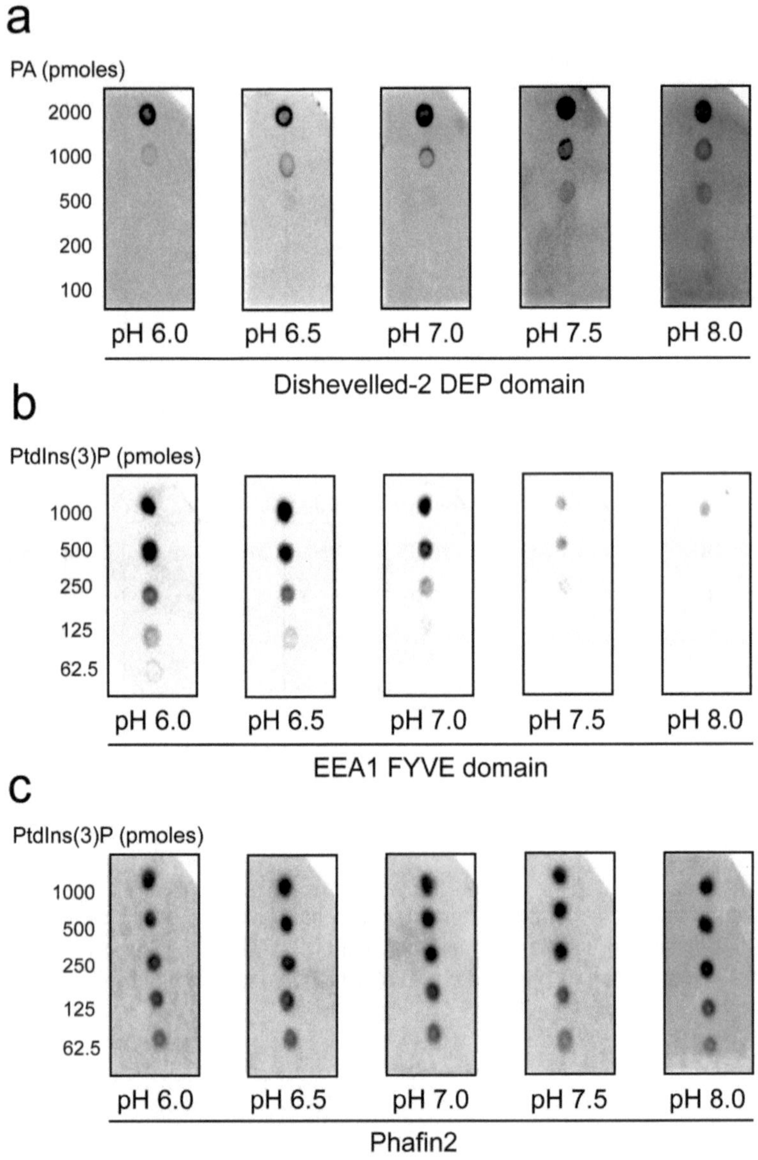

Fig. 2 PLOA showing an opposite pH-dependency of binding of EEA1 FYVE (**a**) and Dishevelled-2 DEP domain (**b**) for PtdIns(3)P and phosphatidic acid, respectively. On the other hand, PtdIns(3)P binding of Phafin2 is pH-independent (**c**)

2. Add 10 mL of blocking buffer to each of the petri dishes and gently block membranes for 1 h at room temperature using a platform shaker. Keep all petri dishes covered with aluminum foil throughout the experiment (*see* **Note** 7).
3. During the blocking incubation time, mix 1–10 μg of the fusion protein of interest without and with increasing molar ratios of the modulator in a final volume of 50–100 μL. Place the mixtures in a rotating mixer for 45 min at room temperature.

4. Discard blocking solution and wash membranes three times with washing buffer.
5. Add 10 mL of washing buffer to each of the Petri dishes (*see* **Note 8**). Add the fusion protein (as control) or the fusion protein-modulator mixtures to each of the Petri dishes and incubate with gentle agitation overnight at 4 °C (*see* **Note 9**). If the effect of pH in lipid binding is assayed (Fig. 2), the fusion protein should be incubated with the washing buffer at the specific range of pH values to be investigated.
6. Discard protein mixtures. Wash membranes three times with washing buffer for 10 min each time at room temperature by shaking on a platform shaker (*see* **Note 10**).
7. Prepare 40 mm petri dishes and add 4 mL of washing buffer containing 3% fatty acid-free BSA. Transfer the lipid membranes to each of these plates and add 2 μL of the rabbit anti-GST antibody commercial product. Incubate membranes gently on a platform shaker for 1 h at room temperature (*see* **Note 11**).
8. Transfer the membranes back to the 100 mm petri dishes. Wash membranes three times with washing buffer for 10 min each at room temperature by gently shaking on a platform shaker.
9. Transfer the membranes back to the 40 mm petri dishes and add 4 mL of washing buffer containing 3% fatty acid-free BSA. Add 0.4 μL of the secondary anti-rabbit-HRP conjugated antibody. Incubate the membranes gently on a platform shaker for 1 h at room temperature.
10. Transfer the membranes back to the 100 mm petri dishes. Wash membranes three times with washing buffer for 10 min each at room temperature by gently shaking on a platform shaker.
11. Wash membranes once with pre-developing buffer for 10 min at room temperature by gently shaking on a platform shaker (*see* **Note 12**).
12. Remove each of the membranes from the Petri dishes and transfer to a clean piece of transparent cellulose acetate sheet. Remove the excess of washing buffer using tissue papers. Do not let the membranes dry. Prepare the working solution of the SuperSignal West Pico Chemiluminescent Substrate mixture in the same cellulose acetate sheet, by mixing equal volumes of the stable peroxide solution and the luminol/enhancer solution. Place all the membranes (facing the lipid complexes down) in the mixture for 1 min. The membranes should be fully in contact with the working solution mixture, which will fluoresce due to the enzymatic activity of HRP.

13. Remove the excess of solution and place another cellulose acetate sheet on the top of the membranes. Immediately expose the reaction at different time points using the chemiluminescence detector instrument.

4 Notes

1. The composition of the solution will depend on the nature of the lipid. For example, phosphoinositides are relatively polar and need water to be soluble. Other more hydrophobic lipids require chloroform only to be soluble. Solutions should be made fresh by the time that the lipid solutions are ready to be spotted. It is recommended to work with a benchtop cooler rack unit to prevent evaporation of the solution.
2. Fatty acid-free BSA (Sigma A7030) is the most suited reagent for blocking nonspecific binding of proteins to the nitrocellulose membrane. However, there are slight variations among lots of this product and, consequently, they affect the efficiency of blocking as well as the intensity of the binding signal. It is, therefore, recommended to compare results using more than one lot if possible. Both dried milk powder or albumin have also been suggested to be used as alternative blocking reagents for nonspecific binding, but results with these reagents are quite inconsistent in our hands.
3. Fusion proteins containing glutathione-*S*-transferase (GST) are the most suitable tag to be employed during PLOA. The major reason for the GST tag use is that it increases the yield of the protein of interest by increasing its solubility. However, other tags, including green fluorescent protein (GFP) and polyhistidine tags, also worked well in our hands. As occurs with most proteins, the location of a tag at either the N- or C-terminus does not have an effect in lipid binding. If this is not the case, it is recommended adding a linker of 4–7 amino acids (preferentially glycine residues), which can extend the distance between the lipid binding site and the tag. Proteins should be checked for optimal quality using SDS-PAGE analysis. Concentration of proteins should be carefully checked using two independent methods.
4. Tris-based buffers are relatively stable. However, large quantities of this buffer are required for this assay. It is, therefore, recommended to prepare a ten-fold concentrated stock of washing buffer (100 mM Tris–HCl, pH 8, 1.5 M NaCl) and store at room temperature. If the effect of the pH is tested, the same buffer is suitable for the assay at a pH range of 6–8.

5. Fatty acid-free BSA containing blocking solutions can be stored at −20 °C.
6. Efficient immobilization of lipids to nitrocellulose membranes occurs when their fatty acid chains are saturated and are at least 16 carbons in length.
7. Do not allow the nitrocellulose membranes to get dry throughout the experiment.
8. We do not recommend using the blocking solution during the overnight incubation of the protein mixtures with the lipid-bound membranes. The initial 1-h blocking step is sufficient to avoid nonspecific binding results. BSA may bind nonspecifically to the macromolecules to be assayed and, therefore, interfere with interpretation of the results.
9. If modulation of lipid binding is studied, it is recommended to have a negative control in the experiment to ensure that the modulator specifically targets the protein under investigation.
10. When using GST fusion proteins and the indicated antibodies, the effect of pH in protein-lipid binding is not critically affected after the incubation of the protein of interest. However, it is important to pay attention to the isoelectric point of the proteins to be studied. We have also carried out pH-dependent PLOA with buffers at the same pH at all times, which led to the same outcomes. This suggests that lowering the pH of the washing buffer does not significantly affect the strength of the antibody binding in later steps.
11. The use of 40 mm petri dishes in **steps 7** and **9** in Subheading 3.2 is suggested in order to minimize the volume of the antibodies used in the assay.
12. Tween-20 can interfere with the HRP-mediated fluorescence reaction.

Acknowledgments

We thank Janet Webster for critical reading and comments on the manuscript. We also thank Tiffany Radle and Morgan Vaughn for the optimization of the purification conditions of the GST-EEA1 FYVE domain. Work in the Capelluto laboratory is supported by the National Institutes of Health (NIAID).

References

1. Wenk MR (2005) The emerging field of lipidomics. Nat Rev Drug Discov 4:594–610
2. McNamara CR, Degterev A (2011) Small-molecule inhibitors of the PI3K signaling network. Future Med Chem 3:549–565
3. Scott JL, Musselman CA, Adu-Gyamfi E et al (2012) Emerging methodologies to investigate lipid-protein interactions. Integr Biol (Camb) 4:247–258

4. Nitulescu GM, Margina D, Juzenas P et al (2016) Akt inhibitors in cancer treatment: the long journey from drug discovery to clinical use (Review). Int J Oncol 48:869–885
5. Busse RA, Scacioc A, Schalk AM et al (2016) Analyzing protein-phosphoinositide interactions with liposome flotation assays. Methods Mol Biol 1376:155–162
6. Saliba AE, Vonkova I, Ceschia S et al (2014) A quantitative liposome microarray to systematically characterize protein-lipid interactions. Nat Methods 11:47–50
7. Saliba AE, Vonkova I, Gavin AC (2015) The systematic analysis of protein-lipid interactions comes of age. Nat Rev Mol Cell Biol 16:753–761
8. Dowler S, Kular G, Alessi DR (2002) Protein lipid overlay assay. Sci STKE 2002:pl6
9. Yu JW, Mendrola JM, Audhya A et al (2004) Genome-wide analysis of membrane targeting by S. cerevisiae pleckstrin homology domains. Mol Cell 13:677–688
10. Gallego O, Betts MJ, Gvozdenovic-Jeremic J et al (2010) A systematic screen for protein-lipid interactions in Saccharomyces cerevisiae. Mol Syst Biol 6:430
11. Narayan K, Lemmon MA (2006) Determining selectivity of phosphoinositide-binding domains. Methods 39:122–133
12. Bonham KS, Orzalli MH, Hayashi K et al (2014) A promiscuous lipid-binding protein diversifies the subcellular sites of toll-like receptor signal transduction. Cell 156:705–716
13. Naguib A, Bencze G, Cho H et al (2015) PTEN functions by recruitment to cytoplasmic vesicles. Mol Cell 58:255–268
14. Murphy JE, Tacon D, Tedbury PR et al (2006) LOX-1 scavenger receptor mediates calcium-dependent recognition of phosphatidylserine and apoptotic cells. Biochem J 393:107–115
15. Zimmermann P, Meerschaert K, Reekmans G et al (2002) PIP(2)-PDZ domain binding controls the association of syntenin with the plasma membrane. Mol Cell 9:1215–1225
16. Klinkenberg D, Long KR, Shome K et al (2014) A cascade of ER exit site assembly that is regulated by p125A and lipid signals. J Cell Sci 127:1765–1778
17. Nakamura Y, Andres F, Kanehara K et al (2014) Arabidopsis florigen FT binds to diurnally oscillating phospholipids that accelerate flowering. Nat Commun 5:3553
18. Lee SA, Eyeson R, Cheever ML et al (2005) Targeting of the FYVE domain to endosomal membranes is regulated by a histidine switch. Proc Natl Acad Sci U S A 102:13052–13057
19. Capelluto DG, Zhao X, Lucas A et al (2014) Biophysical and molecular-dynamics studies of phosphatidic acid binding by the Dvl-2 DEP domain. Biophys J 106:1101–1111
20. Drahos KE, Welsh JD, Finkielstein CV, Capelluto DG (2009) Sulfatides partition disabled-2 in response to platelet activation. PLoS One 4:e8007
21. Alajlouni R, Drahos KE, Finkielstein CV, Capelluto DG (2011) Lipid-mediated membrane binding properties of Disabled-2. Biochim Biophys Acta 1808:2734–2744
22. Xiao S, Brannon MK, Zhao X et al (2015) Tom1 modulates binding of Tollip to phosphatidylinositol 3-phosphate via a coupled folding and binding mechanism. Structure 23:1910–1920

Chapter 14

Resazurin Live Cell Assay: Setup and Fine-Tuning for Reliable Cytotoxicity Results

José Á. Rodríguez-Corrales and Jatinder S. Josan

Abstract

In vitro cytotoxicity tests allow for fast and inexpensive screening of drug efficacy prior to in vivo studies. The resazurin assay (commercialized as Alamar Blue®) has been extensively utilized for this purpose in 2D and 3D cell cultures, and high-throughput screening. However, improper or lack of assay validation can generate unreliable results and limit reproducibility. Herein, we report a detailed protocol for the optimization of the resazurin assay to determine relevant analytical (limits of detection, quantification, and linear range) and biological (growth kinetics) parameters, and, thus, provide accurate cytotoxicity results. Fine-tuning of the resazurin assay will allow accurate and fast quantification of cytotoxicity for drug discovery. Unlike more complicated methods (e.g., mass spectrometry), this assay utilizes fluorescence spectroscopy and, thus, provides a less costly alternative to observe changes in the reductase proteome of the cells.

Key words Resazurin, Cytotoxicity, Cell viability, Optimization, Linear range, Alamar Blue®

1 Introduction

Drug development is a costly (>1 billion dollars per drug) and lengthy (several years) process that utilizes several "filters" to select a lead molecule that can be carried to clinical trials and, finally, be commercialized. Drug screening is performed early in this process to determine cytotoxicity of candidates and identify leads, which are later optimized. Tiered approaches to drug evaluation involving multiple parameters (e.g., cytotoxicity, changes in the proteome) have been proposed to improve our ability to predict the in vivo activity of drugs [1]. The resazurin assay, commercially available as Alamar Blue®, has been extensively used to determine toxicity in human and animal in vitro models since the early 1990s [2, 3]. Resazurin reduction by metabolically active, or "live," cells produces the highly fluorescent resorufin, thus allowing spectroscopic evaluation of cell viability under certain conditions. This reduction process is promoted by NADH reductase and carnitine dehydrogenase in vitro [4, 5]. However, it has been shown that resazurin

Iulia M. Lazar et al. (eds.), *Proteomics for Drug Discovery: Methods and Protocols*, Methods in Molecular Biology, vol. 1647, DOI 10.1007/978-1-4939-7201-2_14, © Springer Science+Business Media LLC 2017

reduction is not carried out by only one specific enzyme, as incubation with different subcellular fractions (cytosolic, microsomal, and mitochondrial) led in all cases to the formation of resorufin [6]. This observation indicates that the results of this assay correlate to the reductase activity of the proteome as a whole, and not to any one enzyme in particular. In this context, resazurin can be used to screen for drugs that target reductases directly or indirectly (e.g., enzymes involved in the same signaling pathway), or promote cell death or growth. Furthermore, reductase activity varies among different cell lines, which influences the assay conditions. Improper or lack of assay validation conditions can generate, therefore, unreliable results and limit reproducibility. Several scientists and organizations have raised concerns due to the irreproducibility of published results that could arise from improper assay validation, among other factors [7, 8]. Herein, we present a method for the optimization of the resazurin assay for cell viability and cytotoxicity measurements, where both analytical and biological parameters are studied and validated. Although the amount of information gathered with this technique is rather limited (resazurin reduction is affected by changes in the reductase proteome and not only one enzyme), it provides a facile, fast, and accurate assessment of drug cytotoxicity in drug discovery, as long as the assay is properly validated. Thus, this inexpensive assay allows the identification of hit molecules in the earlier stages of drug discovery screening. These hits can be further studied through more selective yet cost-intensive proteomics techniques, including SDS-PAGE and MALDI-TOF, to determine their specific biological targets.

2 Materials

Prepare all the solutions in phosphate buffered saline (PBS) or complete medium, i.e., medium supplemented with fetal bovine serum (FBS) and antibiotics, as specified. Use chemical reagents of reagent grade or higher. Sterilize all the solutions by filtering through a sterile 0.22 μm membrane or an equivalent method, or, procure sterile solutions from dedicated manufacturers. Prepare and store all the solutions at 4 °C, except if noted otherwise, and dispose following authorized procedures. Use only sterile equipment and supplies (pipettes, pipette tips, culture flasks, vials, tubes, etc.) for handling the cells. Cell work must be performed under aseptic conditions, in a biosafety cabinet, following all the appropriate safety regulations.

2.1 Reagent Solutions

1. Mammalian cells.
2. Complete cell culture medium: Follow the instructions provided by the supplier of the cells to prepare a cell culture

medium of adequate composition. For example, supplement the base medium (e.g., EMEM or equivalent) with 10% FBS and penicillin/streptomycin (1×). Sterile filter and store in the refrigerator at 4 °C.

3. PBS solution (1×): 0.01 M phosphate buffer, 0.0027 M KCl, and 0.137 M NaCl, pH 7.4 at 25 °C. Dissolve a PBS tablet (e.g., Sigma-Aldrich-100 g) in 200 mL (or volume suggested by the manufacturer) of autoclaved deionized water (*see* **Note 1**). Mix and filter under sterile conditions.
4. Trypsin solution (1×): 0.25% Trypsin, 0.53 mM EDTA in PBS. Use any high quality commercial product.
5. Penicillin/streptomycin solution (100×): 10,000 units/mL penicillin, 10 mg/mL streptomycin. Use any high quality commercial product.
6. Resazurin stock solution: 10 mg/mL. Add 1 g of resazurin into 100 mL of PBS and mix until the reagent has dissolved completely. Sterile filter and store in the refrigerator protected from light at 4 °C (*see* **Note 2**).
7. Resazurin working solution: 15 μM. Add 180 μL of cold resazurin stock solution into 50 mL of complete medium at 37 °C. Prepare fresh immediately before use (*see* **Note 3**).

2.2 Equipment

1. Cell culture supplies: sterile cell culture flasks, pipettes, vials, tubes, etc.
2. Conventional cell culture lab equipment: biosafetycabinet, incubator, microscope, water bath, 4 °C refrigerator, −20 °C and −80 °C freezers, centrifuges, and hemocytometer or flow cytometer.

3 Methods

3.1 Validation of Assay Conditions

1. Grow cells in a T75 flask under the conditions recommended by the supplier. Typical conditions may include EMEM:FBS 9:1 with 1% penicillin/streptomycin, with incubation performed at 37 °C in a 5% CO_2 atmosphere (*see* **Note 4**).
2. Harvest the cells when confluency reaches ~70–90%. Typical detachment conditions include a PBS wash, followed by incubation with 0.25% trypsin/EDTA for 5–10 min at 37 °C. After trypsinization, the trypsin must be inactivated by the addition of complete medium. Next, the cells must be dispersed by aspirating and dispensing gently the cell suspension with a pipette.

3. Transfer the cell suspension to a 15 mL Falcon tube and centrifuge according to supplier's recommendation, e.g., 5 min at 125 × *g* (*see* **Note 5**).
4. Gently aspirate the supernatant without disturbing the cell pellet. Resuspend the pellet in complete medium by careful, repeated pipetting.
5. Determine the cell concentration using a hemocytometer, a flow cytometer, or any other available method (*see* **Note 6**).
6. Prepare a stock solution of 2,000,000 cells/mL (stock A). If the concentration is higher than 2,000,000 cells/mL, dilute with complete medium. If the concentration is lower, centrifuge, remove the supernatant, and resuspend in an appropriate volume.
7. Prepare stocks of 250,000 cells/mL (stock B) and 25,000 cells/mL (stock C) by dilution in cell medium.
8. Seed two 96-well plates with the volumes provided in Table 1, i.e., seed three wells per cell plating density, including blanks, and per incubation time. If five incubation times are tested, the total number of wells per plating density is 3 × 5 = 15, and the total number of wells needed is 3 × 12 × 5 = 180 wells (*see* **Notes 7–9**, and Fig. 1).

Table 1
Volume of stock solutions to be placed per well

Columns	Plating density (cells/well)	Stock A (μL)	Stock B (μL)	Stock C (μL)	Complete medium (μL)
1	300,000	150	–	–	–
2	200,000	100	–	–	50
3	150,000	75	–	–	75
4	100,000	50	–	–	100
5	50,000	25	–	–	125
6	25,000	–	100	–	50
7	10,000	–	40	–	110
8	5000	–	20	–	130
9	1000	–	–	40	110
10	500	–	–	20	130
11	100	–	–	4	146
12 (blanks)	0	–	–	–	150

Note: row H of one of the plates should be used for blanks and, thus, filled with medium only

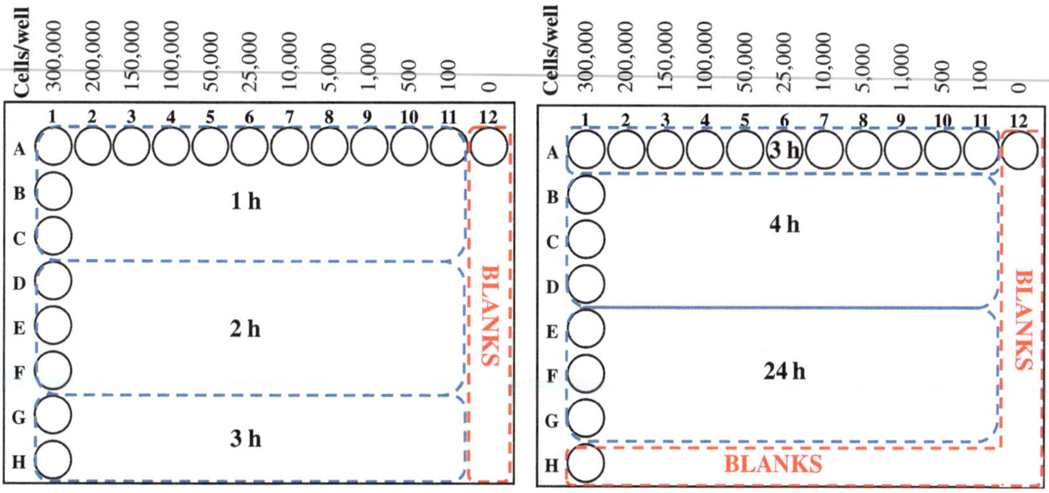

Fig. 1 Sample plating setup. The recommended setup allows triplicate testing of 12 plating densities at 5 resazurin incubation times per triplicate (15 rows/density). **Notes 6** and **7** provide important considerations if the setup is to be modified

9. Place 96-well plates ("culture plates") in the incubator for 24 h to allow for cell attachment (*see* **Note 10**).
10. Aspirate the medium from all wells. Add 250 μL of resazurin working solution per well using a multichannel pipette. Return the plate *to* the incubator (*see* **Note 11**).
11. After 1 h, remove 100 μL from the wells located in wells A1 through C11 and pipette into a separate 96-well plate with black walls and clear bottom ("reader plate"). Also, pipette 100 μL from eight blank wells (column 12 or row-H of one of the plates, *see* Fig. 1). Return the culture plate to an incubator (*see* **Notes 12** and **13**).
12. Place the black reader plate in a fluorescence plate reader and measure the fluorescence intensity using excitation at 560 nm and detection at 590 nm (*see* **Note 14**).
13. Repeat **steps 11** and **12** using a different triplicate of wells after 2, 3, 4, and 24 h incubation with resazurin.
14. For each time point, determine the average fluorescence of the corresponding blanks ($\overline{f_b}$) and the standard deviation (σ_b).
15. Subtract $\overline{f_b}$ from all the samples to obtain the blank-corrected fluorescence intensity. Exclude any blank-corrected values lower than 10 $\times\sigma_b$ from further analysis. These values are considered to be under the limit of quantification.
16. Plot blank-corrected fluorescence intensity vs. number of plated cells (Fig. 2). Identify the linear portion and plateau in the graph (*see* **Note 15**) and determine the limit of linearity (LOL) using an appropriate method (*see* **Note 16**).

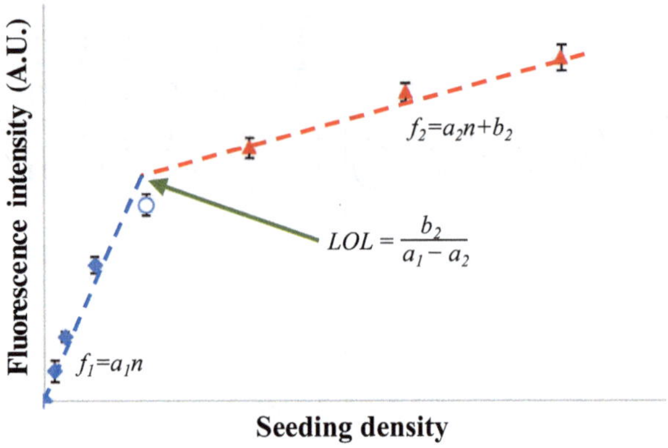

Fig. 2 Sample fluorescence vs. plating density graph, displaying the linear (*blue diamonds*) and plateau (*red triangles*) regions. Best-fit linear equations are provided in the graph, along with the LOL

17. Calculate limit of detection (LOD) and quantification (LOQ), using Eqs. 1 and 2, where a_1 stands for the slope of the best-fit equation for the linear range.

$$\text{LOD} = 3 \times \sigma_b / a_1 \quad (1)$$

$$\text{LOQ} = 10 \times \sigma_b / a_1 \quad (2)$$

18. Under most circumstances, choose the "optimum" incubation time with resazurin ($t_{\text{resazurin}}$) that produces the widest linear range (LOL/LOQ) (*see* **Note 17**).

3.2 Evaluation of Growth Kinetics

1. Perform **steps 1–5** from Subheading 3.1.
2. Seed 1000–5000 cells/well in 12 wells (i.e., three wells per four time-points) of a 96-well plate, which will be used to assess growth kinetics (*see* **Note 18**).
3. Place the 96-well plate in the incubator for 24 h to allow cell attachment (*see* **Note 10**).
4. Carefully aspirate the medium from all the wells.
5. Rinse three wells with 100 μL of PBS (1×), and aspirate.
6. Add 100 μL of trypsin 0.25% with 0.53 mM EDTA to the wells prepared in **step 5** (*see* **Note 19**). Pipette 100 μL of complete medium into all other wells.
7. Return the 96-well plate to the incubator for 10 min.
8. Gently mix the wells with trypsin with a pipette to *improve* cell detachment from the plate.
9. Check cells under the microscope to confirm detachment from the plate. If incomplete, repeat **step 8**.

10. Determine the number of cells/well in the three wells treated with trypsin, using a hemocytometer, a flow cytometer, or any other available method (*see* **Note 6**).
11. Return the 96-well plate to the incubator.
12. Repeat **steps 4–11**, 2, 3, and 4 days after seeding, each day trypsinizing the following three wells.
13. Determine the average number of cells/well for each time point.
14. Normalize all values by using the number of cells/well in day 1 (24 h after seeding) (*see* **Note 20**).
15. Plot natural logarithm of the normalized number of cells/well vs. time and determine the equation for the line of best fit.
16. Determine the population doubling time (τ) using Eq. 3, where a_k is the slope of the line of best fit as determined in **step 15** [9]

$$\tau = \ln(2)/a_k \quad (3)$$

3.3 Cytotoxicity Experiment Design

In order to illustrate the reasoning behind the assay validation and the experiment's design, the general procedure is summarized as a flowchart in Fig. 3.

1. Based on the goals of the experiment, calculate the total number of days that cells will be grown (t_{total}). Take into account the time required for attachment to the plate (t_{attach}), time of incubation with cytotoxic compound (t_{test}), and any other relevant incubations (e.g., toxicity evaluation 24 h after removing cytotoxic compound). Do not add time of incubation of resazurin.

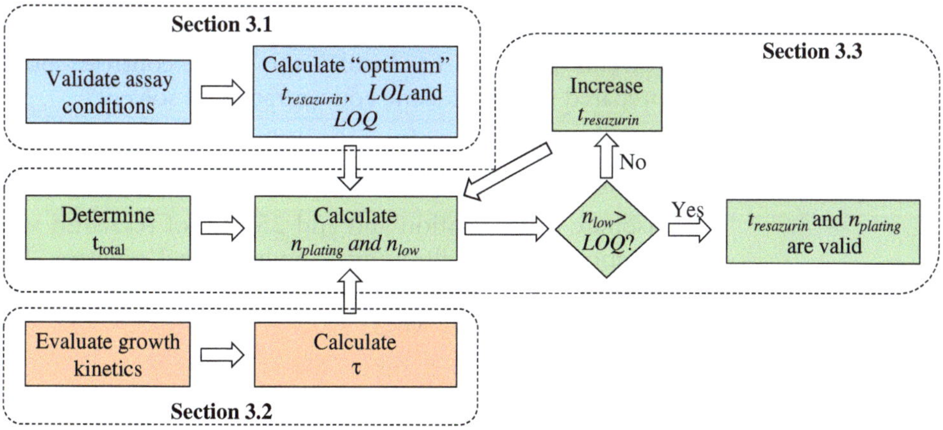

Fig. 3 Flowchart for effective cytotoxicity experiment design

2. Based on the validated resazurin assay conditions, namely incubation time ($t_{resazurin}$) at a given concentration of resazurin (*see* Subheading 3.1, **step 18**), determine the seeding density ($n_{plating}$) required to ensure that the number of cells (n_{final}) at t_{total} remains in the linear range. N_{final} can be estimated as 90% of the LOL, and $n_{plating}$ can be solved using Eq. 4 (*see* **Notes 21** and **22**).

$$n_{plating} = \frac{0.9 \times \text{LOL}}{2^{(t_{total}/\tau)}} \tag{4}$$

3. Assuming ~95% cytotoxicity in the sample treated with the highest concentration of drug, calculate the estimated "lowest" number of live cells after treatment with cytotoxic drug (n_{low}).

$$n_{low} = 0.045 \times \text{LOL} \tag{5}$$

4. If $n_{low} > \text{LOQ}$ at $t_{resazurin}$, all the parameters have been correctly calculated and the resazurin assay conditions should be valid.
5. If $n_{low} < \text{LOQ}$, $t_{resazurin}$ should be extended to increase the sensitivity of the assay and **steps 1–4** of this section should be repeated.

3.4 Cytotoxicity Experiment

1. Perform **steps 1–5** from Subheading 3.1.
2. Pipette enough solution to seed $n_{plating}$ cells per well, as determined in **step 2** from Subheading 3.3. Seed at least two or three wells for each concentration of drug to be tested, and for the controls (growth medium without drug, and growth medium with any solvents used).
3. Incubate cells at 37 °C and 5% CO_2 for 24 h (*see* **Note 10**).
4. Prepare stock solutions of drug to be tested in a complete medium. If dimethyl sulfoxide (DMSO) or other cytotoxic solvents are used to dissolve the drug, controls should be prepared using the same concentration of solvent.
5. Replace the medium from the wells with solutions of drug.
6. Return plate(s) to the incubator for desired period of time.
7. Aspirate drug solution and add 250 μL of resazurin working solution to all wells using a multichannel pipettor. Prepare blanks by pipetting 250 μL of resazurin working solution into at least three empty wells.
8. Incubate with resazurin for the time determined in Subheading 3.3.

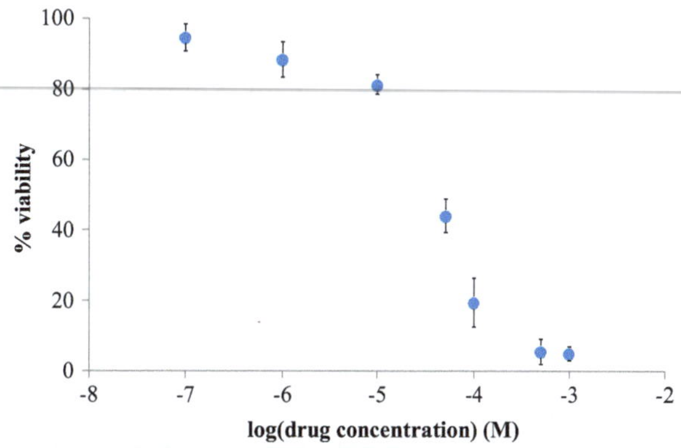

Fig. 4 Sample cytotoxicity graph showing cell viability vs. log of drug concentration. Error bars correspond to standard deviation of samples at each concentration of drug

9. Pipette 100 μL out of each well and into a 96-well plate with black walls and clear bottom.
10. Place the black 96-well plate in a plate reader and measure fluorescence intensity using excitation at 560 nm with detection at 590 nm (*see* **Note 14**).
11. Calculate the average fluorescence of the blanks ($\overline{f_b}$).
12. Subtract $\overline{f_b}$ from all the values to obtain the blank-corrected fluorescence intensity of samples ($\overline{f_s}$) and untreated controls ($\overline{f_c}$) (*see* **Note 23**).
13. Calculate % viability using Eq. 6.

$$\frac{\%\text{viability} = \overline{f_s}}{\overline{f_c} \times 100} \quad (6)$$

14. Determine IC_{50} using an appropriate statistical program. A sample % viability vs. drug concentration plot is shown in Fig. 4.
15. In order to report average IC_{50}, at least two or three different cytotoxicity experiments should be performed. Uncertainty should correspond to the standard deviation of the three experiments.

4 Notes

1. Non-autoclaved deionized water can also be used since the solution will be sterile filtered. However, we recommend use of autoclaved water as an extra cautionary measure.

2. Resazurin stock solution can be stored in 50 mL centrifuge tubes wrapped in aluminum foil to prevent photodegradation of resazurin. Solution stability can be checked periodically by diluting the stock and calculating the concentration based on wavelength of maximum absorption (usually 600 nm).
3. Resazurin stock solution should be kept cold at all times to avoid its thermal degradation. Thus, we recommend warming up the complete medium and then adding the cold resazurin stock solution. Given the small volume used of the latter, the resulting solution will remain warm. For more accurate results, working resazurin solutions should be used immediately to avoid photo- or thermal degradation of resazurin. Never leave working resazurin solutions in a warm water bath, or its equivalent.
4. The growth conditions presented in Subheading 3.1 are representative of procedures recommended by ATCC for several cancer cell lines. Any growth protocols recommended by the supplier of a particular cell line supersede those presented here.
5. Centrifugation allows further removal of trypsin from the cell suspension. Although trypsin should be inactivated by the components of the complete medium, we have observed that cell morphology can be slightly affected if not removed.
6. If possible, perform a viability assay in the cell suspension, such as trypan blue or propidium iodide exclusion for bright field or fluorescence microscopy, respectively. Nonviable cells might skew the results of the experiment.
7. The values presented in Table 1 are based on our experience with several cancer cell lines in 96-well plates. The higher plating densities should be used to ensure that a plateau in resazurin reduction is reached, while the lower densities are especially necessary for long-term (e.g., 24 h) incubations. Plating densities should be changed depending on the specific cell line to be used and the size of the well plate (e.g., 6-, 24-, or 384-well plates).
8. The recommended plating allows for triplicate fluorescence measurements of each plating density at each of five different incubation times. Plating can be modified to include less samples per time point (e.g., duplicate instead of triplicate) or less times points (e.g., two vs. five).
9. If the recommended plating setup is used, only 15 wells per seeding density (5 time points $\times$ 3) are needed. Thus, one row of the two 96-well plates would remain unused. We recommend using this row for blanks (*see* Fig. 1) to perform a more accurate evaluation of the limits of detection and quantification.

10. Some cell lines might require longer time to attach to the plate. If visualization of cells under the microscope reveals a significant proportion of floaters, plates should be returned to the incubator and allowed extra time. Once an appropriate time has been determined, it should be used in ALL experiments, including evaluation of cytotoxicity and kinetics of growth.
11. Alternatively, a lower volume of resazurin working solution can be used at the expense of a shorter linear range for the method, since resazurin will be metabolized faster by the cells. Furthermore, a more concentrated resazurin working solution could be added directly to the cells without need of aspirating the old medium. Although this might reduce the time of analysis, results might be skewed by differences in the volume of medium throughout the wells due to evaporation.
12. Alternatively, fluorescence can be measured directly in the plates containing cells. In our experience, this leads to lower fluorescence intensities perhaps attributed to self-absorption. Moreover, pipetting resazurin solution into black 96-well plates instead of growing cells directly on them might prove more cost effective since "reader plates" can be reused after proper disinfection. Such plates are significantly more expensive than pipette tips or regular 96-well plates ("culture plates"). Periodic inspection for cracks or scratches in the black plates should be performed to ensure acceptable spectroscopic quality.
13. The same blank wells can be used at multiple incubation times (e.g., 1 and 2 h) since no cells are present to reduce resazurin.
14. Excitation and emission wavelengths should be optimized for the instrument. Resazurin working solution should be used as a blank, while a fully reduced sample can be prepared by adding sodium borohydride to resazurin working solution [3]. Alternatively, one of the samples with high plating densities can be used for this purpose. Measure fluorescence at different excitation and emission wavelengths ranging from 540 to 580 nm and 570 to 600 nm, respectively. Wavelengths should be selected as those that yield the highest sample/blank emission ratio. Instrument gain should be adjusted so that no samples surpass the detector's threshold, while keeping the highest possible sample/blank emission ratio.
15. Determination of the linear and plateau portions of the graph is critical for latter steps. It is recommended to start with the linear portion by performing graphical examination based on the lowest plating densities. In some cases, the plateau might be accurately described using a linear equation with a slope of or close to zero.

16. Assuming linear fits are applied to the linear and plateau regions, best-fit equations would be expressed as $f_1 = a_1 n$ and $f_2 = a_2 n + b_2$, respectively, where f stands for blank-corrected fluorescence, n is the number of cells or plating density, a is the slope, and b is the intercept on the y-axis. A sample graph displaying both regions and equations is presented in Fig. 2. Thus, the limit of linearity (expressed in number of cells) would be calculated as:

$$\mathrm{LOL} = \frac{b_2}{a_1 - a_2} \quad (7)$$

17. Although the resazurin assay is moderately nontoxic for most cell lines, inherent cytotoxicity should be tested. In our experience, short exposure (2–3 h) has minimal toxicity, whereas long contact (days) leads to decreased viability.
18. The exact number of cells to be plated depends on their growth kinetics. Seeding density should be low enough to avoid reaching 100% confluency at any time point.
19. Detachment conditions should be validated for each cell line. In our experience, treatment with 0.25% trypsin with 0.53 mM EDTA for 10 min works well for most cell lines. For highly adherent cell lines (e.g., HaCaT), PBS can be substituted with 0.05% EDTA for rinses, and EDTA concentration in trypsin solution can be increased to 0.025%.
20. Although the number of cells seeded in the well is known, it is best to actually measure it 24 h after seeding. This allows correction for incomplete attachment (e.g., cells might die after counting and viability evaluation but before attaching to plate) and incomplete detachment from well plate upon trypsinization.
21. Equation 4 assumes 100% attachment efficiency, which might not be true for all cell lines. Although the equation could be corrected to account for this error, we recommend using as is to ensure that cells do not exceed the linear range of the resazurin assay.
22. Equation 4 was derived from a basic exponential growth function (Eq. 8), where the number of cells (n_i) at a given time (t_i) after plating depends on the doubling time (τ) and initial number of cells (n_0).

$$n_i = n_0 \times 2^{\left(\frac{t_i}{\tau}\right)} \quad (8)$$

Given the assumptions described in **step 2** of Subheading 3.2, n_i at t_{total} corresponds to 90% of the LOL, and n_0 is the plating density.

23. If a "solvent control" was prepared with DMSO or any other potential cytotoxic solvent, the drug samples with the same solvent concentration should be compared to the "solvent control" instead.

References

1. McKim JMJ (2010) Building a tiered approach to in vitro predictive toxicity screening: a focus on assays with in vivo relevance. Comb Chem High Throughput Screen 13(2):188–206. doi:10.2174/138620710790596736
2. Fields RD, Lancaster MV (1993) Dual-attribute continuous monitoring of cell proliferation/cytotoxicity. Am Biotechnol Lab 11(4):48–50
3. Pagé B, Pagé M, Noël C (1993) A new fluorometric assay for cytotoxicity measurements *in vitro*. Int J Oncol 3:473–476
4. Barnes S, Spenney JG (1980) Stoichiometry of the NADH-oxidoreductase reaction for dehydrogenase determinations. Clin Chim Acta 107 (3):149–154. doi:10.1016/0009-8981(80) 90442-8
5. Matsumoto K, Yamada Y, Takahashi M et al (1990) Fluorometric determination of carnitine in serum with immobilized carnitine dehydrogenase and diaphorase. Clin Chem 36 (12):2072–2076
6. Gonzalez RJ, Tarloff JB (2001) Evaluation of hepatic subcellular fractions for Alamar blue and MTT reductase activity. Toxicol In Vitro 15(3):257–259. doi:10.1016/S0887-2333(01) 00014-5
7. Begley CG, Ellis LM (2012) Drug development: raise standards for preclinical cancer research. Nature 483(7391):531–533
8. Prinz F, Schlange T, Asadullah K (2011) Believe it or not: how much can we rely on published data on potential drug targets? Nat Rev Drug Discov 10(9):712–712
9. American Type Culture Collection (2014) ATCC® animal cell culture guide. American Type Culture Collection, Manasas, VA

Chapter 15

Exploring Protein-Protein Interactions as Drug Targets for Anti-cancer Therapy with In Silico Workflows

Alexander Goncearenco, Minghui Li, Franco L. Simonetti, Benjamin A. Shoemaker, and Anna R. Panchenko

Abstract

We describe a computational protocol to aid the design of small molecule and peptide drugs that target protein-protein interactions, particularly for anti-cancer therapy. To achieve this goal, we explore multiple strategies, including finding binding hot spots, incorporating chemical similarity and bioactivity data, and sampling similar binding sites from homologous protein complexes. We demonstrate how to combine existing interdisciplinary resources with examples of semi-automated workflows. Finally, we discuss several major problems, including the occurrence of drug-resistant mutations, drug promiscuity, and the design of dual-effect inhibitors.

Key words Protein-protein interactions, Cancer, Protein interaction inhibitors, Drug resistance mutations

1 Introduction

Protein-protein interactions (PPI) are remarkably complex and diverse. They form intricate interaction networks where some proteins (so-called hubs) make connections to many other proteins. Despite the diversity and complexity, there are recurrent motifs in the PPI networks and protein-protein binding sites [1, 2]. PPI interaction networks can be perturbed by differential gene expression and disease mutations [3–6] and a lot of efforts have been undertaken to study how signal flow in the networks is altered by these factors and how drugs can reestablish the intricate balance. Targeting of PPIs for therapeutic intervention is very challenging because of the size and shape of their binding interfaces, e.g., planar interfaces lacking binding pockets, the difficulty to construct a functional assay to screen out affected interactions. Nevertheless, for the last several years there has been a considerable interest in PPI drug targets. Furthermore, many existing drugs have promiscuous effects on interaction pathways affecting highly connected

Iulia M. Lazar et al. (eds.), *Proteomics for Drug Discovery: Methods and Protocols*, Methods in Molecular Biology, vol. 1647, DOI 10.1007/978-1-4939-7201-2_15,

protein hubs, while PPI inhibitors can be designed to achieve highly specific drug binding to a target protein complex. The success stories include the design of inhibitors of bromo-domains [7], Bax-BclXL [8], p53-MDM2 [9], and VEGF receptor [10]. The up-to-date database 2P2Idb v.2 contains 27 structurally characterized protein-protein complexes and 274 protein-inhibitor complexes with 242 unique small molecule inhibitors [11].

Predicting the phenotypic effects of mutations and molecular mechanisms of drugs cannot be achieved without the detailed knowledge of cellular pathways, interaction networks, and PPI binding interfaces. Studies on PPI binding site properties go back several decades, and some of them reported an extensive similarity between protein interfaces even in the absence of sequence or structure similarity between proteins [12–15]. Since the current coverage of the human proteome with protein structural complexes remains limited, and the number of novel PPI binding sites in the PDB database grows slower than the number of protein complexes [16, 17], different computational methods have been proposed to close this gap. It has been shown that protein interactions and binding sites for human complexes can be predicted from homologous protein structural complexes of another species [18–20]. Such an inference approach may be supported by a recent work that examined the growth dynamics and evolutionary roots of PPI binding sites and found that a majority of binding sites could be traced back to the universal common ancestor of all cellular organisms [16].

There are several challenges in the structural design of PPI inhibitors. First, the inhibitor should bind relatively strongly to the protein target. This is achieved by mimicking the interactions between two proteins, especially those interactions between residues that contribute most to the binding energy, so-called binding hot spots. Second, many anti-cancer therapies are prone to acquired drug resistance which is a major challenge in cancer treatment. Therefore, the designed inhibitor (peptide or small molecule) should retain its properties, even if two interacting target proteins undergo extensive selection in the tumor to eliminate binding to these inhibitors while retaining binding between the two proteins. The latter task can be accomplished by inhibitor design along with in-silico mutagenesis. Finally, the inhibitors should be active with respect to not only the proposed PPI targets but also with respect to their paralogs. We illustrate the PPI inhibitor design principles using the p53-MDM2 interaction as an example.

2 Materials

2.1 Online Resources

1. IBIS https://www.ncbi.nlm.nih.gov/Structure/ibis/ibis.cgi [21, 22]; IBIS API https://www.ncbi.nlm.nih.gov/research/ibis-api/.
2. PubChem BioAssay (https://pubchem.ncbi.nlm.nih.gov) [23, 24].
3. MutaBind (https://www.ncbi.nlm.nih.gov/research/mutabind/) [25, 26].
4. NCI drug dictionary (http://www.cancer.gov/publications/dictionaries/cancer-drug) [27].
5. Drug sensitivity in cancer (http://www.cancerrxgene.org/translation/Drug).
6. 2P2Idb database v2 (http://2p2idb.cnrs-mrs.fr/) [11].
7. Binding DB (https://www.bindingdb.org/bind/) [28].
8. SAAMBE and HotRegion (http://prism.ccbb.ku.edu.tr/hotregion/) [29, 30].

2.2 Software

1. KNIME (https://www.knime.org).

3 Methods

3.1 Practical Challenges in Cancer Drug Design: p53-MDM2 Inhibitors

Cytotoxicity of cancer drugs may lead to genomic instability and consequently to the activation of the p53 tumor suppressor. E3 ubiquitin-protein ligase (MDM2) endogenously inhibits p53, and the disruption of an interaction between the p53 transactivation domain and MDM2 leads to the activation of p53. MDM2 negatively regulates the p53 pathways and is often overexpressed in tumor cells. Different small molecule and stapled peptide-based drugs have been designed to inhibit the p53-MDM2 interaction [31, 32]. One of the very well-studied small molecule p53-MDM2 inhibitors is Nutlin-3 that exhibits anti-cancer effects even in those cells that do not express functional p53 via mechanisms involving p73 and E2F1 activation [33]. Other p53-MDM2 inhibitors are currently under clinical trials. Stapled peptides that inhibit protein-protein interactions are an emerging class of drugs since they can better resist the cancer clonal selection leading to drug resistance. Their effectiveness can be explained by a more extensive interface between the peptides and proteins compared to small molecules, so that a larger number of acquired drug-resistant mutations on the p53-MDM2 interface can be targeted.

3.2 Exploring Binding Interfaces of Homologous Protein Complexes

To find protein complexes that can potentially be targeted by small molecules, one can examine a superposition of protein-protein and protein-small molecule binding interfaces in the existing structural complexes and find if their binding modes overlap. This procedure can help detect druggable PPIs. Several methods have been proposed that explore different types of structurally homologous complexes to detect so-called multibinding interfaces [34, 35]. In the current study, we use the Inferred Biomolecular Interaction Server (IBIS) method [21, 22] that clusters similar binding sites found in homologous proteins based on their sequence and structure conservation, and validates these using various approaches. Analysis of IBIS binding site clusters offers an opportunity to directly compare PPI interfaces with protein-small molecule or with protein-peptide interfaces to identify those interfaces that can be potentially targeted by small molecules or peptides.

According to IBIS, there are three conserved binding site clusters between MDM2 and small molecules, and one conserved binding site cluster between MDM2 and p53 (PDB 1YCR used as a query in IBIS). IBIS allows one to compare binding sites of MDM2-p53 interactions with the binding sites of MDM2-small molecule complexes. As can be seen in Fig. 1, eight sites (sites 38,

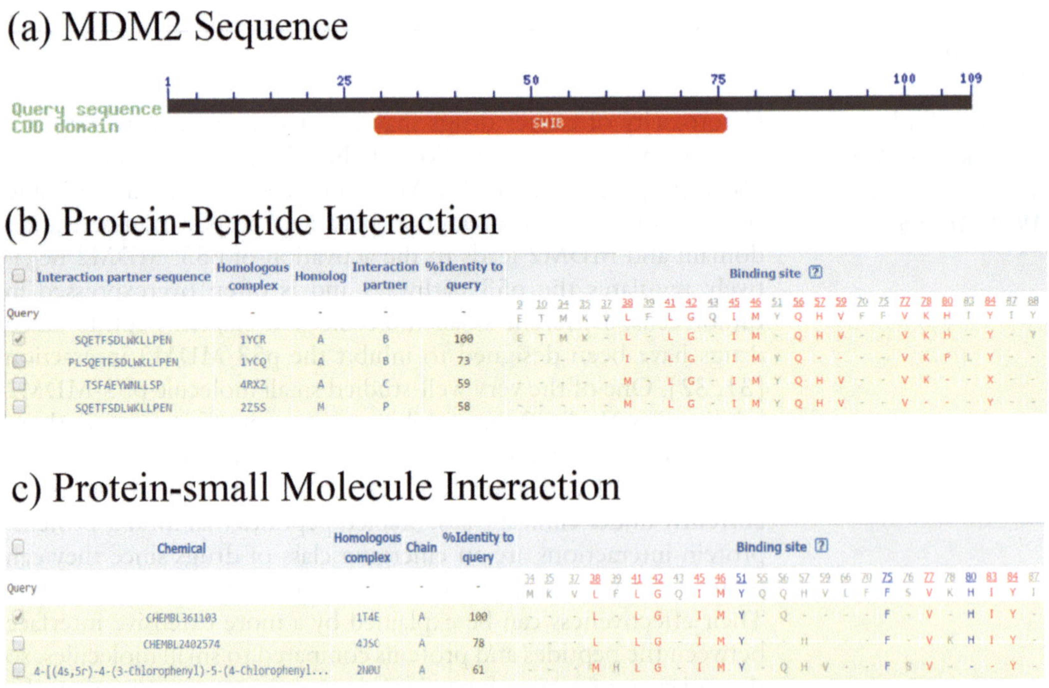

Fig. 1 Analysis of conserved binding sites using IBIS method with PDB 1YCR as a query. (**a**) MDM2 sequence annotation using Conserved Domain Database [36]; (**b**) Conserved protein binding sites between MDM2 and p53; (**c**) Conserved protein binding sites between MDM2 and small molecules. Residues are numbered with respect to the full protein sequence. Colors of binding site residues correspond to their conservation among homologs: conserved residues are colored in *red*, less conserved in *blue* and nonconserved in *gray*

41, 42, 45, 46, 77, 80, and 84) from the second protein-small molecule binding site cluster extensively overlap with the MDM2-p53 binding site cluster. It is also clear that the third small molecule binding site cluster overlaps with MDM2-p53 only partially (by one to two residues) and even without activity assays, one might suggest that the small molecules from this cluster might not be good inhibitors of this interaction and therefore not relevant drugs.

3.3 Identifying Binding Hot Spot Sites

It is challenging to develop modulators of protein–protein interactions because protein–protein interfaces are flat and lack binding pockets [37]. However, binding energy is usually determined by a small number of binding "hot spots" [38]. Targeting binding hot spots in drug design is a simpler and more straightforward strategy. Resistance developed within the tumor cell subpopulations by utilizing preexisting or acquired mutations (drug resistance mutations) may also be predicted from examining binding hot spots.

Binding hot spots can be identified in silico, by calculating the evolutionary conservation of binding sites or by scanning of PPI interfaces and calculating changes in binding affinity upon amino acid substitutions (for example alanine scanning). This task can be achieved by recently developed methods MutaBind and HotRegion. MutaBind uses molecular mechanics force fields, statistical potentials, and fast side-chain optimization algorithms [25]. The MutaBind server maps mutations on a structural protein complex, calculates the associated changes in binding affinity, determines the deleterious effect of a mutation, estimates the confidence of this prediction, and produces a mutant structural model for download. The HotRegion approach uses residue network topology and a statistical pairwise contact energy function [30].

Below is a step-by-step protocol to predict the impact of mutations on binding affinity of MDM2-p53 using MutaBind:

1. Open MutaBind web page (*see* Subheading 2) and use the PDB code 1YCR as an input.
2. Drag and drop "chain A" (MDM2) into the "Partner 1" box and "chain B" (p53) into the "Partner 2" box.
3. Select "Chain A" and binding site residues annotated by IBIS (Fig. 1b) to mutate; mutate them into alanine. Each mutation will be treated independently. Please note the residue numbering difference between PDB (MutaBind uses PDB numbering) and IBIS (IBIS uses the full protein sequence numbering).

Table 1 shows that 12 alanine-scanning mutations are predicted to have deleterious destabilizing effects on the MDM2-p53 interaction (potential binding hot spots). Among these potential binding hot spots, eight are responsible for the interaction with Nutlin-3a inhibitor, whereas nine alanine-scanning mutations are predicted to be deleterious for the paralogous MDMX-p53 interaction (see the next section). Overall, five evolutionarily conserved binding hot

Table 1
Alanine scanning of binding sites annotated on MDM2 (PDB: 1YCR, chain A) or MDMX (PDB: 3DAB, chain A) by IBIS for MDM2-p53 (1YCR) and MDMX-p53 (3DAB) complexes using MutaBind webserver

MDM2/p53			MDMX/p53		
Mutation to Ala	ΔΔ*G* (kcal/mol)	Deleterious	Mutation to Ala	ΔΔ*G*	Deleterious
Glu25	0.56	No			
Thr26	0.44	No			
Met50	1.79	Yes	Val49	1.19	No
Lys51	1.01	No	Lys50	0.86	No
Val53	1.62	Yes			
Leu54#	1.80	Yes	Met53	2.69	Yes
Phe55	0.72	No			
Leu57	1.36	No	L56	1.91	Yes
Gly58#	**1.97**	**Yes**	**Gly57**	**2.00**	**Yes**
Gln59	0.58	No			
Ile61#	**2.12**	**Yes**	**Ile60**	**2.29**	**Yes**
Met62#	1.38	No	Met61	1.04	No
Tyr67	1.16	No	Tyr66	1.42	No
Gln72#	**1.67**	**Yes**	**Gln71**	**1.77**	**Yes**
His73#	1.69	Yes	His72	1.48	No
Val75#	**1.68**	**Yes**	**Val74**	**1.81**	**Yes**
Phe86	1.66	Yes			
Phe91	1.80	Yes			
Val93#	**2.61**	**Yes**	**Val92**	**2.22**	**Yes**
Lys94	0.65	No			
His96#	1.16	No			
Ile99#	2.05	Yes	Leu98	1.95	Yes
Tyr100#	1.12	No	Tyr99	2.68	Yes
Ile103	1.28	No			
Tyr104	0.14	No			

Observed binding sites in MDM2-Nutlin-3a complex (PDB: 4HG7) are labeled with "#." ΔΔ*G* refers to the change in binding affinity upon mutations, positive values correspond to the destabilizing effects. Overlapping binding hot spots (according to MutaBind and HotRegion) among three complexes, MDM2-p53, MDM2-Nutlin-3a, and MDMX-p53, are highlighted in bold fonts. Because of the difference in PDB (used in MutaBind) and full sequence (used in IBIS) numbering, there is a 16 residues offset and Glu25 in MutaBind or PDB corresponds to Glu9 in IBIS

spots overlap between MDM2-p53, MDM2-Nutlin-3a, and MDMX-p53 complexes are highlighted in bold fonts in Table 1.

3.4 Polypharmacology of Dual PPI Inhibitors

Drugs, including PPI inhibitors, may interact with multiple targets, inhibiting multiple PPIs and affecting various pathways. While cytotoxic in some cases, such polypharmacologic effects are desirable in other cases. For example, while targeting the p53-MDM2 interaction, the potency for paralogous interaction with MDMX should also be achieved. MDMX is structurally similar to MDM2 but lacks its p53 ubiquitin-mediated degradation activity. Nevertheless, MDMX can repress p53 and the effectiveness of p53-MDM2 inhibitors would be compromised in cells overexpressing MDMX [39]. Currently, several small-molecule dual p53-MDMX/MDM2 inhibitors are under development [40].

A solution to this problem of finding dual inhibitors requires extensive sampling of protein–small molecule binding modes in the space of similar binding interfaces of homologous complexes. For example, by querying IBIS with the MDMX-p53 structural complex (PDB 3DAB), one can see that MDM2-p53 and MDMX-p53

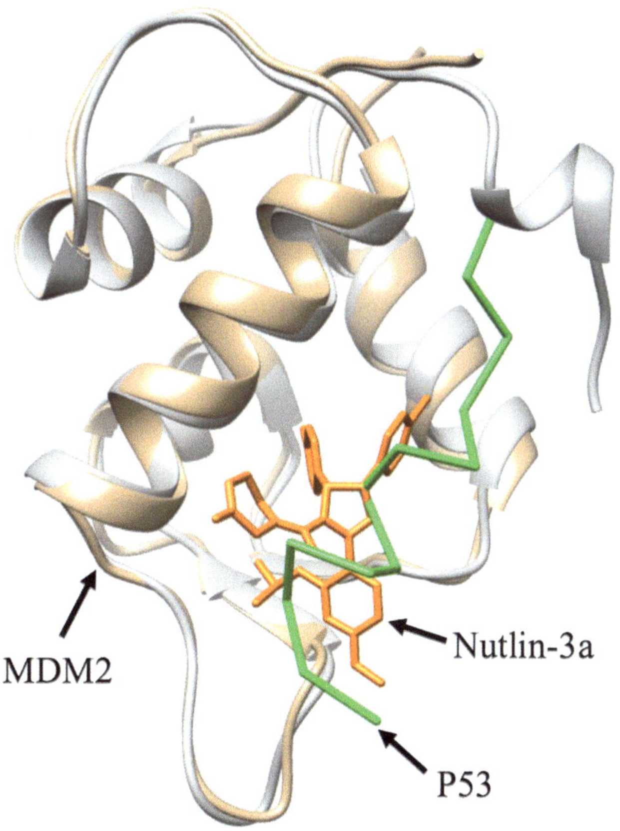

Fig. 2 A superposition of MDM2-p53 (PDB code: 1YCR) and MDM2-Nutlin-3a complex (PDB code: 4HG7) structures

Table 2
Applications program interface (API) to extract and process data on protein-protein and protein-small molecule interactions

URI	Description of retrieved results
/structures/protein/<PROTEIN>	PDB and chain identifiers given a human gene name or a UniProt protein name
/structures/compound/<CID>	PDB identifiers containing a given PubChem compound
/sites/obs/protein/pdb/<PDB>[/<CHAIN>]	Observed *protein-protein* binding site residues given a PDB identifier and an optional chain identifier. Each binding site specifies two interacting chains, the interacting domains and a list of residues
/sites/obs/compound/pdb/<PDB>[/<CHAIN>]	Observed *protein-small molecule* binding sites for each compound in a given protein structure
/sites/obs/peptide/pdb/<PDB>[/<CHAIN>]	Observed *protein-peptide* binding sites given a protein structure. The interacting peptide sequence is returned as well
/sites/obs/compound/compound/<CID>	Proteins and observed *protein-small molecule* binding sites given a compound of interest
/sites/inf/protein/pdb/<PDB>[/<CHAIN>]	Inferred *protein-protein* binding site residues given a protein structure
/sites/inf/peptide/pdb/<PDB ID>[/<CHAIN>]	Inferred *protein-peptide* binding site residues given a PDB and an optional chain identifier
/sites/inf/compound/pdb/<PDB>[/<CHAIN>]	Inferred *protein-small molecule* binding site residues given a protein structure
/compounds/similar/compound/<CID>	Similar *small molecule compounds* given a small molecule. Performs both 2D fingerprint and 3D fast searches

All IBIS API resources have a prefix "https://www.ncbi.nlm.nih.gov/research/ibis-api/rest/v1," followed by more specific parts of the URI. Required parameters are specified in angle brackets, while optional parameters are specified in square brackets. The following parameters specified in angle brackets are used: <CID> PubChem compound identifier; <PDB> <CHAIN> PDB and chain identifiers. Additional instructions are available online at https://www.ncbi.nlm.nih.gov/research/ibis-api/

binding interfaces are similar (explore the first binding site cluster for "protein-peptide" interactions in IBIS, since disordered p53 is classified as peptide). Figure 2 illustrates a structural superposition of MDM2, p53 peptide, and Nutlin-3a small molecule. According to IBIS and MutaBind, the major binding hot spot residues (in bold in Table 1) are retained in both paralogous proteins and can be targeted in designing inhibitors that disrupt both p53-MDM2 and p53-MDMX interactions (Table 1).

In addition to the IBIS webserver, we provide an application program interface (API), discussed in the sections below, to identify binding sites from homologous structures and to find other compounds that would bind homologous complexes (Table 2, Fig. 3). We illustrate a combination of IBIS APIs with PubChem APIs in an example workflow in Figs. 4 and 5.

(a) /sites/inf/protein/pdb/4HG7


```
{
  "response": [
    {
      "CHAIN": "A",
      "INFERRED_PARTNER_CHAIN": "M",
      "INFERRED_PARTNER_PDB": "3DAC",
      "PDB": "4HG7",
      "SDI": 623460,
      "TEMPLATE_CHAIN": "C",
      "TEMPLATE_PARTNER_CHAIN": "",
      "TEMPLATE_PDB": "3DB2",
      "TEMPLATE_SDI": 280697,
      "residues": [
        {
          "id": 42,
          "resi": "53",
          "resn": "VAL"
        },
        {
          "id": 45,
          "resi": "56",
          "resn": "TYR"
        },
        {
          "id": 46,
          "resi": "57",
          "resn": "LEU"
        },
...
```


(b) /compounds/similar/compound/11433190


```
{
  "response": [
    24764418,
    67585029,
    87433224,
    68685834,
    10127376,
    118880883,
    11385877,
    118880884,
    90781721,
    118883353,
    118880885,
    118880886,
    53394469,
...
```


Fig. 3 Examples of results of API calls: (**a**) to find inferred protein-protein binding sites for E3 Ubiquitin-protein Ligase MDM2 (PDB ID 4HG7); JSON structure shows a list of binding site residues from a nutlin-resistant structure of MDM2 (PDB ID 4UMN) with a stapled peptide; (**b**) A list of similar small molecule compounds for nutlin-3a (PubChem compound ID 11433190)

```
import json
from urllib.request import urlopen
from urllib.error import HTTPError

prefix = "https://dev.ncbi.nlm.nih.gov/research/ibis-api/rest/v1"
API_similar_compound = "{}/compounds/similar/compound/{}"
API_structures_compound = "{}/structures/compound/{}"
API_sites = "{}/sites/obs/compound/pdb/{}"

combined_results = []

# Find similar compounds
query_cid = 11433190  # Nultin-3a
with urlopen(API_similar_compound.format(prefix, query_cid)) as f1:
    result_cids = json.loads(f1.read().decode('utf-8'))
    for cid in result_cids['response']:
        # Find structures for each compound
        try:
            with urlopen(API_structures_compound.format(prefix, cid)) as f2:
                result_struct = json.loads(f2.read().decode('utf-8'))
                for pdb in result_struct['response']:

                    # Find small molecules and binding sites for each structure
                    try:
                        with urlopen(API_sites.format(prefix, pdb)) as f3:
                            results_sites = json.loads(f3.read().decode('utf-8'))
                            combined_results.extend(results_sites['response'])
                    except HTTPError:
                        continue  # no small molecule binding sites found
        except HTTPError:
            continue # no structures found for a compound

with open("results.json", 'w') as output:
    output.write(json.dumps(combined_results, indent=4, sort_keys=True))
```

Fig. 4 An example of a programmatic implementation (in Python 3) of a workflow for finding binding sites of similar small molecules in protein structures, given a PubChem Compound identifier. The workflow is using three IBIS API calls described in Table 2

3.5 Incorporating Chemical Similarity and Biological Activity Data

While the biological relevance of binding interfaces via protein similarities can be assessed using the IBIS method, putative drug target alternatives can also be explored using the PubChem database [23]. PubChem is an open archive of chemical biology information including an expansive set of putative drug targets and bioactivity test results. PubChem currently contains over 92M unique chemical compounds and 3.6M substances evaluated for biological activity involving over 10K protein and 20K gene targets. Information can be obtained using the web interface, downloading from the FTP site or using programmatic tools as utilized in this protocol.

Similar compounds can have similar or optimized biological actions, such as the disruption of a protein-protein interaction binding interface. Given a query small molecule, a set of similar compounds can be retrieved via the proposed IBIS API, which, in turn, makes use of the PubChem REST API including both 2D and

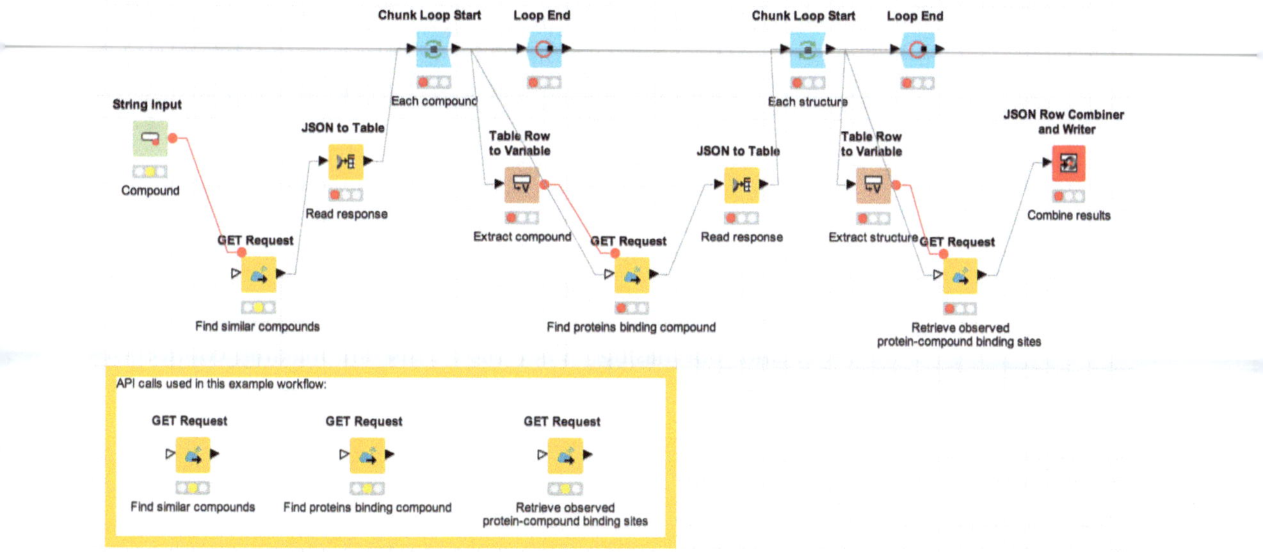

Fig. 5 An example of a workflow diagram built in KNIME using three REST API calls (shown in the figure legend). The workflow identifies binding sites of similar small molecules in protein structures, given a PubChem Compound identifier as an input. Nodes denote services, *while arrows* show direction of information flow in the workflow. *Red lines* show assignment of variables in the flow control structures

3D chemical similarity searches. These two types of similarity searches complement each other. For 2D compound similarity, precomputed sets of binary fingerprints representing the constituent chemical substructures are compared to ensure that the key components of a query molecule are preserved. Using a 3D similarity search, shape features of the molecules can be compared to minimize steric hindrance in binding pockets.

In some cases, as for example in PDB 4HG7, a co-crystallized molecule corresponds to the compound of interest, nutlin-3a. In other cases (e.g., 2NOU, 1T4E, 3VGB, 3U15), the bound small molecules may be different enough to ensure that a compound receives a unique identifier, but their biological activity may be similar. In order to address such cases, an "expansion" of query compounds using similar compounds may be very useful for identifying all relevant hits. We illustrate this approach in an example workflow in Figs. 4 and 5, where an IBIS API (*see* also Table 2) is used to find similar compounds, and then for each compound a list of proteins is found that bind this compound. Finally, all protein-small molecule binding sites are identified for each structure in the list.

In the search for putative drug targets, PubChem offers a wealth of pertinent information to help choose one alternative over another. In combination with the chemical structure search, the large collection of bioactivity test results in PubChem [23, 24]

provide the details required to narrow down a list of candidates. In a first pass, molecules designated "active," as defined by the original experimentalist, can be compared using filters on bioassay types like toxicity studies. If required, detailed analysis can be performed on individual assay experiments which contain large amounts of supporting data and explanations.

Finally, PubChem has collected extensive annotations for many chemical compounds. These include, but are not limited to, safety and hazard, pharmacological, toxicity, patent, vendor, interaction and pathway, classification and synonym annotations. Such annotations impact clinical and practical considerations like adverse reactions and purchasing. They also assist in making connections among existing literature references and categorizations.

3.6 Building Semiautomatic Workflows Using REST Services

Over the years, computational biology and computational chemistry communities have built a large number of services and resources. Some resources are integral parts of infrastructures provided by the large players such as EBI, NCBI, or PDB. However, integrating these resources in data analysis pipelines remains a tedious task that takes up significant amounts of project time and resources.

Some databases and web services can be combined in a modular way in workflows, thereby empowering users to perform complex interdisciplinary analytical processes. A workflow integrates a range of services to perform data-processing tasks. A service can be local or remote, it receives data and parameters as input, processes the request and returns the results. Although each service has a simple interface, complexity of workflows arises from the combination of services. The structure of a workflow is defined by the user and specific tasks. Automated workflows provide a way toward reproducible research. However, reproducibility becomes a complicated issue with distributed services as part of the workflow. Web services typically have a version prefix to make their interfaces and behavior backward-compatible. However, the databases are perpetually being updated and the API calls may return different data the next time a workflow is executed.

Pipelines and automated workflows have a long history in drug discovery, particularly with the advent of high-throughput screening. As robots speed up experiments at the bench, in-silico drug-screening is accelerated by computational pipelines [41]. For instance, the Chemistry Development Toolkit (CDK) has plugins to both KNIME and Taverna [42, 43]. A popular commercial small-molecule drug discovery and protein modeling suite by Schrödinger is automated with KNIME. KNIME may be more suited as a data mining tool [44] and may therefore be more appropriate for drug discovery applications involving data processing in addition to automated execution.

Several primers on connecting proteins with compounds using workflows and distributed web services exist, for example using BindingDB and KNIME [28, 45]. One of our goals here is to provide a primer on data mining in drug discovery for modulating and inhibiting protein-protein interactions. We exemplify an approach implemented with KNIME. Taverna Workflow Management System [46] is an open platform under the Apache project, guaranteeing a long-term development commitment. Taverna offers prebuilt packages with Bioinformatics services and a comprehensive online repository for open exchange of workflows between users—MyExperiment [47]. The latter is a social network for the scientific community facilitating sharing of workflows and snippets. Creators and maintainers of web services may choose to list them in online catalogs, making it easier for the users to find, annotate, and monitor distributed services. Biocatalogue [48] is the largest online registry of SOAP and REST Web services, and all its services can be loaded into Taverna Workbench or into KNIME.

Web services operate on the top of the HTTP protocol, which is used as transport for the web. XML-based web-services, such as SOAP and WSDL specifications, became widespread in computational biology and cheminformatics, and are still well maintained. However, standardization of web browsers and JavaScript—frontend and backend frameworks stimulated a transition to RESTful services using a lightweight JSON-based (JavaScript Object Notation) data exchange. While XML-derived data formats remain heavily used, many services now support JSON in addition to XML. With REST, services behave more like documents or resources that are uniquely identified with a URI address. Not only does this approach simplify the development and scalability of servers but also allows one to create clients in a simple way in any modern scripting language: Perl, Python, or JavaScript.

Table 2 describes a list of IBIS REST resources that we created to facilitate drug discovery workflows aimed at protein-protein interaction inhibitors. All the following resources use HTTPS "GET" to retrieve single entries and "POST" to submit multiple entries. Examples of JSON formatted results for two IBIS API calls (a) requesting inferred protein-protein binding sites for a given protein structure and (b) requesting similar compounds given a PubChem compound ID are shown in Fig. 3. These API calls may be arranged in workflows accepting protein and gene names, protein structure, or small molecule identifier as input. A listing of a Python script (Fig. 4) illustrates the workflow where output of one API call is passed as input to another API call, thus creating a rather complex analysis pipeline. The script in Fig. 4 retrieves a list of small molecule compounds similar to Nutlin-3a, identifies the PDB structures where any of these compounds is cocrystallized with a protein, and uses IBIS to retrieve description of protein-small molecule binding sites. Due to the simplicity of implementation of HTTP clients, the script implementing this pipeline remains relatively short and easy to read. In addition to programmatic access to IBIS REST API, a graphical representation of workflows can be created, for example using KNIME, as we illustrate in Fig. 5. The diagram shown in Fig. 5 implements the same workflow as in the Python script shown in Fig. 4.

Acknowledgments

This work was supported by the Intramural Research Program of the National Library of Medicine. We thank Sunghwan Kim for helpful discussions about PubChem.

References

1. Shoemaker BA, Panchenko AR, Bryant SH (2006) Finding biologically relevant protein domain interactions: conserved binding mode analysis. Protein Sci 15(2):352–361
2. Aloy P, Russell RB (2004) Ten thousand interactions for the molecular biologist. Nat Biotechnol 22(10):1317–1321
3. Rolland T, Tasan M, Charloteaux B et al (2014) A proteome-scale map of the human interactome network. Cell 159(5):1212–1226
4. Yates CM, Sternberg MJ (2013) The effects of non-synonymous single nucleotide polymorphisms (nsSNPs) on protein-protein interactions. J Mol Biol 425(21):3949–3963
5. Teng S, Madej T, Panchenko A, Alexov E (2009) Modeling effects of human single nucleotide polymorphisms on protein-protein interactions. Biophys J 96(6):2178–2188
6. Li M, Goncearenco A, Panchenko AR (2017) Annotating mutational effects on proteins and protein interactions: designing novel and revisiting existing protocols. Methods Mol Biol 1550:235–260
7. Filippakopoulos P, Qi J, Picaud S et al (2010) Selective inhibition of BET bromodomains. Nature 468(7327):1067–1073
8. Smith BJ, Lee EF, Checco JW et al (2013) Structure-guided rational design of alpha/beta-peptide foldamers with high affinity for BCL-2 family prosurvival proteins. Chembiochem 14(13):1564–1572
9. Zhao Y, Aguilar A, Bernard D, Wang S (2015) Small-molecule inhibitors of the MDM2-p53 protein-protein interaction (MDM2 Inhibitors) in clinical trials for cancer treatment. J Med Chem 58(3):1038–1052
10. Haase HS, Peterson-Kaufman KJ, Lan Levengood SK et al (2012) Extending foldamer design beyond alpha-helix mimicry: alpha/beta-peptide inhibitors of vascular endothelial growth factor signaling. J Am Chem Soc 134(18):7652–7655
11. Basse MJ, Betzi S, Morelli X, Roche P (2016) 2P2Idb v2: update of a structural database dedicated to orthosteric modulation of protein-protein interactions. Database (Oxford) 2016
12. Aloy P, Ceulemans H, Stark A, Russell RB (2003) The relationship between sequence and interaction divergence in proteins. J Mol Biol 332(5):989–998
13. Ma B, Elkayam T, Wolfson H, Nussinov R (2003) Protein-protein interactions: structurally conserved residues distinguish between binding sites and exposed protein surfaces. Proc Natl Acad Sci U S A 100(10):5772–5777
14. Valdar WS, Thornton JM (2001) Protein-protein interfaces: analysis of amino acid conservation in homodimers. Proteins 42(1):108–124
15. Guharoy M, Chakrabarti P (2005) Conservation and relative importance of residues across protein-protein interfaces. Proc Natl Acad Sci U S A 102(43):15447–15452
16. Goncearenco A, Shaytan AK, Shoemaker BA, Panchenko AR (2015) Structural perspectives on the evolutionary expansion of unique protein-protein binding sites. Biophys J 109 (6):1295–1306
17. Goncearenco A, Shoemaker BA, Zhang D et al (2014) Coverage of protein domain families with structural protein-protein interactions: current progress and future trends. Prog Biophys Mol Biol 116(2–3):187–193
18. Petrey D, Honig B (2014) Structural bioinformatics of the interactome. Annu Rev Biophys 43:193–210

19. Mosca R, Ceol A, Aloy P (2013) Interactome3D: adding structural details to protein networks. Nat Methods 10(1):47–53
20. Tyagi M, Hashimoto K, Shoemaker BA et al (2012) Large-scale mapping of human protein interactome using structural complexes. EMBO Rep 13(3):266–271
21. Shoemaker BA, Zhang D, Thangudu RR et al (2010) Inferred Biomolecular Interaction Server—a web server to analyze and predict protein interacting partners and binding sites. Nucleic Acids Res 38(Database issue): D518–D524
22. Shoemaker BA, Zhang D, Tyagi M et al (2012) IBIS (Inferred Biomolecular Interaction Server) reports, predicts and integrates multiple types of conserved interactions for proteins. Nucleic Acids Res 40(Database issue): D834–D840
23. Kim S, Thiessen PA, Bolton EE et al (2016) PubChem Substance and Compound databases. Nucleic Acids Res 44(D1): D1202–D1213
24. Wang Y, Bryant SH, Cheng T (2017) PubChem BioAssay: 2017 update. Nucleic Acids Res 45(D1):D955–D963
25. Li M, Simonetti FL, Goncearenco A, Panchenko AR (2016) MutaBind estimates and interprets the effects of sequence variants on protein-protein interactions. Nucleic Acids Res 44(W1):W494–W501
26. Li M, Petukh M, Alexov E, Panchenko AR (2014) Predicting the impact of missense mutations on protein? Protein binding affinity. J Chem Theor Comput 10(4):1770–1780
27. Yang W, Soares J, Greninger P (2013) Genomics of Drug Sensitivity in Cancer (GDSC): a resource for therapeutic biomarker discovery in cancer cells. Nucleic Acids Res 41(Database issue):D955–D961
28. Gilson MK, Liu T, Baitaluk M, Nicola G et al (2016) BindingDB in 2015: a public database for medicinal chemistry, computational chemistry and systems pharmacology. Nucleic Acids Res 44(D1):D1045–D1053
29. Petukh M, Dai L, Alexov E (2016) SAAMBE: webserver to predict the charge of binding free energy caused by amino acids mutations. Int J Mol Sci 17(4):547
30. Cukuroglu E, Gursoy A, Keskin O (2012) HotRegion: a database of predicted hot spot clusters. Nucleic Acids Res 40(Database issue): D829–D833
31. Estrada-Ortiz N, Neochoritis CG, Domling A (2016) How to design a successful p53-MDM2/X interaction inhibitor: a thorough overview based on crystal structures. ChemMedChem 11(8):757–772
32. Shangary S, Wang S (2009) Small-molecule inhibitors of the MDM2-p53 protein-protein interaction to reactivate p53 function: a novel approach for cancer therapy. Annu Rev Pharmacol Toxicol 49:223–241
33. Cinatl J, Speidel D, Hardcastle I, Michaelis M (2014) Resistance acquisition to MDM2 inhibitors. Biochem Soc Trans 42(4):752–757
34. Thangudu RR, Bryant SH, Panchenko AR, Madej T (2012) Modulating protein-protein interactions with small molecules: the importance of binding hotspots. J Mol Biol 415 (2):443–453
35. Davis FP, Sali A (2010) The overlap of small molecule and protein binding sites within families of protein structures. PLoS Comput Biol 6(2):e1000668
36. Marchler-Bauer A, Derbyshire MK, Gonzales NR et al (2015) CDD: NCBI's conserved domain database. Nucleic Acids Res 43(D1): D222–D226
37. Wells JA, McClendon CL (2007) Reaching for high-hanging fruit in drug discovery at protein-protein interfaces. Nature 450 (7172):1001–1009
38. Bogan AA, Thorn KS (1998) Anatomy of hot spots in protein interfaces. J Mol Biol 280 (1):1–9
39. Graves B, Thompson T, Xia M, Janson C (2012) Activation of the p53 pathway by small-molecule-induced MDM2 and MDMX dimerization. Proc Natl Acad Sci U S A 109 (29):11788–11793
40. Ribeiro CJ, Rodrigues CM, Moreira R, Santos MM (2016) Chemical variations on the p53 reactivation theme. Pharmaceuticals (Basel) 9(2)
41. Perez-Sanchez H, Rezaei V, Mezhuyev V et al (2016) Developing science gateways for drug discovery in a grid environment. Spring 5 (1):1300
42. Beisken S, Meinl T, Wiswedel B et al (2013) KNIME-CDK: workflow-driven cheminformatics. BMC Bioinformatics 14:257
43. Truszkowski A, Jayaseelan KV, Neumann S et al (2011) New developments on the cheminformatics open workflow environment CDK-Taverna. J Cheminform 3:54
44. Mazanetz MP, Marmon RJ, Reisser CB, Morao I (2012) Drug discovery applications for KNIME: an open source data mining platform. Curr Top Med Chem 12(18):1965–1979
45. Nicola G, Berthold MR, Hedrick MP, Gilson MK (2015) Connecting proteins with drug-like compounds: open source drug discovery

workflows with BindingDB and KNIME. Database (Oxford) 2015
46. Wolstencroft K, Haines R, Fellows D et al (2013) The Taverna workflow suite: designing and executing workflows of Web Services on the desktop, web or in the cloud. Nucleic Acids Res 41(Web Server issue):W557–W561
47. Goble CA, Bhagat J, Aleksejevs S et al (2010) myExperiment: a repository and social network for the sharing of bioinformatics workflows. Nucleic Acids Res 38(Web Server issue): W677–W682
48. Bhagat J, Tanoh F, Nzuobontane E et al (2010) BioCatalogue: a universal catalogue of web services for the life sciences. Nucleic Acids Res 38(Web Server):W689–W694

Chapter 16

Method to Identify Silent Codon Mutations That May Alter Peptide Elongation Kinetics and Co-translational Protein Folding

Ronald Worthington, Elijah Ball, Brentsen Wolf, and Gregory Takacs

Abstract

Due to the redundancy of the protein genetic code, mutational changes in the second or third nucleotide of an existing codon may not change the amino acid specification of the resulting modified codon. When peptide primary sequence is unchanged by mutation, that mutation is assumed to have no functional consequences. However, for one key gene involved in drug transport, MDR-1, several silent, synonymous mutations have been shown to alter protein structure and substrate affinity (Kimchi-Sarfaty et al., Science 315:525–528, 2007). The mechanism of these changes, in the absence of primary amino acid sequence changes, appears to be the change in abundance of the transfer RNA molecules complementary to the mutated, although synonymous, new codon. Transfer RNA abundance is proportional to the frequency of each codon as found in human protein coding DNA (Sharp et al., Nucleic Acids Res 14(13):5125–5143, 1986). These frequencies have been mapped for many thousands of human proteins (Nakamura et al., Nucleic Acids Res 28:292, 2000). This method analyzes silent codon mutations in whole genome data. Where there are large changes in codon frequency resulting from codon sequence mutation, the affected proteins are mapped to potential disease pathways, in the context of clinical phenotypes associated with the patient genome data.

Key words Co-translational protein folding, Peptide elongation kinetics, Synonymous codon mutation, Human whole genome sequence data

1 Introduction

Protein folding, which contributes significantly to ultimate protein functionality, has been shown to occur during translation, as ribosomes read codons on messenger RNA (mRNA) and accept the cognate aminoacyl-transfer RNA (tRNA) to elongate the nascent peptide chain by the tRNA's cargo of a single amino acid [1–4]. The protein genetic code maps DNA codons to tRNA. There is usually redundancy in tRNA and thus amino acid specification by codons, yet each codon has only one cognate tRNA, and each cognate tRNA has its own gene, whose expression is regulated.

Iulia M. Lazar et al. (eds.), *Proteomics for Drug Discovery: Methods and Protocols*, Methods in Molecular Biology, vol. 1647, DOI 10.1007/978-1-4939-7201-2_16, © Springer Science+Business Media LLC 2017

The abundance of cognate tRNA molecules varies by species according to the regulation of expression of these genes and by gene copy number [5]. Thus, the process of translation can show variations in kinetics—elongation kinetics—depending on the abundance of each cognate tRNA molecule at the ribosome, and the specific codon sequence appearing in the exon. Protein folding is influenced by the rate of translation, and translation will slow down as the ribosome encounters codons that call for tRNA molecules in low abundance, or will speed up when reading codons whose cognate tRNA molecules are in high abundance [4]. The discovery that the abundance of cognate tRNA molecules varies by species was a sea change in molecular biology that enabled efficient expression of human proteins in bacterial, yeast, and mammalian cells [6].

Mutations in the second or third nucleotide of an ancestral codon often do not change the amino acid specified by the "new" codon. These "silent" or synonymous mutations are generally thought to be inconsequential to protein function, since the peptide primary structure does not change because of the mutation. However, that assumption ignores the possibility that the abundance of the cognate tRNA for the new codon may differ, sometimes dramatically, from that of the ancestral codon. For example, the codons CUA and CUG both code for the amino acid leucine. In the case of codon usage rates for *Homo sapiens*, the CUA rate is 5%, while that of CUG is 46%. If the codon frequency change is large enough, as in the CUA<->CUG example, then effects on translational kinetics, and resulting co-translational protein folding patterns, are possible. The ratio (calculated as ancestral codon usage rate ÷ mutant codon usage rate) of these two usage rates can range from small fractions to values greater than one, with a ratio of one signifying no change in codon usage rate between ancestral and mutant codons. If the codon usage rate ratio of the ancestral (e.g., 46%) to mutant (e.g., 5%) is greater than 1, that may predict a stalling effect on the rate of translation, because the ribosome must wait longer for the cognate tRNA to appear at the docking site. If the ratio is less than 1, e.g., ancestral 5%, mutant 46%, the translation rate may be accelerated because of the greater abundance of the cognate tRNA molecule for the mutant codon. In either case, the co-translational folding process may be altered, resulting in folding conformations that may alter protein function.

This hypothesis was confirmed when Kimchi-Sarfaty et al. [1] studied the reason that three silent codon mutations in the MDR1 gene (also known as P-glycoprotein) cause a change in substrate specificity for drug transport by this important pharmacokinetic protein. Using several physical techniques that detect conformational changes in protein structure, they used site-directed mutagenesis of monkey kidney cells to show that changing synonymous codons from higher to lower usage rate changes both the final protein conformation and substrate specificity (protein function).

Geneticists typically ignore silent mutations because of the synonymy of protein sequence between the ancestral and new codons. The Kimchi-Sarfaty study teaches us that silent mutations can have profound functional consequences. The purpose of this project is to annotate silent codon mutations that result in codon frequency ratio shifts smaller or greater than 1.0 in human whole genome data sets, and map these changes to disease pathways relevant to clinical phenotypes of the respective individuals. The effect has been confirmed in a large variety of studies [6, 7].

2 Materials

1. Individual whole genome or exome data files were obtained from the Harvard Personal Genomes Project (PGP) (*see* **Note 1**). All the files were produced by Complete Genomics Inc. (CGI) after performing next-generation DNA sequencing of participant blood samples, and were generated by the PGP using the cgatools program to produce VAR type files in VCF4.1 format. Informed consent was obtained by the Harvard Medical School IRB, and all genome data, along with clinical phenotype information as provided by the subject, are available for research use without local IRB approval. In this study, we only used genome data referenced to the GRCh37/hg19 build (release) of the human genome by the National Center for Biotechnology Information of the United States National Institutes of Health (*see* **Note 2**).
2. SnpEff software [8] was downloaded from SourceForge (*see* **Note 3**). SnpEff accesses human genome databases based on genome build, which must be locally installed in addition to compiling the SnpEff program itself. For our studies, the SnpEff reference build was GRCh37/hg19.
3. Text parsing programs were written by the authors in Python and Perl and are available for download under the GNU creative license (*see* **Note 4**).
4. Computation was performed primarily in the Unix (or Linux) operating system on a server farm operated by Penguin on Demand (*see* **Note 5**). We used a Penguin Relion server with Intel Haswell-EP Xeon microarchitecture running at 2.66 GHz, using 2 nodes, 20 cores per node, and 128 GB RAM per node. With this configuration, a single CGI whole genome variant file typically used about 7 min of walltime.

3 Methods

While some of our programming and testing was performed on local Darwin Unix and various Linux environments, using current releases of the PERL and PYTHON interpreters for those platforms, most of program execution was carried out on our account at the Penguin on Demand cloud server farm located at Indiana University. Our computational pipeline is shown in Fig. 1.

1. First, we had to manually download the individual variant genome files, because the PGP does not provide an application programming interface to access data. The files are compressed, and in binary format, are around 250 MB each. Following bzip2 decompression, the text files are about 1 GB in size.
2. We then used a wrapper script written in PYTHON to submit and receive data through a series of PERL programs, and to run SnpEff. The input file for the first wrapper script (CGI-var.-ann-2-SnpEff-input.pl) is the uncompressed CGI VAR file. Since the CGI VAR type files contain only the following information for each variant nucleotide: (a) chromosome ID; (b) chromosome coordinates, reference allele, variant allele, and quality of the nucleotide read (pass or fail), the first script in the

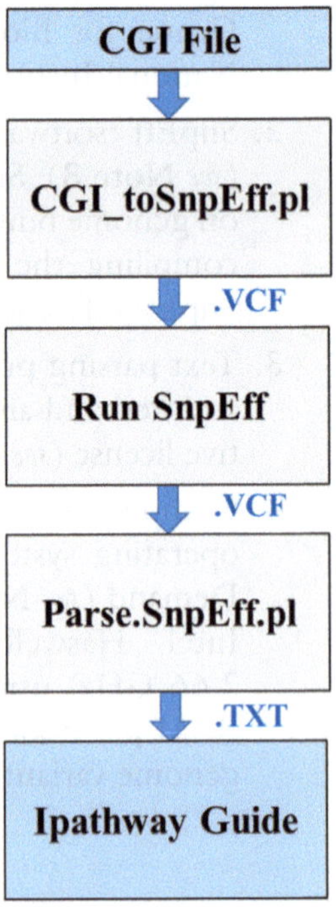

Fig. 1 Pipeline for identifying and visualizing silent codon mutations with potential functional effects

wrapper pipeline filters out all failed reads, and then formats the necessary data as a text VCF4.1 file for input to SnpEff.

3. Because SnpEff must sift through hundreds of gigabits of binary data to perform variant effect prediction, a high-performance computing environment is required (*see* Subheading 2). The SnpEff output file is a VCF text file with extensive annotation of likely functional effects. The second script in the wrapper, parse.snpeff.pl, identifies all variants annotated as silent codon mutations, the gene symbol of the affected protein, and captures the DNA sequence of both the ancestral and mutant codons. The script then looks up the human codon usage rate for the ancestral and mutant codons, and calculates the $\log_2$ ratio of ancestral rate ÷ mutant rate. The gene symbol and $\log_2$ rate are written to a text file that serves as the input data into the AdvaitaBio iPathwayGuide server. As of this writing, the AdvaitaBio servers do not have an application programming interface to allow our wrapper to directly upload the data to our account, and so we must do that using an html interface. However, the company plans to implement the API in the first quarter of 2017.
4. Since the data of interest for us are ratios, we can utilize analysis software designed for gene expression studies. In place of using a ratio of differential gene expression signals, we ratio the ancestral to the mutant codon usage rates. We chose the AdvaitaBio iPathwayGuide platform (*see* **Note 6**), because it uses systems biology algorithms [9] to predict emergent biological relationships among proteins and their pathways. An example pathway is shown in Fig. 2. The data are from a PGP subject diagnosed with Barrett's Esophagus, which is an epithelial dysplasia that can lead to malignancy. In this case, we are particularly interested in pathways associated with malignancies, and Fig. 2 shows codon usage rate changes in genes belonging to the pathway of the tumor suppressor transcription factor gene TP53. The intensity of the red color signifies the absolute value of the change from higher to lower (ancestral ÷ mutant) codon usage rates in silent mutations, while the intensity of the blue color signifies lower to higher rate changes. Genes of high intensity may produce proteins of altered function [1], and could be studied using in vitro protein expression models [1]. If functional effects are demonstrated, these gene products could become therapeutic targets for disease.

4 Notes

1. Downloaded from: https://my.pgp-hms.org/public_genetic_data. Participants provided Informed Consent for research use

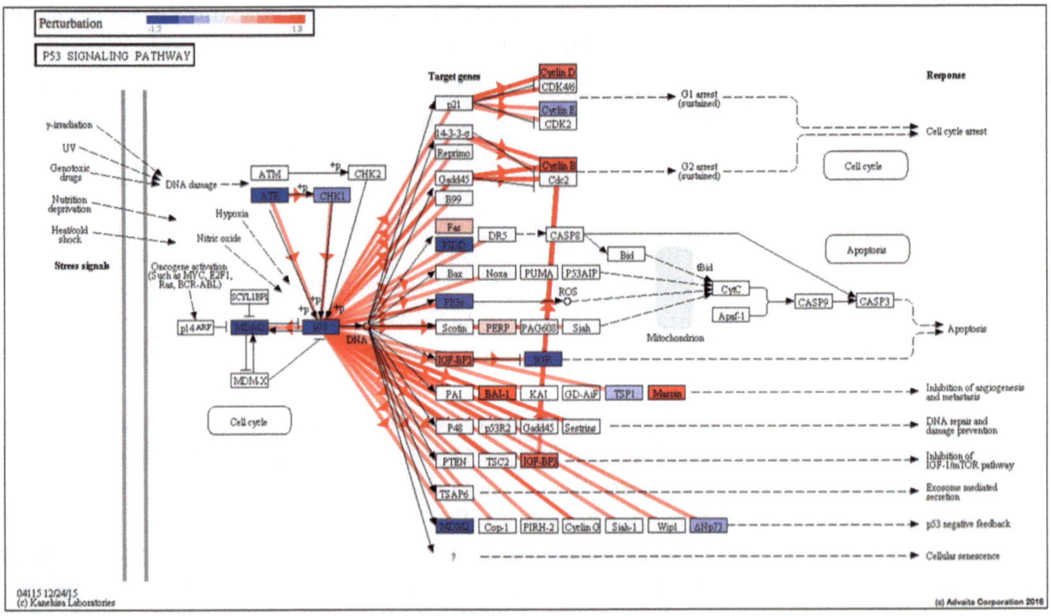

Fig. 2 Silent codon mutations of high codon usage rate change visualized in AvaitaBio iPathwayGuide software for the TP53 signaling pathway. Predicted protein folding effects are proportional to color intensity. The results suggest perturbation of the cell cycle arrest or apoptosis pathway (cell type dependent) regulated by tumor protein p53. Key proteins in this pathway with large codon frequency changes include p53, MDM2, ATR, PIDD, IGF-BP3, IGF, Cyclin B, and Cyclin D. Note: The original TP53 signaling pathway was obtained from the freely available Kegg Pathway Database (http:/www.genome.jp/kegg/pathway.html)

of their DNA sequence data, and for voluntarily provided clinical data, under a protocol approved by the Harvard Medical School Institutional Review Board.

2. Build 37 of the human reference genome is available at: ftp://ftp.ncbi.nlm.nih.gov/genomes/Homo_sapiens/ARCHIVE/BUILD.37.3.
3. Downloaded from: http://snpeff.sourceforge.net; *See* ref. 3.
4. Available for download at: http://www.siue.edu/~rworthi/software.
5. https://pod.penguincomputing.com/hpc_on_demand.
6. https://www.advaitabio.com/support-ipathwayguide.html.

Acknowledgment

This work was supported by the SIUE School of Pharmacy Research Grant program for FY16 and FY17, along with additional student stipend support from the SIUE FY17 Undergraduate Research and Creative Activities Program.

References

1. Kimchi-Sarfaty C, Oh J, Saune Z et al (2007) A "silent" polymorphism in the MDR1 gene changes substrate specificity. Science 315:525–528
2. Sharp PM, Tuohy TMF, Mosurski KR (1986) Codon usage in yeast: cluster analysis clearly differentiates highly and lowly expressed genes. Nucleic Acids Res 14(13):5125–5143
3. Nakamura Y, Gojobori T, Ikemura T (2000) Codon usage tabulated from the international DNA sequence databases: status for the year 2000. Nucleic Acids Res 28:292
4. Purvis IJ, Bettany AJ, Santiago TC et al (1987) The efficiency of folding of some proteins is increased by controlled rates of translation in vivo. A hypothesis. J Mol Biol 193:413–417
5. Frydman J (2001) Folding of newly translated proteins in vivo: the role of molecular chaperones. Annu Rev Biochem 70:603–606
6. Komar A (2008) A pause for thought along the co-translational folding pathway. Trends Biochem Sci 34(1):16–24
7. Komar A (2007) NPs, Silent but not invisible. Science 315:456–457
8. Cingolani P, Platts A, Wang le L et al (2012) A program for annotating and predicting the effects of single nucleotide polymorphisms, SnpEff: SNPs in the genome of Drosophila melanogaster strain w1118; iso-2; iso-3. Fly (Austin) 6(2):80–92
9. Draghici S, Khatri P, Tarca A et al (2007) A systems biology approach for pathway level analysis. Genome Res 17(10):1537–1545

Chapter 17

In Silico Design of Anticancer Peptides

Shailesh Kumar and Hui Li

Abstract

In the past few years, small peptides having anticancer properties have emerged as a potential avenue for cancer therapy. Compared to current anti-cancer chemotherapeutic drugs (or small molecules), anticancer peptides (ACPs) have numerous advantageous properties, such as high specificity, low production cost, high tumor penetration, ease of synthesis and modification. However, in wet lab setups, identification and characterization of novel ACPs is a time-consuming and labor-intensive process. Therefore, in silico designing of anticancer peptides is beneficial, prior to their synthesis and characterization. This approach is less time consuming and more cost-effective. In this chapter, we discuss a web-based tool, AntiCP (http://crdd.osdd.net/raghava/anticp/), for designing ACPs.

Key words Anti-cancer peptides, Machine learning, Support vector machine

1 Introduction

Cancer is a major health problem for both developed and developing countries [1]. Despite advancements in cancer treatment, the mortality rate due to cancer is still very high. In year 2012 alone, a total of 8.2 million patients died of the disease worldwide [2], and this number is expected to increase to approximately 11.5 million in 2030 [3]. There are several challenges associated with the treatment of cancer. Conventional chemo- and radiotherapy mainly kill the cancer cells, but without high specificity, and cause several side effects and toxicity [4–6]. The development of drug resistance is another important issue in the treatment of cancer. Therefore, there is an urgent need for the development of novel tumor-targeted therapies that can precisely and efficiently target tumor cells, with less toxicity to normal tissues and cells [6]. In the last decade, researchers explored small peptides that have anticancer properties as a potential alternative approach for cancer therapy [7, 8]. The main advantages of anticancer peptides include high specificity, low production cost, high tumor penetration, and ease of synthesis and modification, etc. [9].

Iulia M. Lazar et al. (eds.), *Proteomics for Drug Discovery: Methods and Protocols*, Methods in Molecular Biology, vol. 1647, DOI 10.1007/978-1-4939-7201-2_17, © Springer Science+Business Media LLC 2017

Anticancer peptides (ACPs) are small peptides having 5–30 amino acids in length. They are cationic peptides mainly derived from antimicrobial peptides (AMPs) [10–12] and their source may include animals [13–18] or plants [19–21]. The exact mechanism of action of ACPs is still unknown. Mader et al. [10] describe that the selective killing of cancer cells by ACPs is due to the electrostatic interactions between the cationic amino acids of ACPs and the anionic components of cancer cell membranes. Some studies also show that the high membrane fluidity and high cell-surface area [22, 23] of cancer cells, compared to normal cells, lead to enhanced lytic activity of ACPs and binding of an increased number of ACPs, respectively. Induction of apoptosis by disrupting the mitochondrial membrane was also noticed by researchers [24]. There are several studies involving ACPs that have been used for various kinds of cancer therapies and that are currently in preclinical and clinical trials [25–28].

In general, experimental identification and characterization of novel ACPs is a time-consuming and labor-intensive process. Therefore, prior to synthesis, it is beneficial to design the appropriate ACPs by in silico methods, to save time, cost and manpower. Raghava and co-workers have developed several in silico methods [29–33] for the prediction, screening and design of small peptides. They also developed an in silico method for the design and discovery of novel anticancer peptides, i.e., AntiCP [34]. AntiCP is a Support Vector Machine (SVM)-based in silico method, freely available for the scientific community (*see* **Note 1**). It can design ACPs and their mutants with different physicochemical properties. This chapter describes the various modules of AntiCP for the prediction and design of ACPs.

2 Materials

For this study, Tyagi et al. [34] collected a total of 225 antimicrobial peptides having anticancer properties, with appropriate negative datasets (*see* **Notes 2** and **3**), and developed in silico models for the identification and design of the anticancer peptides. Based on these models, they also developed a web-based platform called AntiCP (*see* **Notes 4** and **5**).

3 Methods

3.1 AntiCP: Web Based Tool for the Design of Anticancer Peptides

AntiCP is a webserver for the prediction and design of anticancer peptides. Tyagi et al. [34] developed several SVM based models by considering various properties such as amino acid composition, dipeptide composition and binary profile of pattern for the discrimination of ACPs from non-ACPs. In this study, the performance of binary-based models performed reasonably well. Binary profile models were developed for each peptide, where each amino acid is represented by a vector of dimensions of 20 (e.g., Ala by 1,0,0,0,0,0,0,0,0,0,0,0,0,0,0,0,0,0,0,0). In the AntiCP webserver, the best models were implemented (*see* **Note 6**). AntiCP enables users to generate ACPs and ACP mutants with different physicochemical properties. The prediction server's major features include: "Peptide Design," "Virtual Screening," "Protein Scan," and "Motif Scan".

3.2 Prediction and Designing of ACPs

The AntiCP "Peptide Design" module enables users to design desired single query peptides. The user is prompted to enter a query peptide into the given text box. After the peptide sequence is entered, the user can choose the desired peptide model (*see* **Note 1**), SVM threshold (*see* **Note** 7), and specific physicochemical properties to be displayed when the peptide design is generated (Fig. 1). This module also returns all possible mutant peptides and predicts the anti-cancer activity. We used **'DIVKCGLSPAI'** as an example sequence to design ACPs by using 'Model 1' at SVM threshold of '0' (Fig. 1). The server predicted this peptide to have anti-cancer properties with a descent SVM score of 0.77 (*see* **Note 8**). The server also generated >200 mutants for this mother peptide with all selected properties (Fig. 1). The mutants were sorted on the basis of the SVM score. Out of these mutants, a mutant **'DIVKCGLSP*KI*'** with SVM score of 1.06 was the best ACP that was found. Next, several mutants of **'DIVKCGLSP*KI*'** were generated, and **'DI*G*KCGLSPKI'** was found to be the best ACP with an SVM score of 1.35. After several cycles of mutant generation, a list of ~23 mutant peptides, predicted ACPs, was found, with a maximum SVM score of 2.04. These mutant peptides had different physiological properties.

3.3 Virtual Screening of ACPs

The AntiCP "Virtual Screening" module allows users to design multiple peptides in batch mode simultaneously. This tool provides the prediction of anti-cancer properties and displays the desired physicochemical properties as chosen by the user. The user inputs the peptide sequences in FASTA format in the given text box, and then is prompted to select the type of model to be returned, the threshold, and specific physicochemical properties (Fig. 2). Users can also generate mutants of one peptide by selecting the desired peptide on this server. In this module, five example sequences already available on the web page (Fig. 2) were used with 'Model 1'. Out of those, the server predicted four sequences as ACPs, with the best ACP (i.e., seq4, **LTAEHYAAQATS**) having an SVM score of 0.95 (Fig. 2). This peptide can be used for further generation of mutants with best SVM scores and physiological properties.

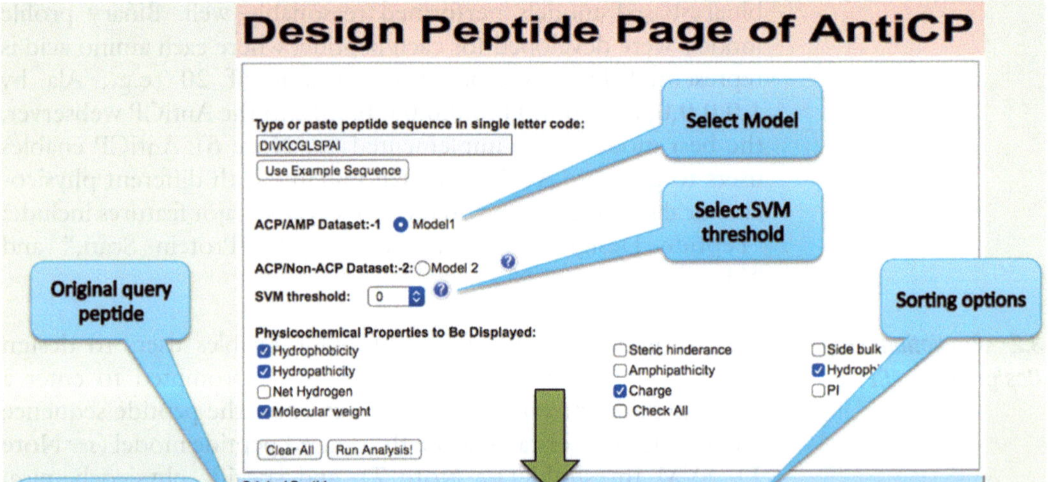

Peptide Sequence	Mutation Position	SVM score	Prediction	Hydrophobicity	Hydropathicity	Hydrophilicity	Charge	Mol wt
DIVKCGLSPAI	No	0.77	Anticp	0.08	1.01	-0.19	0.00	1115.51
AIVKCGLSPAI	1	0.74	Anticp	0.16	1.49	-0.51	1.00	1071.50
CIVKCGLSPAI	1	0.78	Anticp	0.14	1.55	-0.55	1.00	1103.56
EIVKCGLSPAI	1	0.76	Anticp	0.08	1.01	-0.19	0.00	1129.54
FIVKCGLSPAI	1	0.85	Anticp	0.20	1.58	-0.69	1.00	1147.60
GIVKCGLSPAI	1	-0.13	Non-Anticp	0.16	1.29	-0.46	1.00	1057.48
HIVKCGLSPAI	1	0.62	Anticp	0.10	1.04	-0.51	1.50	1137.57
IIVKCGLSPAI	1	0.81	Anticp	0.21	1.74	-0.63	1.00	1113.59
KIVKCGLSPAI	1	0.85	Anticp	0.04	0.97	-0.19	2.00	1128.60
LIVKCGLSPAI	1	0.77	Anticp	0.19	1.67	-0.63	1.00	1113.59
MIVKCGLSPAI	1	0.73	Anticp	0.16	1.50	-0.58	1.00	1131.62
NIVKCGLSPAI	1	0.71	Anticp	0.08	1.01	-0.45	1.00	1114.53
PIVKCGLSPAI	1	0.72	Anticp	0.13	1.18	-0.46	1.00	1097.54
QIVKCGLSPAI	1	0.71	Anticp	0.08	1.01	-0.45	1.00	1128.56
RIVKCGLSPAI	1	0.78	Anticp	-0.02	0.92	-0.19	2.00	1156.61
SIVKCGLSPAI	1	0.75	Anticp	0.12	1.25	-0.44	1.00	1087.50
TIVKCGLSPAI	1	0.78	Anticp	0.12	1.26	-0.50	1.00	1101.53
VIVKCGLSPAI	1	0.74	Anticp	0.19	1.71	-0.60	1.00	1099.56
WIVKCGLSPAI	1	0.71	Anticp	0.17	1.25	-0.77	1.00	1186.64
YIVKCGLSPAI	1	0.74	Anticp	0.14	1.21	-0.67	1.00	1163.60
DAVKCGLSPAI	2	0.94	Anticp	0.03	0.76	-0.07	0.00	1073.42
DCVKCGLSPAI	2	0.95	Anticp	0.01	0.83	-0.12	0.00	1105.48
DDVKCGLSPAI	2	0.96	Anticp	-0.06	0.28	0.25	-1.00	1117.43
DEVKCGLSPAI	2	1.02	Anticp	-0.05	0.28	0.25	-1.00	1131.46

Fig. 1 Schematic representation of the "Peptide Design" module and its output

3.4 Protein Scanning

"Protein Scanning" is an AntiCP module that enables users to scan a protein in order to detect novel anti-cancer peptides. Once the user inputs a protein sequence in the text box, the user will select the peptide fragment length, the SVM threshold, model, and physicochemical properties, in order for the server to generate and display information from the resulting overlapping peptides (Fig. 3). This module is responsible for generating fragments of length chosen by the user, and predicting the anti-cancer and physicochemical properties of the fragments. For this module, the example sequence shown in Fig. 3 was used, and the peptide fragment length of 5 with 'Model 1' was selected. The server generated >100 peptide fragments with associated SVM scores and physiochemical properties. Out of these, peptide **'IRTVR'** had the best SVM score (i.e., 1.46) (Fig. 3). This peptide can be used to generate better peptide mutants.

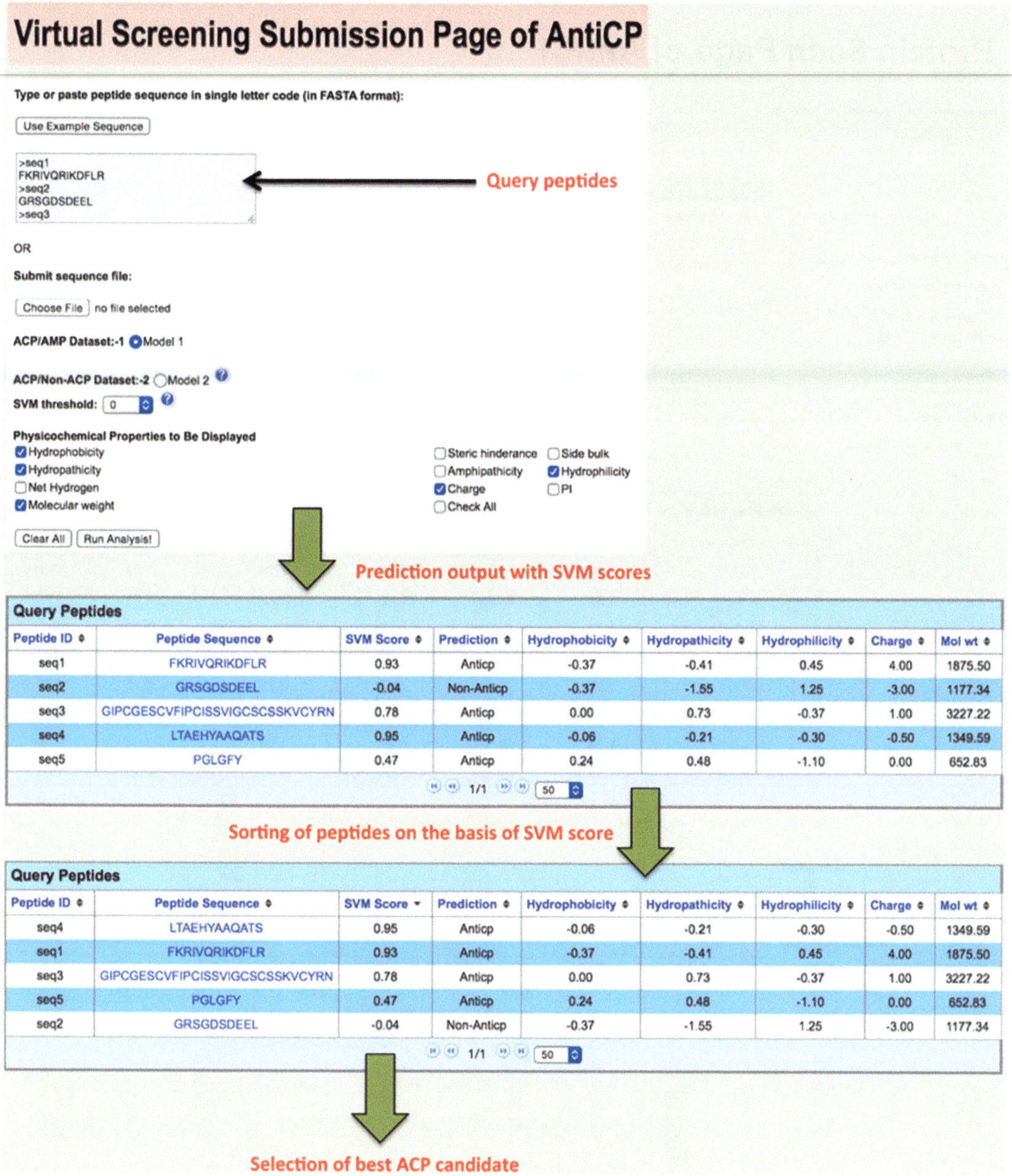

Fig. 2 Schematic representation of the "Virtual Screening" module of AntiCP

3.5 Motif Scanning

The AntiCP "Motif" module provides users with the ability to scan anti-cancer and non-anti cancer peptides. The user is prompted to enter a peptide or protein sequence, and will then select the desired model for the output (Fig. 4). The server scans the sequence for the presence of all ACP motifs and returns the result in tabular format with the number of positive and negative motifs in the query sequence. In this module, a total of five example sequences were used, already available on the web page, with the model option of

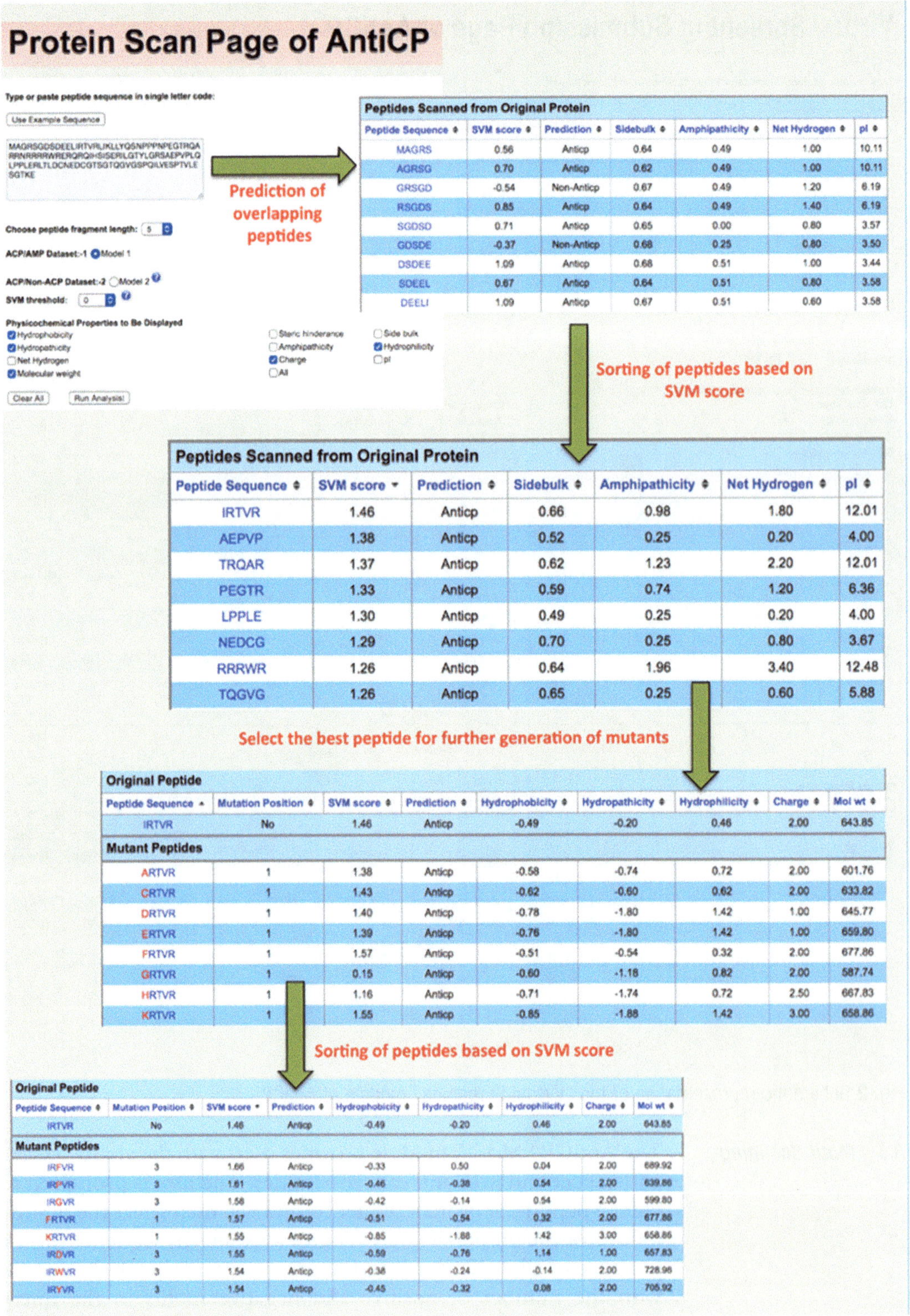

Fig. 3 Schematic representation of the "Protein Scan" module of AntiCP

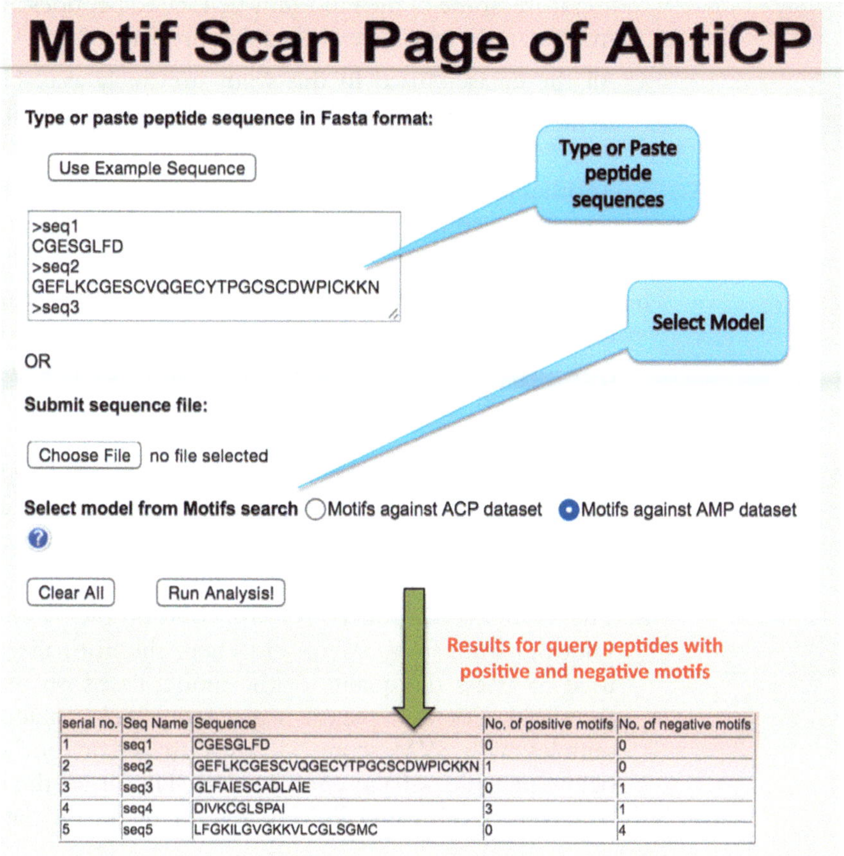

Fig. 4 Schematic representation of the "Motif Scan" module of AntiCP

'**Motifs against AMP dataset**'. The result table is shown in Fig. 4 with the number of positive and negative motifs present in each sequence. On the basis of this result, a user can select the best ACP.

4 Notes

1. All the modules of AntiCP are based on two types of SVM based prediction models: (a) model based on ACP/AMP dataset and (b) model based on ACP/Non-ACP dataset.
2. The ACP/AMP model was developed from 225 ACPs and 1372 peptides with antimicrobial properties but no anti-cancer properties.
3. The ACP/Non-ACP model was developed from 225 ACPs and 2250 random peptides derived from SwissProt proteins, considered as non-ACPs (i.e., negative dataset). Since experimentally validated non-ACPs were not reported in the

literature, some of the randomly generated peptides in the non-ACPs dataset may have anti-cancer properties.

4. All the datasets used in this study are freely available for the scientific community at the datasets section of the AntiCP web-server (http://crdd.osdd.net/raghava/anticp/datasets.php).
5. Another server for ACPs prediction is ACPP [35] (http://acpp.bicpu.edu.in). Users can explore and compare the results generated by the two servers.
6. Since, classification techniques, particularly machine learning techniques, perform best on balanced datasets, both models performed best when developed with an equal number of positive and negative datasets (i.e., 225 ACPs and 225 AMP or non-ACPs).
7. All the modules produced results on the default SVM threshold value of '0'. The default threshold is the value at which the learning of the prediction model is best. A user can select different SVM thresholds values ranging from −1 to +1.
8. The SVM score produced is normalized on the SVM threshold values ranging from −1 to +1. Then, the normalized score is used to assess the quality of the model based on the selected threshold. For example, at a given threshold value 0, all the samples with SVM score >0 are classified in one class, and below that (i.e. <0) in another class. The larger the difference between the calculated SVM scores, the better the classification.

Acknowledgement

AntiCP server was developed by Raghava's group (http://www.imtech.res.in/raghava/) at CSIR-IMTECH, India.

References

1. Siegel RL, Miller KD, Jemal A (2016) Cancer statistics, 2016. CA Cancer J Clin 66:7–30. doi:10.3322/caac.21332
2. Mayne ST, Playdon MC, Rock CL (2016) Diet, nutrition, and cancer: past, present and future. Nat Rev Clin Oncol. doi:10.1038/nrclinonc.2016.24
3. Wu D, Gao Y, Qi Y et al (2014) Peptide-based cancer therapy: opportunity and challenge. Cancer Lett 351:13–22. doi:10.1016/j.canlet.2014.05.002
4. Peer D, Karp JM, Hong S et al (2007) Nanocarriers as an emerging platform for cancer therapy. Nat Nanotechnol 2:751–760. doi:10.1038/nnano.2007.387
5. Amit D, Hochberg A (2010) Development of targeted therapy for bladder cancer mediated by a double promoter plasmid expressing diphtheria toxin under the control of H19 and IGF2-P4 regulatory sequences. J Transl Med 8:134. doi:10.1186/1479-5876-8-134
6. Kang TH, Mao C-P, He L et al (2012) Tumor-targeted delivery of IL-2 by NKG2D leads to accumulation of antigen-specific CD8+ T cells in the tumor loci and enhanced anti-tumor effects. PLoS One 7:e35141. doi:10.1371/journal.pone.0035141
7. Bidwell GL (2012) Peptides for cancer therapy: a drug-development opportunity and a drug-delivery challenge. Ther Deliv 3:609–621

8. Thundimadathil J (2012) Cancer treatment using peptides: current therapies and future prospects. J Amino Acids 2012:967347. doi:10.1155/2012/967347
9. Vlieghe P, Lisowski V, Martinez J, Khrestchatisky M (2010) Synthetic therapeutic peptides: science and market. Drug Discov Today 15:40–56. doi:10.1016/j.drudis.2009.10.009
10. Mader JS, Hoskin DW (2006) Cationic antimicrobial peptides as novel cytotoxic agents for cancer treatment. Expert Opin Investig Drugs 15:933–946. doi:10.1517/13543784.15.8.933
11. Gaspar D, Veiga AS, Castanho MARB (2013) From antimicrobial to anticancer peptides. A review. Front Microbiol 4:294. doi:10.3389/fmicb.2013.00294
12. Chu H-L, Yip B-S, Chen K-H et al (2015) Novel antimicrobial peptides with high anticancer activity and selectivity. PLoS One 10:e0126390. doi:10.1371/journal.pone.0126390
13. Okumura K, Itoh A, Isogai E et al (2004) C-terminal domain of human CAP18 antimicrobial peptide induces apoptosis in oral squamous cell carcinoma SAS-H1 cells. Cancer Lett 212:185–194. doi:10.1016/j.canlet.2004.04.006
14. Risso A, Braidot E, Sordano MC et al (2002) BMAP-28, an antibiotic peptide of innate immunity, induces cell death through opening of the mitochondrial permeability transition pore. Mol Cell Biol 22:1926–1935
15. Lehmann J, Retz M, Sidhu SS et al (2006) Antitumor activity of the antimicrobial peptide magainin II against bladder cancer cell lines. Eur Urol 50:141–147. doi:10.1016/j.eururo.2005.12.043
16. Hui L, Leung K (2002) Chen HM The combined effects of antibacterial peptide cecropin A and anti-cancer agents on leukemia cells. Anticancer Res 22:2811–2816
17. Kim S, Kim SS, Bang YJ et al (2003) In vitro activities of native and designed peptide antibiotics against drug sensitive and resistant tumor cell lines. Peptides 24:945–953
18. Ohtake T, Fujimoto Y, Ikuta K et al (1999) Proline-rich antimicrobial peptide, PR-39 gene transduction altered invasive activity and actin structure in human hepatocellular carcinoma cells. Br J Cancer 81:393–403. doi:10.1038/sj.bjc.6690707
19. Fang X-Y, Chen W, Fan J-T et al (2013) Plant cyclopeptide RA-V kills human breast cancer cells by inducing mitochondria-mediated apoptosis through blocking PDK1-AKT interaction. Toxicol Appl Pharmacol 267:95–103. doi:10.1016/j.taap.2012.12.010
20. Cao Q, Lin Z-B (2006) Ganoderma lucidum polysaccharides peptide inhibits the growth of vascular endothelial cell and the induction of VEGF in human lung cancer cell. Life Sci 78:1457–1463. doi:10.1016/j.lfs.2005.07.017
21. Kim SE, Kim HH, Kim JY et al (2000) Anticancer activity of hydrophobic peptides from soy proteins. Biofactors 12:151–155
22. Kozłowska K, Nowak J, Kwiatkowski B, Cichorek M (1999) ESR study of plasmatic membrane of the transplantable melanoma cells in relation to their biological properties. Exp Toxicol Pathol Off J 51:89–92. doi:10.1016/S0940-2993(99)80074-8
23. Sok M, Sentjurc M, Schara M (1999) Membrane fluidity characteristics of human lung cancer. Cancer Lett 139:215–220
24. Ellerby HM, Arap W, Ellerby LM et al (1999) Anti-cancer activity of targeted pro-apoptotic peptides. Nat Med 5:1032–1038. doi:10.1038/12469
25. Hariharan S, Gustafson D, Holden S et al (2007) Assessment of the biological and pharmacological effects of the alpha nu beta3 and alpha nu beta5 integrin receptor antagonist, cilengitide (EMD 121974), in patients with advanced solid tumors. Ann Oncol 18:1400–1407. doi:10.1093/annonc/mdm140
26. Gregorc V, De Braud FG, De Pas TM et al (2011) Phase I study of NGR-hTNF, a selective vascular targeting agent, in combination with cisplatin in refractory solid tumors. Clin Cancer Res 17:1964–1972. doi:10.1158/1078-0432.CCR-10-1376
27. Khalili P, Arakelian A, Chen G et al (2006) A non-RGD-based integrin binding peptide (ATN-161) blocks breast cancer growth and metastasis in vivo. Mol Cancer Ther 5:2271–2280. doi:10.1158/1535-7163.MCT-06-0100
28. Deplanque G, Madhusudan S, Jones PH et al (2004) Phase II trial of the antiangiogenic agent IM862 in metastatic renal cell carcinoma. Br J Cancer 91:1645–1650. doi:10.1038/sj.bjc.6602126
29. Sharma A, Kapoor P, Gautam A et al (2013) Computational approach for designing tumor homing peptides. Sci Rep 3:1607. doi:10.1038/srep01607
30. Gupta S, Kapoor P, Chaudhary K et al (2013) In silico approach for predicting toxicity of peptides and proteins. PLoS One 8:e73957. doi:10.1371/journal.pone.0073957
31. Gautam A, Chaudhary K, Kumar R et al (2013) In silico approaches for designing highly

effective cell penetrating peptides. J Transl Med 11:74. doi:10.1186/1479-5876-11-74
32. Sharma A, Singla D, Rashid M, Raghava GPS (2014) Designing of peptides with desired half-life in intestine-like environment. BMC Bioinformatics 15:282. doi:10.1186/1471-2105-15-282
33. Kumar R, Chaudhary K, Singh Chauhan J et al (2015) An in silico platform for predicting, screening and designing of antihypertensive peptides. Sci Rep 5:12512. doi:10.1038/srep12512
34. Tyagi A, Kapoor P, Kumar R et al (2013) In silico models for designing and discovering novel anticancer peptides. Sci Rep 3:2984. doi:10.1038/srep02984
35. Vijayakumar S, PTV L (2014) ACPP: a Web server for prediction and design of anti-cancer peptides. Int J Pept Res Ther 21:99–106. doi:10.1007/s10989-014-9435-7

Chapter 18

Docking and Virtual Screening in Drug Discovery

Maria Kontoyianni

Abstract

Stages in a typical drug discovery organization include target selection, hit identification, lead optimization, preclinical and clinical studies. Hit identification and lead optimization are very much intertwined with computational modeling. Structure-based virtual screening (VS) has been a staple for more than a decade now in drug discovery with its underlying computational technique, docking, extensively studied. Depending on the objective, the parameters for VS may change, but the overall protocol is very straightforward. The idea behind VS is that a library of small compounds are docked into the binding pocket of a protein (e.g., receptor, enzyme), a number of solutions per molecule, among the top-ranked, are being returned, and a choice is made on the fraction of compounds to be moved forward for testing toward hit identification. The underlying principle of VS is that it differentiates between active and inactive compounds, thus reducing the number of molecules moving forward and possibly offering a complementary tool to high-throughput screening (HTS). Best practices in library selection, target preparation and refinement, criteria in selecting the most appropriate docking/scoring scheme, and a step-wise approach in performing Glide VS are discussed.

Key words Virtual screening, High-throughput screening, Docking, Scoring, Structure-based drug design, GOLD, Glide, Drug discovery

1 Introduction

VS is a complementary tool to high-throughput screening (HTS) that attempts to find hits in the early stages of drug discovery. Specifically, once a macromolecular target is selected, compounds are needed to initiate efforts toward a clinical candidate. The goal of VS is to identify these early "hits" among a library of compounds. What differentiates HTS from VS is that HTS is an experimental approach, while VS is a theoretical one. HTS tests large numbers of compounds for their ability to affect the activity of target molecules by addressing whether a compound reacts biochemically with the target. For example, questions such as "does it bind to the target protein?," "does it trigger enzymatic reactions?," "does it activate signaling pathways?" are explored by HTS assays. Compounds showing positive results are passed onto a more rigorous assay. It

Iulia M. Lazar et al. (eds.), *Proteomics for Drug Discovery: Methods and Protocols*, Methods in Molecular Biology, vol. 1647, DOI 10.1007/978-1-4939-7201-2_18,

cannot be emphasized enough that positive results must be reconfirmed, because if false positives are being pursued, the investment detriment down the road will be high. On the other hand, negative results can mean that a potentially valuable compound is not considered. The latter could be an issue if no hits are found. However, the goal of HTS is not to find ***all*** possible hits in a library collection, but a sufficiently enough set to use as starting scaffolds for initial discovery efforts. Therefore, while false positives could be costly if pursed and thus, re-confirmation of the results is critical, false negatives are not and should not be a cause for worry. In the following step, chemists need to intelligently select two to three classes of compounds that show the most promise for potential clinical candidates. HTS is time consuming, requires an infrastructure, and has a low success rate (<5%); nonetheless, it has been the method of choice for the last 20 years in the pharmaceutical sector.

VS, an in-silico HTS method, consists of virtually placing (docking) collections of millions of compounds into a biological target, followed by an evaluation of the tightness of the fit (scoring). VS offers a quick assessment of huge libraries and reduces the number of compounds that need testing in order to identify early hits. Thus, the basic requirements for VS are:

- A compound collection, which highly depends on the objective of the project
- The structure of the biological target
- An appropriate docking/scoring scheme

The choices that are made for each of these requirements come with a set of questions that need to be addressed in order to make the process as efficient and accurate as possible. However, depending on the objective and the target investigated, the protocol varies. The steps of the protocol are discussed in detail in the following sections.

2 Materials

2.1 Compound Collections

2.1.1 Public Collections

Table 1 provides a representative set of diverse private and public compound collections. Most public databases have grown out of the need of academics to have access to chemicals. **PubChem** [1, 2] was launched in 2004 and to date it includes three primary databases: Substance, Compound, and Bioassay. While the Substance database may have redundant records of the same molecule from different contributors, the Compound database extracts all records for a specific molecule into an aggregate record, thus making searching more efficient. The Bioassay database provides descriptions of biological experiments on the tested compounds, particularly from HTS. It should be pointed out that PubChem represents

Table 1
List of available databases for VS

Database	Website	Availability	Size
PubChem	https://pubchem.ncbi.nlm.nih.gov/	Public	92,345,074
ChEMBL	https://www.ebi.ac.uk/chembl/	Public	2,036,512
BindingDB	https://www.bindingdb.org/bind/index.jsp	Public	600,622
ZINC	http://zinc.docking.org/	Public	35 million
ChemSpider	http://www.chemspider.com/	Public	57 million
DrugBank	http://www.drugbank.ca/	Public	8,261 drugs
GRAC	http://www.guidetopharmacology.org/about.jsp	Public	8,674
ChemBridge	http://www.chembridge.com/index.php	Commercial	1 million
Maybridge	http://www.maybridge.com	Commercial	53,000
ChemDiv	http://www.chemdiv.com/products/screening-libraries/	Commercial	1.5 million
Life Chemicals	http://www.lifechemicals.com/	Commercial	1.2 million
Specs	http://www.specs.net/	Commercial	1.5 million
Enamine	http://www.enamine.net/	Commercial	2 million

the largest body of molecular structures. The **ChEMBL** [3] database holds to date about 1.5 million distinct compounds, with accompanying information on functional assays, binding data, and ADMET (absorption, distribution, metabolism, excretion, and toxicity) assays. The bioassay data are derived from the literature, multiple screening resources, PubChem bioassays, GSK (Glaxo SmithKline) deposited data, and the BindingDB database. The **BindingDB** [4–6] collection focuses on small molecules interacting with proteins. As of May 2017, it contained 1,346,745 binding findings for 7,100 protein targets and 600,622 small molecules. BindingDB reports binding data stemming from the literature, PubChem confirmed assays, and those ChEMBL entries that are associated with a clearly defined protein target. It includes findings from enzyme inhibition and kinetics, isothermal titration calorimetry, NMR, and binding and competition assays. Details about experimental conditions accompany the data extracted from the BindingDB. **ZINC15** [7–9] is a database of over 120 million drug-like compounds that can be purchased. Besides being compound-centric, the latest version of ZINC [9] links compounds to biological targets or processes, using data from other databases, while maintaining its original capability to provide information on purchasing of reagents. **ChemSpider** provides access to 57 million chemical structures from 500 data sources. The entire database is not available for a free download, however, without permission, while one download can include up to 5,000 structures along with respective properties. For downloading a bigger dataset, it is necessary to contact the **ChemSpider** team. The **DrugBank** [10] database is a bioinformatics and cheminformatics resource with detailed drug information and relevant drug target data (i.e., sequence, structure, and pathway). As of today, the database contains 8,261 drug entries, including 2,012 FDA-approved small molecules, 233 FDA-approved protein/peptide drugs, 94 nutraceuticals, and more than 6,000 experimental drugs. Furthermore, 4,338 non-redundant protein sequences are linked to these drug entries. Selected text components and sequence data can be downloaded. Similarly, a smaller database, **GRAC** [11], includes about 8,611 ligands. These are approved drugs, phase I and beyond candidates, monoclonal antibodies, compounds from repurposing initiatives, representative compounds directed against reported Alzheimer's disease targets, new human Protein Data Bank ligand structures, and ligands disclosed in papers with ligand-protein relationships.

2.1.2 Commercial Collections

ChemBridge and **Maybridge** are commercially available screening libraries complementary to one another, with little overlap between the two. ChemBridge includes the CORE library with about 640,000 compounds, covering unique chemical spaces produced from 810 scaffolds, and another non-overlapping library, the EXPRESS-Pick collection. The latter consists of 460,000 novel and drug-like chemical structures, covering a broad chemical space, and offering diversity for initial structure-activity relationship efforts. EXPRESS-Pick compounds are selected using novelty, diversity, drug-like properties, and chemical structure analyses. The Maybridge library is much smaller containing approximately 53,000 compounds. **ChemDiv** offers a collection of over 1.5 million compounds, which have been validated in biological assays and are mostly focused libraries. **Life Chemicals** consists of 431,000 diverse compounds based on 2,800 distinct scaffolds. The **Specs** compounds can be downloaded at specs.net in "sd" format. **Enamine** is another commercial library with the largest collection of screening compounds (2,000,000). It includes the Advanced Collection with 294,000 compounds for targeted library design, an HTS Collection of 1,757,000 diverse screening compounds, and the Premium Collection of 120,000 compounds having favorable physicochemical properties.

2.1.3 Considerations Regarding the Library Selection

The choice of the database should not be driven by the number of compounds that it contains, but by the existing knowledge-base of the target regarding already known actives or information pertaining to the binding pocket. For example, are there known active compounds in the library? If yes, perform ligand-based similarity searching and eliminate the compounds that are similar to the already known actives (binders). If not, then the use of filters to eliminate potentially dead-end-leads or the generation of a diverse set may be the answer to selecting the most appropriate library. In regards to filters, much debate exists on whether to use the Lipinski's rule of five (RO-5) [12], which was developed based on the molecular properties of orally bioavailable compounds. This is due to the fact that in attempting to find early hits and using the RO-5 approach, one might eliminate scaffolds that could otherwise be optimized in later lead optimization cycles to improve oral bioavailability. On the other hand, filtering based on the Pan Assay Interference Compounds (PAINS) [13] is advisable in order to eliminate promiscuous binders. Thus, PAINS would be the filter of choice, while RO-5 is a filter used in the hit-to-lead stage. The size of the binding pocket can also be used as a guide in that if it is small, larger compounds may be eliminated from the library. Finally, a diverse subset is generated by identifying compounds of similar structure and choosing one representative of each scaffold class. The resultant library will then be employed in VS diversity sets.

Another critical aspect of library selection refers to the ligand geometries. Even though most libraries are in an "sd" format, attention should be given to bond lengths, rings, chirality, protonation and charges, and proper atom types for the program to be used. Docking programs consider the ligands as flexible at the torsional angle level. There is no optimization for lengths and bonds, thus the researcher needs to make sure that these parameters are accurate before proceeding. There are several excellent programs, freely available to academics, that generate conformations [14].

2.2 The Structure of the Biological Target

2.2.1 Crystal Structures

Typically, crystal structures of the biological targets are used. All the experimentally resolved receptor structures (NMR or X-ray) can be found in the Protein Databank (www.rcsb.org) as PDB IDs (for example, 3nxu corresponds to the cytochrome P450 3A4 isozyme complexed with the inhibitor ritonavir) [15]. If the PDB ID is not known, in the keyword window one should enter the name of the target of interest, and a list of resolved crystal or NMR structures will appear with their respective PDB IDs. However, even if the crystal structure is available, its quality needs to be carefully evaluated, because crystal structures are static snapshots not accounting for the dynamic behavior of macromolecules. Another point to note is that the crystal structures many times have water molecules attached. Common practice is to remove all waters. This is a point of contention, however, because some water molecules are very critical, i.e., form hydrogen-bonding networks with the ligand, or are involved in the mechanism of action of the enzyme. If there are water molecules involved in ligand binding, there are programs that can include these molecules either as part of the grid that defines the active site (i.e., Glide, however, the water molecules remain static) [16–18] or enable the user to select the water molecules that move during docking (i.e., GOLD) [19, 20]. In this chapter, we will not discuss the methods used to identify which water molecules are critical and/or are needed.

2.2.2 Target Considerations Prior to VS

The following parameters are evaluated prior to performing VS: (a) The resolution of the structure (the higher, the better); (b) The R-factor (or residual factor or agreement factor) which is indicative of the agreement between the crystallographic model and the experimental X-ray diffraction data, and (c) The B factor that is reflective of the true static or dynamic mobility of an atom, and therefore it shows where the errors are in the structure (the lower the B factor, the better). Also, crystal structures do not have hydrogens, so hydrogen atoms must be added prior to doing any VS experiments. Finally, we have to consider if the crystal structure is a complex (receptor-small molecule) or not. In the first case, the binding pocket, which needs to be defined for docking, represents the area around the small molecule. However, if the crystal structure is an apo-structure (containing no small molecule within it), then programs must be used that find the binding pockets prior to VS [21]. If the crystal structure is not available, homology modeling can be used to generate a receptor structure, provided there is reasonable homology with template(s). The reader is advised to refer to CASP (Critical Assessment of Protein Structure Prediction) papers regarding advances in the field of model building or homology modeling [22]. Caution should be given to metal ions and cofactors. If the receptor is a metalloenzyme, docking with programs that have been parameterized in their scoring functions for the specific ion is critical, or VS will not work.

2.2.3 Other Considerations

Biological targets are dynamic in nature, however, experimentally resolved macromolecular structures with or without a bound ligand are isolated snapshots and not reflective of flexibility. Some docking programs attempt to address flexibility by including rotamers of side chains of the binding site residues [23] or by using an induced fit (IF) methodology or ensemble docking. In IF, a ligand is docked giving rise to multiple receptor-ligand complexes; each ligand is subsequently re-docked into each of these receptor conformations, as is the case with Schrodinger's Induced Fit Docking protocol. Ensemble docking employs multiple protein structures stemming from the PDB, if available, or from molecular dynamics simulations, or from normal mode analyses. It is not within the scope of the present chapter to delve into alternative approaches and methodologies that incorporate receptor flexibility. It should be, however, emphasized that not all receptors undergo conformational changes, which means that the success of a VS experiment is not always determined by the use of a flexible receptor structure.

2.3 An Appropriate Docking/Scoring Scheme

Most investigators use a docking program that they are familiar with or one that is available in the laboratory. However, this is not good practice. If the receptor-small molecule complex structure is available (*see* Subheading 2.2), several docking algorithms must be used to see which one places the small molecule in the same orientation as in the PDB ID of that target. Toward that goal, a number of poses should be generated and visually inspected (typically 30–60 poses/ligand). It is advisable to visually inspect all poses, as the top-ranked ones with the native scoring function are often not accurate. Once the docking program that reproduces the experimental pose (i.e., what is seen in the crystal structure) is identified, the researcher should check how the pose is ranked by the native scoring function (docking programs come with their own scoring functions). If it is ranked at the top, then this docking/scoring scheme should be used in the VS experiment. If not, poses should be rescored with different scoring functions. This will lead to a docking/scoring combination and provide information on how close to being top-ranked the observed modes are.

Finally, it is advisable that a pilot VS experiment with a smaller subset of the chosen library, seeded with the known active compounds against the target of interest as decoys, is performed using the docking/scoring scheme identified in the preceding paragraph. This is performed to ensure optimal parameter choices, robustness, and to hopefully get a sense of accuracy in ranking the known hits high(er). An excellent list of available docking programs can be found on the Swiss Institute of Bioinformatics website (https://www.click2drug.org), whereas Table 2 shows a representative list of the most widely used programs.

3 Methods

VS with the Glide program is described below. The main steps that must be performed include: (1) Prepare the target after downloading it from the PDB (Glide has its own preparation routine, called "protein preparation wizard"); (2) Generate a GRID, which will define the area that the docking algorithm has to search (binding pocket) to place/dock the ligands; (3) Run the docking algorithm with the generated GRID.

3.1 Target Preparation

1. Under Project, select "get pdb auto," enter the PDB entry of the receptor of interest, and download. This downloads the PDB ID that the investigator is interested in.
2. Under tasks, the "protein preparation wizard" should be selected.
3. Then "add hydrogens". Hydrogens may not be displayed (even if they are added) if the button "none" for "display hydrogens"

Table 2
Representative docking programs used in VS

Program	Free for academics	Source
Autodock	Yes	http://autodock.scripps.edu/
Autodock Vina	Yes	http://vina.scripps.edu/
DOCK	Yes	http://dock.compbio.ucsf.edu/
RosettaLigand	Yes	http://rosettadock.graylab.jhu.edu/
iGEMDOCK	Yes	http://gemdock.life.nctu.edu.tw/dock/igemdock.php
SLIDE	Yes	http://www.kuhnlab.bmb.msu.edu/
rDOCK	Yes	http://rdock.sourceforge.net/
iDOCK	Yes	http://istar.cse.cuhk.edu.hk/idock/
FlipDock	Yes	http://flipdock.scripps.edu/
paraDOCKs	Yes	https://github.com/cbaldauf/paradocks
DAIM/SEED	Yes	http://www.biochem-caflisch.uzh.ch/download/
GlamDock	Yes	http://www.chil2.de/Glamdock.html
BetaDock	Yes	http://voronoi.hanyang.ac.kr/software.htm
FRED	Yes	http://www.eyesopen.com/oedocking
ICM	No	http://www.molsoft.com/docking.html
FlexX(E)	No	https://www.biosolveit.de/FlexX/
Glide	No	https://www.schrodinger.com/glide
GOLD	No	http://www.ccdc.cam.ac.uk/solutions/csd-discovery/components/gold/
MOW	No	http://www.chemcomp.com/

is selected. Other options for hydrogen display include polar only, mixed, or all. The default is "Mixed" (Fig. 1 shows the 1ubq receptor structure before and after this step).

4. Next, we assign bond orders (double bonds, single bonds, and aromaticity are corrected in this step).
5. Select "delete waters" within 5 Å. If some water molecules are thought to be critical, we can keep them in the pdb, and delete the rest with an editor. In the latter case, we do not delete any waters in the protein preparation wizard.
6. Proceed with "preprocess."
7. In the "review and modify" button, the option to delete a chain, in the case that the crystal structure has more than one chain and they are all the same, is available. Keeping only one chain saves time.

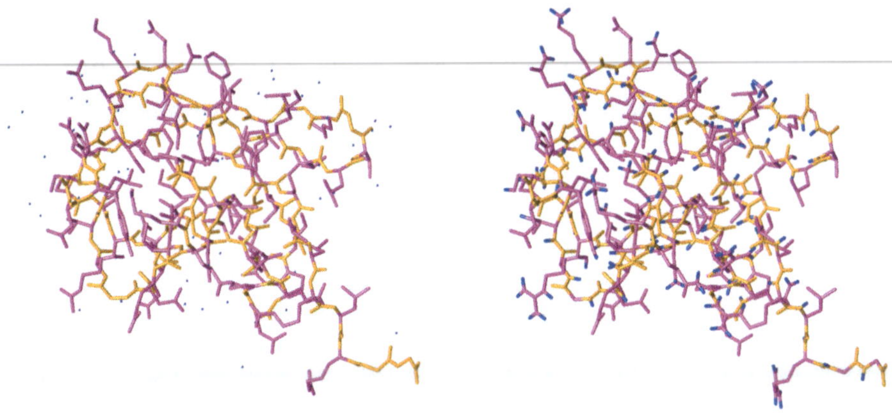

Fig. 1 PDB ID 1ubq, without hydrogens (*left panel*) and with hydrogens (*right panel*). Backbone is *orange*, side chains are *purple*, and water molecules are depicted in *blue* in the *left panel*, while the hydrogens are blue in the *right panel*

8. Under "refine," hydrogen-bonds are assigned at pH 7.0, while the water orientations are sampled, if this option is checked. This is done via the "optimize" button. The next step should be minimization, which refines the positions of hydrogens and heavy atoms.

3.2 Grid Generation

1. The grid will define the space within which the docking program will place the ligands (binding pocket). Under tasks, select "docking grid generation."
2. To generate the grid, one needs to click on an atom of the ligand ("pick to identify the ligand" is checked), thus defining it as the center of the grid. In the case that the crystal structure does not have a ligand bound to it, the button that identifies the ligand should be left unchecked. The van der Waals (vdw) scaling factor is set to 1.0 (default), unless docking does not result in any poses; in the latter case, the scaling factor should be lowered to 0.8 in order to allow for more tolerance for closer contacts (*see* **Note 1**).
3. Under "site," either the centroid of the workspace ligand or the centroid of selected residues should be checked. In the first case, the ligand is part of the structure (*see* **step 2** above), while in the latter, an amino acid needs to be selected, provided we know from experiments (i.e., mutagenesis) which amino acids are critical for binding and/or for the enzymatic mechanism. Once this is done, a purple rectangular box surrounding the binding pocket appears on the workspace.

3.3 Docking/VS

1. Under "tasks," glide docking should be selected.
2. Indicate the name of the grid that was just generated.

3. Select the filename of the library collection that will be used for screening.
4. The vdw scaling factor and partial charge cutoff should be left at default values, unless softening of the potential is thought to be necessary (*see* **Note 2**).
5. Under "settings," HTVS (high-throughput virtual screening) should be selected. Ligand sampling is flexible, which means that unlike the receptor which remains rigid throughout the experiment, the ligands are flexible so that different conformations are identified and docked.
6. In the output tab, "write pose viewer" should be enabled because this is the output file. Depending on the library, "write out at most 5 poses per ligand" means that five poses per ligand will be generated. This may not be ideal, if the library has millions of compounds. It is up to the researcher to decide how many poses per ligand are wanted.
7. Click Run.

3.4 Analysis

1. The results come in a project table that will be opened once the poseviewer file (output from the run) is imported (*see* **Note 3**).
2. Open the project table, and under "entry," "view poses" should be selected. The arrows on the top can be used to step through each of the docked ligands.
3. To view interactions, in the Pose Viewer menu, one can click on what he/she is interested, i.e., contacts, hydrogen-bonds, etc. which will be marked on the screen for each pose.

4 Notes

1. The docking programs consider the receptor as a rigid structure that is not moving during docking. With the scaling of van der Waals radii of nonpolar atoms, the investigator has the option to decrease the penalties for close contacts, and in turn allow for a slight "give" in the receptor and/or the ligand. This is by no means an approach to generate a flexible receptor, but a way of allowing poses that would be otherwise rejected due to close contacts to active site residues.
2. Successful docking requires sometimes that the ligand or the receptor "give" a bit in order to bind. To model this behavior, Glide can scale the vdw radii of nonpolar atoms (where nonpolar is defined by a partial charge threshold that the researcher can set), thereby decreasing penalties for close contacts. By default, scaling is performed for qualifying atoms in the ligand, but not those in the receptor. Ligand atom radii scaling settings can be changed using the options in this section.

3. Before proceeding to visual inspection of the results, one needs to address the following questions: (1) Which percentage of the database should be considered in order to identify a sufficient number of actives for initial scaffolding? It is suggested that 2.5% of the library collection should be examined [24]; (2) Which criteria to use, besides scoring, to select compounds? The reader should be reminded that the objective of VS is to find actives among inactives. Ideally, all actives are scored on the top, but realistically that is never the case. Consequently, knowledge regarding interactions with key residues, if experimental data is available, structural diversity, and possibly fingerprints are the most common criteria used for aiding compound selection.

References

1. Kim S, Thiessen PA, Bolton EE, Chen J, Fu G, Gindulyte A, Han L, He J, He S, Shoemaker BA, Wang J, Yu B, Zhang J, Bryant SH (2016) PubChem Substance and Compound databases. Nucleic Acids Res 44(D1): D1202–D1213. doi:10.1093/nar/gkv951
2. Wang Y, Xiao J, Suzek TO, Zhang J, Wang J, Zhou Z, Han L, Karapetyan K, Dracheva S, Shoemaker BA, Bolton E, Gindulyte A, Bryant SH (2012) PubChem's BioAssay Database. Nucleic Acids Res 40(Database issue): D400–D412. doi:10.1093/nar/gkr1132
3. Gaulton A, Bellis LJ, Bento AP, Chambers J, Davies M, Hersey A, Light Y, McGlinchey S, Michalovich D, Al-Lazikani B, Overington JP (2012) ChEMBL: a large-scale bioactivity database for drug discovery. Nucleic Acids Res 40(Database issue):D1100–D1107. doi:10.1093/nar/gkr777
4. Liu T, Lin Y, Wen X, Jorissen RN, Gilson MK (2007) BindingDB: a web-accessible database of experimentally determined protein-ligand binding affinities. Nucleic Acids Res 35(Database):D198–D201. doi:10.1093/nar/gkl999
5. Gilson MK, Liu T, Baitaluk M, Nicola G, Hwang L, Chong J (2016) BindingDB in 2015: A public database for medicinal chemistry, computational chemistry and systems pharmacology. Nucleic Acids Res 44(D1): D1045–D1053. doi:10.1093/nar/gkv1072
6. Nicola G, Liu T, Gilson MK (2012) Public domain databases for medicinal chemistry. J Med Chem 55(16):6987–7002. doi:10.1021/jm300501t
7. Irwin JJ, Shoichet BK (2005) ZINC–a free database of commercially available compounds for virtual screening. J Chem Inf Model 45 (1):177–182. doi:10.1021/ci049714+
8. Irwin JJ, Sterling T, Mysinger MM, Bolstad ES, Coleman RG (2012) ZINC: a free tool to discover chemistry for biology. J Chem Inf Model 52(7):1757–1768. doi:10.1021/ci3001277
9. Sterling T, Irwin JJ (2015) ZINC 15–Ligand Discovery for Everyone. J Chem Inf Model 55 (11):2324–2337. doi:10.1021/acs.jcim.5b00559
10. Wishart DS, Knox C, Guo AC, Cheng D, Shrivastava S, Tzur D, Gautam B, Hassanali M (2008) DrugBank: a knowledgebase for drugs, drug actions and drug targets. Nucleic Acids Res 36(Database issue):D901–D906. doi:10.1093/nar/gkm958
11. Pawson AJ, Sharman JL, Benson HE, Faccenda E, Alexander SP, Buneman OP, Davenport AP, McGrath JC, Peters JA, Southan C, Spedding M, Yu W, Harmar AJ, Nc I (2014) The IUPHAR/BPS Guide to PHARMACOLOGY: an expert-driven knowledgebase of drug targets and their ligands. Nucleic Acids Res 42(Database issue):D1098–D1106. doi:10.1093/nar/gkt1143
12. Lipinski CA (2000) Drug-like properties and the causes of poor solubility and poor permeability. J Pharmacol Toxicol Methods 44 (1):235–249
13. Baell JB, Holloway GA (2010) New substructure filters for removal of pan assay interference compounds (PAINS) from screening libraries and for their exclusion in bioassays. J Med Chem 53(7):2719–2740. doi:10.1021/jm901137j
14. Ebejer JP, Morris GM, Deane CM (2012) Freely available conformer generation methods: how good are they? J Chem Inf Model 52(5):1146–1158. doi:10.1021/ci2004658

15. Berman HM, Westbrook J, Feng Z, Gilliland G, Bhat TN, Weissig H, Shindyalov IN, Bourne PE (2000) The Protein Data Bank. Nucleic Acids Res 28(1):235–242
16. Friesner RA, Banks JL, Murphy RB, Halgren TA, Klicic JJ, Mainz DT, Repasky MP, Knoll EH, Shelley M, Perry JK, Shaw DE, Francis P, Shenkin PS (2004) Glide: a new approach for rapid, accurate docking and scoring. 1. Method and assessment of docking accuracy. J Med Chem 47(7):1739–1749
17. Friesner RA, Murphy RB, Repasky MP, Frye LL, Greenwood JR, Halgren TA, Sanschagrin PC, Mainz DT (2006) Extra precision glide: docking and scoring incorporating a model of hydrophobic enclosure for protein-ligand complexes. J Med Chem 49(21):6177–6196
18. Halgren TA, Murphy RB, Friesner RA, Beard HS, Frye LL, Pollard WT, Banks JL (2004) Glide: a new approach for rapid, accurate docking and scoring. 2. Enrichment factors in database screening. J Med Chem 47(7):1750–1759
19. Jones G, Willett P, Glen RC, Leach AR, Taylor R (1997) Development and validation of a genetic algorithm for flexible docking. J Mol Biol 267(3):727–748
20. Verdonk ML, Chessari G, Cole JC, Hartshorn MJ, Murray CW, Nissink JW, Taylor RD, Taylor R (2005) Modeling water molecules in protein-ligand docking using GOLD. J Med Chem 48(20):6504–6515
21. Ghersi D, Sanchez R (2011) Beyond structural genomics: computational approaches for the identification of ligand binding sites in protein structures. J Struct Funct Genomics 12 (2):109–117. doi:10.1007/s10969-011-9110-6
22. Moult J, Fidelis K, Kryshtafovych A, Schwede T, Tramontano A (2014) Critical assessment of methods of protein structure prediction (CASP)–round x. Proteins 82(Suppl 2):1–6. doi:10.1002/prot.24452
23. Gaudreault F, Chartier M, Najmanovich R (2012) Side-chain rotamer changes upon ligand binding: common, crucial, correlate with entropy and rearrange hydrogen bonding. Bioinformatics 28(18):i423–i430. doi:10.1093/bioinformatics/bts395
24. Kontoyianni M, Sokol GS, McClellan LM (2005) Evaluation of library ranking efficacy in virtual screening. J Comput Chem 26 (1):11–22

Chapter 19

Bioinformatics Resources for Interpreting Proteomics Mass Spectrometry Data

Iulia M. Lazar

Abstract

Developments in mass spectrometry (MS) instrumentation have supported the advance of a variety of proteomic technologies that have enabled scientists to assess differences between healthy and diseased states. In particular, the ability to identify altered biological processes in a cell has led to the identification of novel drug targets, the development of more effective therapeutic drugs, and the growth of new diagnostic approaches and tools for personalized medicine applications. Nevertheless, large-scale proteomic data generated by modern mass spectrometers are extremely complex and necessitate equally complex bioinformatics tools and computational algorithms for their interpretation. A vast number of commercial and public resources have been developed for this purpose, often leaving the researcher perplexed at the overwhelming list of choices that exist. To address this challenge, the aim of this chapter is to provide a roadmap to the basic steps that are involved in mass spectrometry data acquisition and processing, and to describe the most common tools that are available for placing the results in biological context.

Key words Proteomics, Mass spectrometry, Bioinformatics, Data interpretation

1 Introduction

Proteomic and mass spectrometry technologies have been extensively used in recent years for assessing the dynamic composition of the protein complement of cells, tissues, and organisms. Proteomic profiling and differential expression analysis approaches have been advanced for establishing correlations between gene and protein expression levels, determining the spatial and temporal distribution of proteins in a cell, identifying mutations and posttranslational modifications (PTMs), and in a more complex systems biology context, mapping entire signaling pathways or biological processes that describe the dynamic behavior of a living cell or organism. Proteomic protocols involve a large number of both experimental and computational steps, a generic workflow being provided in Fig. 1. Once a particular problem is identified and a hypothesis launched, the workflow can be divided into three major modules:

Iulia M. Lazar et al. (eds.), *Proteomics for Drug Discovery: Methods and Protocols*, Methods in Molecular Biology, vol. 1647, DOI 10.1007/978-1-4939-7201-2_19, © Springer Science+Business Media LLC 2017

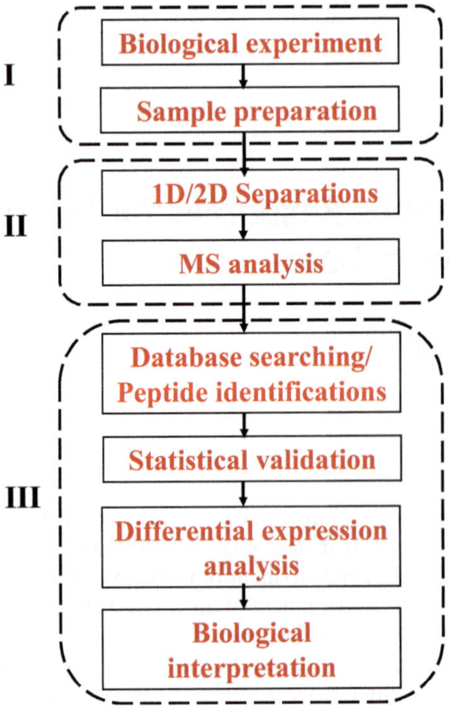

Fig. 1 Generic proteomics workflow. Module I: biological experiment and sample preparation; Module II: 1D/2D separations with MS detection; Module III: mass spectrometry data processing and biological interpretation

(I) experimental, (II) MS analysis, and (III) data processing and interpretation. After a brief description of the processes that are involved in the first two modules, the forthcoming discussion will center on the computational and bioinformatics tools that are available for module III, with focus on processing proteomic data generated through workflows that build on various implementations of nano-high performance liquid chromatography (HPLC) electrospray ionization (ESI) MS technologies. Over the years, a vast list of such computational and bioinformatics tools and resources have been developed, many still active, while some abandoned or not maintained any longer. In Table 1, only the active and most referenced tools and resources are summarized [1–106]. The abbreviations of these tools and/or brief descriptions are included in the table.

2 Experimental Module

With a particular end-goal in mind, the biological experiment must be designed to generate proteomic samples at relevant time-points and of sufficient size. While modern mass spectrometers enable the

Table 1
List of computational and bioinformatics tools for processing mass spectrometry proteomics data

Software/database	Type	Description	Link	Reference
Protein sequence databases				
UniProt	Freeware	Universal Protein Resource	http://www.uniprot.org	[1, 2]
SwissProt	Freeware	Swiss Protein Sequence Database	http://www.uniprot.org	[3]
TrEMBL	Freeware	Translated European Molecular Biology Laboratory	http://www.uniprot.org	[4]
PIR	Freeware	Protein Information Resource	http://pir.georgetown.edu	[5]
PDB	Freeware	Protein Data Bank		[6]
RefSeq	Freeware	Reference Sequence Database	https://www.ncbi.nlm.nih.gov/refseq	[7, 8]
Peptide mass fingerprinting				
MASCOT	Commercial	Search engine for the identification, characterization and quantitation of proteins using mass spectrometry data	http://www.matrixscience.com	[9]
MS-FIT/ ProteinProspector	Freeware	Correlation of Mass Spectrometry data (parent masses only) with a protein in a sequence database which best FITs the data	http://prospector.ucsf.edu/prospector	[10]
PROWL/ ProFound	Freeware	Tool for searching protein sequence collections with peptide mass maps	http://prowl.rockefeller.edu/prowl-cgi/profound.exe	[11, 12]
Tandem MS data interpretation				
MASCOT	Commercial	Search engine for the identification, characterization, and quantitation of proteins using mass spectrometry data	http://www.matrixscience.com	[9]
SEQUEST	Commercial	Correlates uninterpreted tandem mass spectra of peptides with amino acid sequences from protein and nucleotide databases	https://www.thermofisher.com	[13]
PEAKS	Commercial	Software for LC-MS/MS data analysis to provide identification and quantification	http://www1.bioinfor.com/peaks/features/peaksdb.html	[14]

(continued)

Table 1
(continued)

Software/database	Type	Description	Link	Reference
Bionic	Commercial	Software for peptide and protein identification from tandem MS data	http://www.proteinmetrics.com	–
Spectrum Mill	Commercial	Comprehensive suite of software tools for high-throughput proteomics	www.agilent.com	–
MassLynx	Commercial	Software for acquiring, analyzing, managing, and sharing MS data	www.waters.com	–
PLGS	Commercial	ProteinLynx Global SERVER, Advanced features for peptide identification and de novo sequencing	www.waters.com	
SWATH-MS	Commercial	Sequential Windowed Acquisition of All Theoretical Fragment Ion Mass Spectra	https://sciex.com/swath-acquisition	–
OpenSWATH	Open-source	Targeted analysis of DIA data in an automated, high-throughput fashion	http://www.openswath.org	[15, 16]
X!Tandem	Open-source	Software for matching tandem mass spectra to peptide sequences in a database	http://www.thegpm.org/tandem	[17]
OMSSA	Open-source	Open Mass Spectrometry Search Algorithm	http://massqc.proteomesoftware.com/help/faq/omssa.php	[18]
MyriMatch	Open-source	Search engine that scores peptide matches with greater emphasis on matching intense peaks	https://medschool.vanderbilt.edu/msrc-bioinformatics/software	[19]
Andromeda	Freeware	Peptide search engine based on probabilistic scoring	http://www.coxdocs.org	[20]
MS Amanda	Freeware	Database search engine developed for high-resolution tandem mass spectrometry data	http://ms.imp.ac.at/?goto=msamanda	[21]
MS-Product/Protein Prospector	Freeware	Tool for generating hypothetical fragment ions from a given peptide sequence	http://prospector.ucsf.edu/prospector	–

(continued)

Table 1
(continued)

Software/database	Type	Description	Link	Reference
MS-Tag/ Protein Prospector	Freeware	Tool for matching ions from tandem mass spectra to proteins in a database by making use of tags (series of fragment ions that correspond to coherent sequences of amino acids)	http://prospector.ucsf.edu/prospector	–
PepFrag/ Prowl	Freeware	Tool for identifying proteins from a collection of sequences that match a single tandem mass spectrum	http://prowl.rockefeller.edu/prowl/pepfrag.html	–
MS data validation				
PeptideProphet	Open-source	Statistical model that evaluates the confidence of peptide identifications from tandem mass spectra, as returned by a database search	http://peptideprophet.sourceforge.net	[22, 23]
ProteinProphet	Open-source	Statistical model that validates the protein identifications based on peptide assignments to tandem mass spectra	http://proteinprophet.sourceforge.net	[24]
Percolator	Open-source	Tool that discriminates correct from incorrect peptide-spectrum matches by using semi-supervised learning	http://percolator.ms	[25]
De novo sequencing				
PEAKS	Commercial	Software platform for peptide identification, quantification, PTM analysis, and de novo sequencing	http://www.bioinfor.com	[26]
Spectrum Mill	Commercial	Comprehensive suite of software tools for high-throughput proteomics	http://www.agilent.com	–
PLGS	Commercial	ProteinLynx Global SERVER, Advanced features for peptide identification and de novo sequencing	www.waters.com	–
ProSight PTM	Freeware	Identification and characterization of intact proteins and their PTMs (top-down approach)	https://prosightptm.northwestern.edu	[27]
InsPecT	Freeware	Identification of posttranslationally modified peptides from tandem mass spectra	http://proteomics.ucsd.edu/Software/Inspect	[28]

(continued)

Table 1
(continued)

Software/database	Type	Description	Link	Reference
Proteomic data quantitation				
PEAKSQ	Commercial	Stable isotope label and label-free quantification	http://www.bioinfor.com/denovo.php	[26]
MASCOT	Commercial	Stable isotope label and label-free quantification	http://www.matrixscience.com	–
Proteome Discoverer	Commercial	Stable isotope label quantification	https://www.thermofisher.com	–
SIEVE	Commercial	Software for label-free, semiquantitative differential analysis of complex LC-MS data sets	https://www.thermofisher.com	–
PLGS	Commercial	ProteinLynx Global SERVER, Advanced features for label-free quantitation from MS^E data	www.waters.com	–
Spectrum Mill	Commercial	Stable isotope label and label-free quantification	http://www.agilent.com	
ProteinPilot	Commercial	Stable isotope label-based quantitation	https://sciex.com/products/software/proteinpilot-software	–
ProteinScape	Commercial	Stable isotope label and label-free quantification	https://www.bruker.com	–
TPP	Open-source	Trans Proteomics Pipeline, Multiple quantification tools	http://tools.proteomecenter.org/wiki/index.php?title=Software:TPP	[29]
Skyline	Open-source	SRM/MRM/PRM quantification	https://skyline.ms/project/home/software/Skyline/begin.view	[30]
ASAP Ratio/TPP	Freeware	Automated Statistical Analysis on Protein Ratio, Stable isotope label-based quantification	http://tools.proteomecenter.org/wiki/index.php?title=Software:ASAPRatio	[31]
XPRESS/TPP	Freeware	Stable isotope label-based quantification	http://tools.proteomecenter.org/wiki/index.php?title=Software:XPRESS	[32]

(continued)

Table 1
(continued)

Software/database	Type	Description	Link	Reference
Libra/ TPP	Freeware	Stable isotope label-based quantification	http://tools.proteomecenter.org/wiki/index.php?title=Software:Libra	[31, 33]
ProteoSuite	Open-source	Stable isotope label and label-free quantification	http://www.proteosuite.org	[34]
TOPP	Open-source	The OpenMS Proteomics Pipeline, Stable isotope label and label-free quantification	http://www.openms.de	[35–37]
MaxQuant	Freeware	Stable isotope label and label-free quantification	http://www.biochem.mpg.de/5111795/maxquant	[38]
MSQuant	Open-source	Stable isotope label and label-free quantification	http://msquant.alwaysdata.net	[39]
APEX	Open-source	Absolute Protein Expression, Label-free quantification (spectral counting)	https://sourceforge.net/projects/apexqpt	[40]
Pipelines for processing proteomic data				
TPP	Open-source	Trans Proteomic Pipeline, Open-source software to support tandem MS-based shotgun proteomics research	http://tools.proteomecenter.org/wiki/index.php?title=Software:TPP	[29]
Skyline	Open-source	Open-source application for building SRM/MRM/PRM-targeted MS/MS, DIA/SWATH and targeted DDA quantitative methods	https://skyline.ms/project/home/software/Skyline/begin.view	[30]
TOPP	Open-source	The OpenMS Proteomics Pipeline, Compilation of a broad range of tools for analyzing proteomic datasets	http://www.openms.de	[35–37]
MaxQuant	Freeware	Suite of algorithms for peak detection, scoring, mass calibration, database searching, peptide/protein identification, quantification, and statistical analysis	http://www.coxdocs.org/doku.php?id=maxquant:start	[38]
Scaffold	Commercial	Advanced bioinformatics platform for protein identification and quantitation	http://www.proteomesoftware.com	[41]

(continued)

Table 1
(continued)

Software/database	Type	Description	Link	Reference
Sorcerer	Commercial	Desktop or cloud-based server for high-throughput protein identification and characterization	https://www.sagenresearch.com	[42]
IP2/IPA	Commercial	Integrated Proteomics Pipeline (IP2)/ Integrated Proteomics Applications (IPA)	www.integratedproteomics.com	–
Proteomics data repositories				
PRIDE	Freeware	PRoteomics IDEntifications	http://www.proteomexchange.org https://www.ebi.ac.uk/pride/archive	[43]
Peptide Atlas	Freeware	Peptide Atlas, SRM Atlas, SWATH Atlas	http://www.peptideatlas.org	[44]
CrossTalkDB	Freeware	Database for proteins with multiple PTMs (histone modifications)	http://crosstalkdb.bmb.sdu.dk	[45]
GPM	Freeware	Global Proteome Machine	http://www.thegpm.org/	–
NIST	Freeware	National Institutes of Standards	https://www.nist.gov/programs-projects/peptide-mass-spectral-libraries	[46, 47]
Biological interpretation of proteomics data				
Annotation and enrichment				
GO	Freeware	Gene Onthology Consortium	http://www.geneontology.org	[48–50]
GoMiner	Freeware	GO Miner	https://discover.nci.nih.gov/gominer	[51]
DAVID/ EASE	Freeware	Database for Annotation, Visualization, and Integrated Discovery/ Expression Analysis Systematic Explorer	https://david.ncifcrf.gov	[52–54]
GSEA	Freeware Open-source	Gene Set Enrichment Analysis	https://www.broadinstitute.org/gsea	[55, 56]
Protein-protein interactions databases and tools				
Cytoscape	Open-source	Platform for network data integration, analysis, and visualization	http://www.cytoscape.org	[57]

(continued)

Table 1
(continued)

Software/database	Type	Description	Link	Reference
STRING	Freeware	Search Tool for the Retrieval of Interacting Genes/Proteins	http://string-db.org	[58, 59]
IIS	Freeware	Integrated Interactome System	http://bioinfo03.ibi.unicamp.br/lnbio/IIS2	[60]
PPI Spider	Freeware	Protein-Protein Interaction Spider	http://www.bioprofiling.de/PPI_spider.html	[61]
PIPs	Freeware	Human Protein-Protein Interaction Prediction	http://www.compbio.dundee.ac.uk/www-pips	[62, 63]
BioGRID	Freeware	Biological General Repository for Interaction Datasets	https://thebiogrid.org	[64, 65]
HPRD	Freeware	Human Reference Protein Database	http://www.hprd.org	[66]
IntAct/MIntAct	Freeware Open-source	Database system and analysis tools for molecular interaction data	http://www.ebi.ac.uk/intact	[67, 68]
MINT	Freeware	Molecular INTeraction database	http://mint.bio.uniroma2.it	[69]
MIPS	Freeware	Mammalian Protein-Protein Interaction Database	http://mips.helmholtz-muenchen.de/proj/ppi	[70]
DIP	Freeware	Database of Interacting Proteins	http://dip.doe-mbi.ucla.edu/dip/Main.cgi	[71]
ChEMBL/ChEMBLdb	Freeware	Chemical database of bioactive molecules with drug-like properties (manually curated and maintained by EMBL)	https://www.ebi.ac.uk/chembl	[72]
DrugBank	Freeware	Bioinformatics and cheminformatics resource that combines detailed drug data with comprehensive drug target information	https://www.drugbank.ca	[73]
Pathway analysis				
Reactome	Freeware Open-source	Free, open-source, curated, peer-reviewed pathway database	http://www.reactome.org	[74]

(continued)

Table 1
(continued)

Software/database	Type	Description	Link	Reference
Cytoscape	Freeware	Platform for network data integration, analysis, and visualization	http://www.cytoscape.org	[57, 75, 76]
KEGG	Freeware	Kyoto Encyclopedia of Genes and Genomes	http://www.genome.jp/kegg	[77]
NCI-PID	Freeware	National Cancer Institute—Pathway Interaction Database	http://www.home.ndexbio.org	[78]
Wiki Pathways	Freeware	Open, public platform dedicated to the curation of biological pathways by and for the scientific community	http://www.wikipathways.org/index.php/WikiPathways	[79, 80]
PathVisio	Open-source	Pathways editing, analysis, visualization	http://www.pathvisio.org	[81]
Biocarta	Freeware	Resource for biological pathways, displayed in a graphical format, part of NCI Cancer Genome Anatomy project	https://cgap.nci.nih.gov/Pathways	–
PathwayCommons	Freeware	Database of publicly available biological pathway information	http://www.pathwaycommons.org	[82]
Kinome Explorer/NetPhorest	Freeware	Integrated framework for modeling kinase-substrate interactions	http://www.netphorest.info http://kinomexplorer.info	[83]
IPA	Commercial	Ingenuity Pathways Analysis	www.qiagen.com/ingenuity	–
Metacore	Commercial	Integrated software suite for functional analysis of next-generation sequencing, variant, CNV, microarray, metabolic, SAGE, proteomics, siRNA, microRNA, and screening data	https://lsresearch.thomsonreuters.com/pages/solutions/1/metacore	–
Protein PTMs, motifs, and domains				
ExPASy	Freeware	EXpert Protein Analysis SYstem	https://www.expasy.org	[84]
InterPro	Freeware	The integrative protein signature database, Functional analysis of protein families and domains	https://www.ebi.ac.uk/interpro	[85]

(continued)

Table 1
(continued)

Software/database	Type	Description	Link	Reference
PFAM	Freeware	Protein Families Database	http://pfam.xfam.org	[86]
ProSite	Freeware	Database of protein domains, families, and functional sites	http://prosite.expasy.org	[87]
SMART	Freeware	Simple Modular Architecture Research Tool (protein domains)	http://smart.embl-heidelberg.de	[88]
CORUM	Freeware	The comprehensive resource of mammalian protein complexes	http://mips.helmholtz-muenchen.de/corum	[89]
PhosphoSite Plus	Freeware	Systems biology resource providing comprehensive information and tools for the study of protein PTMs	http://www.phosphosite.org/homeAction.action	[90]
Phosida	Freeware	PHOsphorylation SIte DAtabase	http://141.61.102.18/phosida/index.aspx	[91]
Phospho.ELM	Freeware	Phospho-Eukaryotic Linear Motif; Database of experimentally verified phosphorylation sites in eukaryotic proteins	http://phospho.elm.eu.org	[92]
KinasePhos	Freeware	Tool for computational prediction of phosphorylation sites within given protein sequences	http://kinasephos.mbc.nctu.edu.tw	[93]
PhosphoMotif Finder	Freeware	Kinase/phosphatase substrate and binding motifs curated from the published literature	http://www.hprd.org/PhosphoMotif_finder	[94]
GlycoMod	Freeware	Tool for predicting possible oligosaccharide structures on proteins	http://web.expasy.org/glycomod	[95, 96]
UniCarbDB/GlycoSuiteDB	Freeware	Glycomic MS database repository, Relational database of glycoprotein glycan structures and their biological sources	http://unicarb-db.expasy.org	[97–99]
UniPep	Freeware	Resource for serum protein biomarkers and glycopeptides	http://www.unipep.org	[100]
PEAKS PTM	Commercial	Posttranslational modification identifications	http://www1.bioinfor.com	[101]

(continued)

Table 1
(continued)

Software/database	Type	Description	Link	Reference
Scientific data workflow management platforms				
Galaxy	Open-source	Platform for data intensive biomedical research	https://galaxyproject.org	[102–104]
Taverna	Open-source	Suite of tools for designing and executing scientific workflows and in-silico experiments	http://www.taverna.org.uk	[105]
KNIME	Open-source	Konstanz Information Miner	https://www.knime.org	[106]

identification of >2000 proteins from <1 μg sample [107], starting protein amounts of 50–100 μg are typically necessary for ensuring the availability of sufficient quantities for sample manipulation, pre-fractionation, processing through complementary protocols, or enrichment in various protein subfractions. Once the biological experiment is completed, the proteins are extracted from the biological source, either as a whole or as organelle subfractions, and prepared for analysis [108]. To generate peptides that fit the operational m/z range of most mass spectrometers, the protein extract is subjected to proteolytic digestion with an enzyme. The disulfide bonds in proteins are reduced first with a reducing agent [(e.g., DTT (1,4-dithiothreitol), TCEP-HCl (Tris (2-carboxyethyl) phosphine hydrochloride), TBP (tributylphosphine)] in the presence of a denaturant (e.g., urea), and alkylated with, e.g., iodoacetamide, to block the newly freed cysteine residues. The proteins are then subjected to proteolytic digestion typically with trypsin, LysC, or a combination of the two, to cleave the proteins after the C-terminal of Lys and Arg, or of Lys-only residues, respectively. Alternatively, to increase protein coverage, other enzymes have been suggested, as well (LysN, ArgC, AsnN, GluC, or chymotrypsin [109]). The samples are further processed for the removal of salts, detergents, or reagents that have been introduced during sample processing, and then subjected to 1D or 2D chromatographic separations. For the case of 2D separations, the first dimension consists of a pre-fractionation step commonly involving strong cation exchange (SCX), basic reversed phase (RP), or hydrophilic interaction chromatography. Alternatively, the first dimension of separation can be performed at the protein level, prior to proteolytic digestion, by using polyacrylamide gel electrophoresis (PAGE). The second dimension is most often a nano-reversed phase (C18) HPLC separation step that is interfaced to the electrospray ionization source of the mass spectrometer for delivering the separated peptide mixture for analysis. For workflows that involve the analysis of PTMs, specialized protein or peptide enrichment steps (e.g., based on immobilized-metal affinity chromatography-IMAC or TiO_2 beads for phosphorylation, or lectin chromatography for glycosylation) must be introduced in the protocol [110, 111].

Proteome-level differential expression analysis for quantitative studies is enabled by following stable isotope label or label-free approaches [112]. For cell culture experiments that are amenable to metabolic labeling, the SILAC (metabolic labeling by using stable isotope labeled amino acids in cell culture) or $^{14}N/^{15}N$-labeling protocols have become the most common choice. For samples that cannot be labeled in cell culture, e.g., tissues, chemical labeling is performed at the peptide level, after proteolytic digestion, by using various stable isotope-labeled tags. While a variety of chemical tags have been developed, the use of ICAT (isotope-coded affinity tag), iTRAQ (isobaric tags for relative and absolute quantitation), or TMT (tandem mass tag) reagents has become routine. All stable isotope-based approaches rely on measuring and comparing the heavy-to-light peptide intensities either in the MS (ICAT) or MS/MS (iTRAQ, TMT) modes of data acquisition. For label-free approaches, the chromatographic peak intensities or areas, the MS ion intensities, or the total number of spectral counts (tandem mass spectra per protein) are measured and compared. For targeted quantitation, selected reaction monitoring (SRM) or multiple reaction monitoring (MRM) methods have been implemented. In this case, the samples are spiked with stable isotope-labeled peptides, representative of the proteins of interest, and only the selected peptides (precursors) and their associated ion fragments (products) are measured by MS and compared [113]. The precursor-product combinations are known as "transitions," and are specific to a given amino acid sequence. Contamination is unlikely, as for a particular transition the contaminant should have the same precursor and fragment m/z, as well as the same LC elution time.

3 Mass Spectrometry Analysis Module

Proteomic data are acquired by mass spectrometers comprising one or more analyzers (quadrupole, ion trap/linear ion trap, time-of-flight, FTICR, or orbitrap [114]). Various ion fragmentation techniques have been developed for enabling the accurate identification of the amino acid sequence of proteins/peptides and their modifications, with modern mass spectrometers making use of a combination of conventional collision-induced dissociation (CID), electron transfer dissociation (ETD), and/or higher energy collisional dissociation (HCD). In contrast to CID [115] which generates mainly **b/y**-type ion pairs with loss of the labile PTMs, ETD favors peptide backbone fragmentation of multiple charged precursor ions preferentially at the N-C_α bond, with the generation of **c/z**-type ion pairs [116]. The site of the PTMs can be identified, as the modification is retained during fragmentation. HCD, used with orbitrap instruments, is performed in a collision cell which is separate from the analyzer. It enables a broader range of fragmentation pathways and a more complete fragmentation of peptides at higher energies, with no low-mass cutoff [117]. Accordingly, a range of MS scanning functions have been developed for orchestrating the analysis of precursor or fragment ions, and of their PTMs (i.e., full MS, MS^n with an assortment of fragmentation choices, zoom, neutral loss, etc.).

Mass analysis on advanced MS instruments with multiple analyzers involves the development of a method that comprises a combination of scanning functions that enable the collection of MS and MS^n spectra generated by the choice of fragmentation method, in a specific order and for specific peptides. By using such complex methods, the newest Orbitrap Fusion Lumos instrument has reached unprecedented performance in terms of resolution (up to 500,000), mass accuracy (<3 ppm), *m/z* range (up to 4000–6000), dynamic range (>5000), scan speed (20 Hz), and multiplexing/throughput ability. Manufacturers have developed instrument-specific software packages for mass spectrometer operation and data acquisition (e.g., XCalibur/Thermo Fisher, MassLynx/Waters, Analyst/Sciex, MassHunter/Agilent, etc.), specialized functions being described for data-dependent analysis (DDA), data-independent analysis (DIA), or targeted (MRM) acquisition of mass spectra. DDA and MRM are complementary techniques that facilitate global and targeted proteomic profiling studies, respectively. Data-independent acquisition is performed on high-resolution/high mass accuracy instruments that enable the simultaneous fragmentation of multiple peptides within a broad precursor *m/z* range (e.g., 25 *m/z*), the caveat being related to the difficulty in interpreting the complex tandem mass spectra comprised of fragment ions belonging to multiple precursors. Such methods are optimized to match the ion activation/fragmentation sequence of operations to the LC separation time-line. The output of an LC-MS proteomic analysis is a chromatogram comprised of tens of thousands of mass or tandem mass spectra that are subjected to further processing (Fig. 2).

4 Mass Spectrometry Data Processing and Biological Interpretation Module

Raw MS data are subjected to multiple levels of processing that can be grouped into two major categories: (1) processing of mass spectra for the identification, validation, and quantitation of

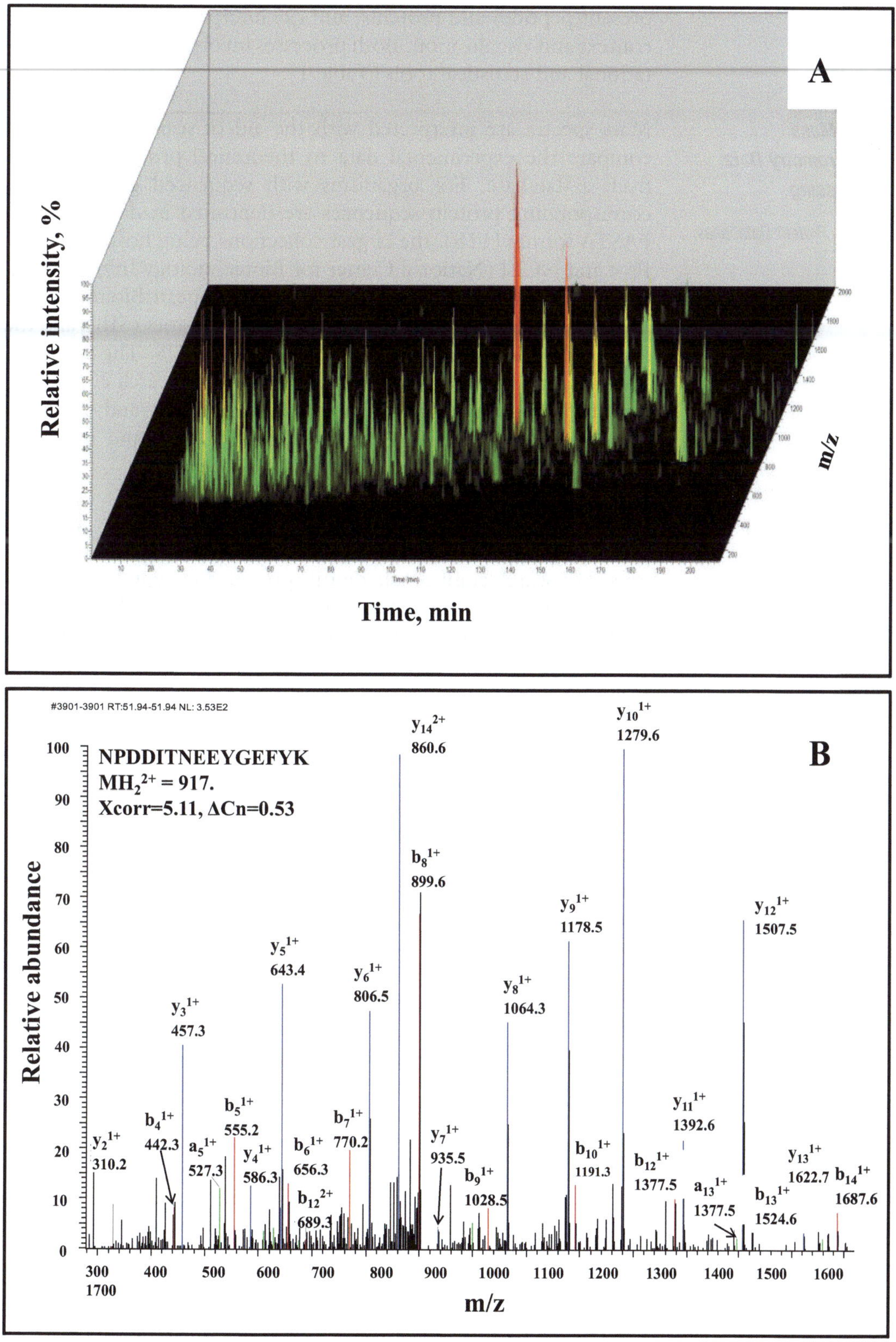

Fig. 2 (**a**) Nano-HPLC separation; (**b**) Tandem MS of a peptide acquired through DDA-MS/MS analysis

present peptides and proteins, and (2) interpretation in biological context and visualization. Both processes involve the use of computational and statistical tools (Table 1).

4.1 Mass Spectrometry Data Processing

4.1.1 Protein Databases

Mass spectra are interpreted with the aid of search engines that compare the experimental data to theoretical protein sequences from a database. For organisms with sequenced genomes, the corresponding protein sequences are deposited in databases in a FASTA format [118], the largest collections being hosted by UniProt and NCBI (National Center for Biotechnology Information). UniProt is a consortium formed by the European Bioinformatics Institute (EBI), the Swiss Institute of Bioinformatics (SIB), and the Protein Information Resource (PIR) from the US. The two UniProt databases that are accessed routinely by MS search engines are SwissProt (a nonredundant, manually annotated) and TrEMBL (automatically annotated). Likewise, NCBI hosts and maintains the RefSeq database.

4.1.2 Search Engines

Based on enzyme specificity, the proteins in a database are processed in silico into peptides with known m/z and known fragment ions. The search engines can identify protein and peptide sequences from experimental data either based on the m/z of the precursor peptides alone, an approach called peptide mass fingerprinting (PMF), or from a combination of precursor and fragment ions from MS^n data. The experimental spectra are: (a) preprocessed by background subtraction/noise removal and isotope and charge state deconvolution, (b) normalized based on intensities, (c) merged if presenting similar features, and (d) searched against the in silico generated peptide database. For this purpose, the experimental mass spectra are compared to all theoretical spectra from the database that have matching parent/fragment ions, typical search parameters including the enzyme specificity, mass accuracy and mass tolerance, and the presence of PTMs (permanent/static or changing/dynamic). PTMs can be identified from precursor and fragment ion mass shifts associated with a particular PTM (e.g., +80 for Ser/Thr/Tyr phosphorylation, +42 for Lys/N-terminal acetylation, +16 for Met oxidation, +43 for Lys carbamylation, etc.). Both commercial or proprietary (e.g., SEQUEST/Thermo-Fisher, MassLynx/Waters, Spectrum Mill/Agilent) and free or open-source (e.g., X!Tandem, OMSSA) search engines have been developed. Batch-interpretation of large proteomic datasets is frequently performed by the commercial SEQUEST and MASCOT engines. SEQUEST runs within the Thermo Proteome Discoverer™ platform, while the MASCOT server, licensed from Matrix Science, supports the processing of raw data generated by most major MS platforms (Thermo, Waters, Agilent, Sciex, Bruker, or Shimadzu) and fragmentation methods (CID, ETD), as well as the handling of PTMs. Free or open-source search engines handle best the analysis of only simple protein mixtures, while platforms such as ProteinProspector (UCSF) and PROWL (Rockefeller University) offer a suite of tools that can help manual PMF and tandem MS data interpretation. Recently, however, advanced search engines that are capable of handling complex proteomic samples have been developed and made freely available to the entire proteomics community either as standalone packages or as integrated within other data processing platforms (e.g., MS Amanda and Andromeda).

Many search engines enable peptide/protein identifications by using either PMF or tandem MS. Due to the complexity of the data, and the fact that PMF is feasible only when the data are collected by high mass-accuracy instruments, the vast majority of proteomic experiments rely on the use of tandem MS. For peptide identification, the search engines generate peptide spectrum matches (PSMs) between experimental tandem mass spectra and theoretical peptide sequences. The search engine results are then scored (e.g., see the cross-correlation score-Xcorr in SEQUEST, or the probability-based Mascot Identity Threshold (MIT)—a similarity measure—in MASCOT) and statistically validated to generate probabilities of peptide identifications and protein inferences. Multiple algorithms can be used to calculate the false discovery rates (FDRs) associated with each peptide and protein. Commonly, FDRs are evaluated by searching the raw data against decoy databases with randomized or reversed sequences. Decoys by reversing the sequence of each protein in the target database are preferred due to the similarities between features such as the number of proteins and protein length distribution, as well as amino acid composition [119]. FDR values are then used to set a cutoff threshold for the peptide or protein scores. The final output of such a search process is a list of proteins and matching peptides with associated scores and FDR values (Fig. 3). For cases when databases do not exist to allow comparisons between experimental and theoretical data, e.g., the case of organisms with un-sequenced genomes, a different set of computational algorithms have been developed, collectively coined de novo sequencing. Both commercial and open-source platforms exist (Table 1). A useful feature of the de novo sequencing tools is that they can be used as a complementary means for the interpretation of tandem mass spectra with noise or un-assigned ions that receive poor scores from conventional search engines.

4.1.3 Postprocessing of Database Search Results

Database search results can be further processed and rescored for providing more reliable significance measures. PeptideProphet and ProteinProphet are statistical approaches, relying on the expectation maximization algorithm, that were among the first to be developed for improving the confidence in peptide and protein

2	Protein					Probability, -10lg(P)	Score	Sequence
3						Protein		coverage
4	Peptide sequence	Charge, z	Spectral count	MH+	DeltaM	Peptide	Xcorr	DeltaCn
5	Q15149|PLEC1_HUMAN Plectin-1 - Homo sapiens (Human)					171	1050.4	31.66
6	K.AGVVGPELHEQLLSAEK.A	2	3	1777.0	-0.56	104	4.99	0.45
7	K.AKLEQLFQDEVAK.A	2	2	1518.8	-0.51	55	4.94	0.53
8	K.AVTGYRDPYTGQSVSLFQALKK.G	3	2	2429.3	0.2	125	6.04	0.49
9	K.AYSDPSTGEPATYGELQQR.C	2	12	2069.9	0.01	133	5.07	0.54
10	K.DLLPSDMAVALLEAQAGTGHIIDPATSAR.L	3	3	2933.5	-0.14	300	3.73	0.6
11	K.GIIRPGTAFELLEAQAATGYVIDPIK.G	3	1	2743.5	-0.04	38	3.32	0.27
12	K.GIIRPGTAFELLEAQAATGYVIDPIKGLK.L	3	1	3041.7	1.1	102	7.89	0.63
13	K.GYYSPYSVSGSGSTAGSR.T	2	1	1782.8	0.4	128	5.13	0.71
14	K.LSIYNALKK.D	2	1	1049.6	-0.48	41	2.46	0.14
15	K.QYINAIKDYELQLVTYK.A	2	1	2102.1	-0.64	35	4.06	0.56
16	K.VLALPEPSPAAPTLRSELELTLGKLEQVR.S	3	1	3127.8	0	108	3.66	0.57
17	K.VQSGSESVIQEYVDLR.T	2	1	1808.9	0.33	37	4.64	0.34
18	R.AALAHSEEVTASQVAATK.T	2	4	1783.9	-0.57	114	5.84	0.69
19	R.AALAHSEEVTASQVAATK.T	3	6	1783.9	-0.69	36	4.19	0.48
20	R.AGEVERDLDKADSMIR.L	2	1	1804.9	-0.46	54	3.34	0.37
21	R.AHEEQLKEAQAVPATLPELEATK.A	3	2	2503.3	0.17	55	3.63	0.48
22	R.ALQALEELRLQAEEAER.R	3	1	1969.0	-0.89	109	4.7	0.41
23	R.ALQALEELRLQAEEAERR.L	3	1	2125.1	0.64	64	5.08	0.38
24	R.APVPASELLASGVLSR.A	2	1	1566.9	-0.49	89	5.07	0.61
25	R.ARQEELYSELQAR.E	2	4	1592.8	-0.49	67	5.13	0.47
26	R.AVTGYKDPYTGQQISLFQAMQK.G	3	2	2474.2	0.06	36	4.72	0.39
27	R.AVTGYRDPYTEQTISLFQAMKK.E	3	2	2547.3	0.27	84	6.08	0.45

Fig. 3 Output of a database search process: list of proteins with matching peptides and associated scores

identifications. Another re-scoring tool, the Percolator (works with Sequest or Mascot), uses a support vector machine (SVM) learning approach that extracts and processes features from the search results to better discriminate between correct and incorrect spectrum identifications. Scaffold combines multiple search engine results and PeptideProphet probabilities to create a single aggregated score with increased confidence of protein identifications.

4.1.4 Quantitative Proteomics

Quantitative proteomic comparisons can be performed by either stable isotope labeling (metabolic and chemical labeling) or label-free (spectral counting, intensity or area-based) methods, at the peptide or protein level, in an absolute or relative fashion, and using a targeted or nontargeted approach. Due to the complexity of the sample, the measurements are typically relative in nature and are performed at the peptide level. Comparisons of heavy-to-light peptides can be performed in MS mode (ICAT, cleavable ICAT), but also in MS2, such as in the case of iTRAQ and TMT labeling methods. Specialized computational modules have been developed for extracting quantitative information from the mass spectral data generated by labeled peptides. The challenges faced by developers when developing software for addressing problems posed by quantitative LC-MS proteomics have been summarized by Cappadona *et al.* [120], and include the ability to handle: (a) data complexity, (b) denoising/peak picking, (c) feature finding (isotopes and charge states), (d) isobaric interference, (e) peptide identifications, (f) protein inference, (g) retention time alignments, (h) normalization, (i) quantification, and (j) calculation of statistical significance.

The quantitation of certain PTMs is particularly difficult due to their short biological lifetime or labile nature, and, as a result, additional scoring schemes are necessary for their accurate identification. Furthermore, when interpreting label-free/intensity-based quantitative LC-MS data, noise reduction, peak picking, and retention time alignment are particularly challenging tasks that have to be addressed. A vast number of commercial (Mascot Distiller, Proteome Discoverer, SIEVE, PEAKSQ, QuanLynx, ProteinPilot, ProteinScape, Spectrum Mill) and free or open-source (APEX, ASAP Ratio, Max Quant, MSQuant, OpenMS, ProteoSuite, XPRESS) computational approaches have been developed for handling peptide and protein quantitation [121], the most advanced packages being capable of processing data produced by a variety of metabolic/chemical labeling and label-free experimental approaches (Table 1). Open-source platforms such as the Trans Proteomic Pipeline (TPP) and Skyline provide advanced tools for global and targeted SRM/MRM quantitation, respectively. While stable isotope label-based approaches have been demonstrated to be more accurate and reproducible than the label-free counterparts [122], label-free methods that rely on using ion intensities (e.g., iBAQ-Intensity-based Absolute Quantification [123]), peptide counting (e.g., PAI-Protein Abundance Index [124], emPAI-exponentially modified Protein Abundance Index [125]), or spectral counting [126] have progressively attracted more interest due to simplicity, cost-effectiveness, as well as ability to compare protein abundances in a large number of samples.

4.1.5 Proteomics Pipelines

To support the integration of results generated by various search engines with tools developed for the interpretation of proteomics data, several free or open-source workflows with broad scope such as the TPP, OpenMS Proteomics Pipeline (TOPP), MaxQuant, or Skyline have been developed. The TPP is an open-source software package that comprises an entire suite of tools for interpreting tandem MS data including data conversion from vendor-specific to open file formats, database searching, protein inference, statistical validations of proteins, quantification, visualization, and reporting. It can handle data from a variety of MS platforms and accepts the output of all commonly used search engines. TPP itself is bundled with the open-source X!Tandem search engine, uses the Prophet line of tools for the validation of peptide and protein identifications, and integrates the XPRESS, ASAPRatio, and LIBRA tools for quantitation. TOPP performs similar operations, MaxQuant is a free quantitative software package that incorporates the Andromeda search engine for analyzing large-scale MS data, while Skyline is an open-source editor that facilitates the analysis, quantification, and sharing of SRM/MRM/PRM (parallel reaction monitoring) proteomics experiments and results. Skyline supports the use of publicly available spectral libraries, as well as the building of custom libraries from user data including libraries that incorporate PTMs. It also allows for building quantitative methods with DIA/SWATH (Sequential Windowed Acquisition of all THeoretical fragment ion mass spectra) and targeted DDA with MS1. Commercial packages for streamlining the proteomics data processing have been developed by many MS instrument manufacturers (e.g., ProteinLynx/Waters, ProteinScape/Bruker, Scaffold/Proteome Software) or data analytics companies (e.g., Sorcerer/Sage-N Research, Integrated Proteomics Pipeline/Integrated Proteomics Applications) and accept data produced by all major mass spectrometry platforms.

With year-to-year improvements in MS instrument performance, new data processing platforms are continuously developed. For example, the recently released PatternLab 4.0 handles shotgun proteomic data processing from peptide spectrum matching to differential expression, de novo sequencing and assessment of biological significance [127]. Packages that make use of proteogenomics data have received special attention. For example, the Spectrogene software tool uses top-down proteomics tandem mass spectra and ORFs of different lengths from the six-frame translation of a genome (the ORFeome) to identify novel genes or exons, and to better annotate the genome of that particular organism [128]. Data that pass certain quality thresholds, as described in the recommendations provided by the HUPO (Human Proteome Organization) and the MIAPE (The minimum information about a proteomics experiment) documents [129, 130], can be deposited in public repositories (e.g., PRIDE, PeptideAtlas, NIST, GPM).

4.2 Biological Interpretation of MS Proteomics Data

High quality proteomics data can be interpreted and visualized in biological context, a process that can involve a number of steps: biological annotation, identification of enriched functional categories, identification of protein-protein interactions and networks, mapping to biological pathways, and, ultimately, assessing dynamical behavior in a systems biology framework.

4.2.1 Biological Annotation and Enrichment Analysis

The biological annotation and identification of enriched protein categories are the steps that will provide a first (and important) appraisal of the biological quality of the data in terms of cellular location, biological function, and biological processes that can be associated with the list of identified proteins, as well as of the categories that are enriched relative to a given background list of genes (often represented by the whole genome of the organism under study). A very large number of bioinformatics Gene Ontology (GO) and enrichment resources exist. Lempicki and coworkers compiled a list of 68 available tools with the associated key statistical methods [53]. They classified the enrichment tools in three categories: class I/singular enrichment analysis (SEA) for identifying individually enriched genes; class II/gene set enrichment analysis (GSEA) for identifying enriched pathways from entire lists of genes, and class III/modular enrichment analysis (MEA) for identifying genes enriched in a network context based on common functional annotation. The most commonly used tools in proteomics include the ones provided by GO, GoMiner, DAVID, and EASE from class I, GSEA from class II, and DAVID from class III. GoMiner classifies genes from a user-provided list into biologically relevant categories, displaying the results in the form of a tree, a "Directed Acyclic Graph," or a clustered image map. It also performs statistical and quantitative data analysis. DAVID provides access to a broad range of annotation and enrichment tools that enable: (a) the identification of functional categories, (b) GO-categorization based on cellular location, function, and biological process, (c) identification of disease-related genes, (d) mapping of pathways (e.g., via KEGG, Biocarta, Reactome), (e) exploration of protein domains and motifs (e.g., via PFAM, SMART, InterPro), (f) identification of protein-protein interactions, and (g) tissue expression analysis. It also provides combined results in the form of functional annotation tables, charts, and clusters. Comparisons can be performed to the whole genome of an organism or to a user-defined gene background, and gene enrichment is measured based on a modified Fisher's exact method that generates an EASE score. Unlike the case of class I single-gene enrichment analysis tools, the GSEA strategy enables the selection of enriched sets of genes with common function, chromosomal location, or regulation. GSEA builds on prior knowledge related to gene co-expression and pathways, enabling the identification of entire cellular processes and pathways that may be altered in a particular biological state. Based on GSEA, a new annotation tool PSEA (Protein Set Enrichment Analysis [131]) was developed for assessing whole protein quantitative profiles.

4.2.2 Protein-Protein Interaction Networks and Biological Pathways

Once that enriched sets of proteins are identified, the next steps involve the identification of the nature of relationships between these proteins, e.g., of co-expression or co-localizations patterns or of protein-protein interactions. Corresponding data mining tools have been made available through multiple platforms such as Cytoscape, STRING, PPI Spider, or IIS. These platforms offer a variety of choices for enabling data integration, visualization, analysis of complex biological protein-protein interaction networks, or even of protein-gene, protein-metabolite, or protein-drug networks (IIS, ChEMBL). The platforms also use or have developed their own protein-protein interaction databases (e.g., BioGRID, HRPD, IntAct, MINT, MIPS, DIP). Pathways Commons offers single point access to many of these tools.

Mapping to biological pathways is performed by another suite of computational packages [132], commercial (e.g., Ingenuity Pathway Analysis-IPA, Metacore toolbox for pathway analysis and data mining) or freely available to the scientific community (e.g., Reactome, Cytoscape, KEGG, NCI Pathways, Wiki Pathways, PathVisio, BioCarta, KinomeExplorer). Two Cytoscape plugins, BiNGO and GeneMANIA, have received particular attention from the proteomics community. BiNGO calculates and displays statistically overrepresented GO terms in a network, while the GeneMANIA plugin enables fast gene (protein) function prediction. The necessary information for enabling pathway analysis is retrieved from databases such as KEGG, Ingenuity Pathway Knowledge Base, Human Database/Metacore, Reactome, BioCarta, NCI, or Wiki pathways. In addition, a variety of tools have been developed for mining the data for the presence of protein functional domains and motifs (IntePpro, ProSite, SMART, PFAM), PTMs (PhopshositePlus, Phosida, PhosphoMotif Finder, Phospho. ELM, GlycoMod, UniCarbDB, etc.), assembly in protein complexes (CORUM), or presence of amino acid mutations [133]. The ExPASy portal provides access to a vast range of software resources and scientific databases that enable a detailed analysis of proteomics, genomics, and transcriptomics data. Data interpretation is ultimately visualized through plots (scatter, profile), heat maps, and protein-protein interaction maps and pathways (Fig. 4) [134].

With the advent of omics technologies and burst of biological data generated by high-throughput platforms, the need for workflow management systems that integrate multi-omics data has become clear. Several systems that handle genomics, epigenomics, transcriptomics, proteomics, and metabolomics data, and that help establish relationships between the various levels of gene expression (e.g., proteome vs. genome or transcriptome, proteome vs. interactome), have been developed (Galaxy, Taverna, KNIME). The ultimate goal of these platforms is to provide the users with the necessary computational, statistical, and bioinformatics tools for developing streamlined, multi-step, custom workflows that enable data interpretation, visualization, reporting, sharing, storage, archiving, and reuse. The platforms can be installed locally or operated on the cloud, and accept data in various formats. KNIME, for example, has over 1000 modules that include math, statistical, machine learning, workflow control, data viewing, and reporting functions, among many others. To further facilitate data sharing and increase the impact of valuable omics data, new, peer-reviewed data journals (e.g., *Scientific Data*/Nature, *Data in Brief*/Elsevier) have been launched.

Altogether, the ability to combine an unprecedented amount of information, generated through expression and functional

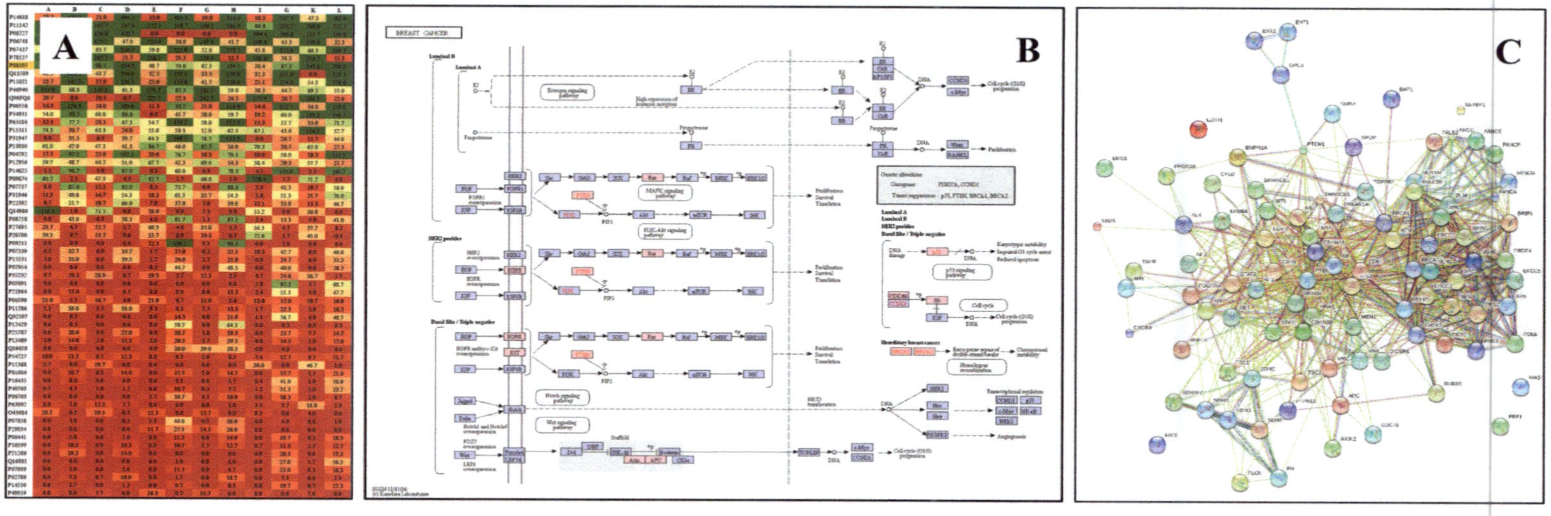

Fig. 4 Data interpretation in biological context. (**a**) Heat map showing the expression level of a set of putative cancer marker proteins in various cell lines; (**b**) Germline cancer proteins (http:/cancer.sanger.ac.uk/cosmic) mapped to the KEGG breast cancer pathway (pink boxes); (**c**) STRING protein-protein interaction diagram of a subset of germline cancer proteins. The nodes represent actual proteins, while the multicolored edges represent various types of protein-protein associations: known (experimentally measured or from curated databases), predicted (gene fusions, co-occurrence, or neighborhood), or of other types (co-expression, text mining, or protein homology)

proteomics studies, with progressively improved computational approaches for data analysis, will enable a detailed characterization of biological systems and of their dynamic behavior. This, in turn, will support the development of novel mathematical models that describe the molecular mechanisms that lead to healthy or diseased cell states. The capability to experimentally assess a systems-level response to artificially induced perturbations will further enable the advance of new hypotheses that support the identification of novel drug targets and the rational design of more effective drugs [135, 136].

Acknowledgment

This work was supported by grant NSF/DBI-1255991 to I.M.L.

References

1. UniProt Consortium (2015) UniProt: a hub for protein information. Nucleic Acids Res 43:D204–D212
2. Apweiler R, Bairoch A, CH W et al (2004) UniProt: the universal protein knowledgebase. Nucleic Acids Res 32 (90001):115D–1119
3. Alpi E, Griss J, da Silva AW et al (2016) UniProtKB/Swiss-Prot, the manually annotated section of the UniProt KnowledgeBase: how to use the entry view. Methods Mol Biol 1374:23–54
4. Boeckmann B, Bairoch A, Apweiler R et al (2003) The SWISS-PROT protein knowledgebase and its supplement TrEMBL. Nucleic Acids Res 31(1):365–370
5. CH W, Yeh L-SL, Huang H et al (2003) The protein information resource. Nucleic Acids Res 31(1):345–347
6. Westbrook J, Feng Z, Jain S et al (2002) The Protein Data Bank: unifying the archive. Nucleic Acids Res 30:245–248
7. Pruitt KD, Tatusova T, Maglott DR (2005) NCBI Reference Sequence (RefSeq): a curated non-redundant sequence database of genomes, transcripts and proteins. Nucleic Acids Res 33:D501–D504
8. Suzek BE, Huang H, McGarvey P et al (2007) UniRef: comprehensive and non-redundant UniProt reference clusters. Bioinformatics 23(10):1282–1288
9. Perkins DN, Pappin DJ, Creasy DM et al (1999) Probability-based protein identification by searching sequence databases using mass spectrometry data. Electrophoresis 20:3551–3567
10. Jacob RJ, Baker PR, Huang L et al (2000) Maximinizing proteomic information from MS data: enhancements to protein prospector, a suite of programs for mining genomic databases. Paper presented at the 48th ASMS conference of mass spectrometry and allied topics, Long Beach, 27–31 May 2000
11. Zhang W, Chait BT (2000) ProFound: an expert system for protein identification using mass spectrometric peptide mapping information. Anal Chem 72(11):2482–2489
12. Beavis R, Fenyö D (2004) Finding protein sequences using PROWL. Curr Protoc Bioinform Chapter 13:Unit 13.2
13. Eng J, McCormack AL, Yates JR III (1994) An approach to correlate tandem mass spectral data of peptides with amino acid sequences in a protein database. J Am Soc Mass Spectrom 5(11):976–989
14. Zhang J, Xin L, Shan B et al (2012) PEAKS DB: de novo sequencing assisted database search for sensitive and accurate peptide identification. Mol Cell Proteomics 11:M111 010587
15. Gillet LC, Navarro P, Tate S et al (2012) Targeted data extraction of the MS/MS spectra generated by data-independent acquisition: a new concept for consistent and accurate proteome analysis. Mol Cell Proteomics 11(6):O111.016717
16. Röst HL, Rosenberger G, Navarro P et al (2014) OpenSWATH enables automated, targeted analysis of data-independent acquisition MS data. Nat Biotechnol 32(3):219–223
17. Craig R, Beavis RC (2004) TANDEM: matching proteins with tandem mass spectra. Bioinformatics 20:1466–1467

18. Geer LY, Markey SP, Kowalak JA et al (2004) Open mass spectrometry search algorithm. J Proteome Res 3:958–964
19. Tabb DL, Fernando CG, Chambers MC (2007) MyriMatch: highly accurate tandem mass spectral peptide identification by multivariate hypergeometric analysis. J Proteome Res 6:654–661
20. Cox J, Neuhauser N, Michalski A et al (2011) Andromeda: a peptide search engine integrated into the MaxQuant environment. J Proteoem Res 10:1794–1805
21. Dorfer V, Pichler P, Stranzl T et al (2014) MS Amanda, a universal identification algorithm optimized for high accuracy tandem mass spectra. J Proteome Res 13:3679–3684
22. Keller A, Nesvizhskii AI, Kolker E, Aebersold R (2002) Empirical statistical model to estimate the accuracy of peptide identifications made by MS/MS and database search. Anal Chem 74:5383–5392
23. Ma K, Vitek O, Nesvizhskii AI (2012) A statistical model-building perspective to identification of MS/MS spectra with PeptideProphet. BMC Bioinformatics 13(Suppl 16):S1
24. Nesvizhskii AI, Keller A, Kolker E, Aebersold R (2003) A statistical model for identifying proteins by tandem mass spectrometry. Anal Chem 75:4646–4658
25. The M, Noble WS, MacCoss MJ, Kall L (2016) Fast and accurate protein false discovery rates on large-scale proteomics data sets with Percolator 3.0. J Am Soc Mass Spectrom 27(11):1719–1727
26. Ma B, Zhang K, Hendrie C et al (2003) PEAKS: powerful software for peptide de novo sequencing by tandem mass spectrometry. Rapid Commun Mass Spectrom 17 (20):2337–2342
27. LeDuc RD, Taylor GK, Kim YB et al (2004) ProSight PTM: an integrated environment for protein identification and characterization by top-down mass spectrometry. Nucleic Acids Res 32:W340–W345
28. Tanner S, Shu H, Frank A et al (2005) InsPecT: identification of posttranslationally modified peptides from tandem mass spectra. Anal Chem 77:4626–4639
29. Deutsch EW, Mendoza L, Shteynberg D et al (2010) A guided tour of the trans-proteomic pipeline. Proteomics 10:1150–1159
30. MacLean B, Tomazela DM, Shulman N et al (2010) Skyline: an open source document editor for creating and analyzing targeted proteomics experiments. Bioinformatics 26 (7):966–968
31. Li X-J, Zhang H, Ranish JR, Aebersold R (2003) Automated statistical analysis of protein abundance ratios from data generated by stable isotope dilution and tandem mass spectrometry. Anal Chem 75:6648–6657
32. Han DK, Eng J, Zhou H, Aebersold R (2001) Quantitative profiling of differentiation-induced microsomal proteins using isotope-coded affinity tags and mass spectrometry. Nat Biotechnol 19:946–951
33. Pedrioli PGA, Eng JK, Hubley R et al (2004) A common open representation of mass spectrometry data and its application to proteomics research. Nat Biotechnol 22:1459–1466
34. Gonzalez-Galarza FF, Lawless C, Hubbard SJ et al (2012) A critical appraisal of techniques, software packages and standards for quantitative proteomic analysis. OMICS 16 (9):431–442
35. Sturm M, Bertsch A, Gropl C et al (2008) OpenMS – an open-source software framework for mass spectrometry. BMC Bioinformatics 9:163
36. Kohlbacher O, Reinert K, Gröpl C et al (2007) TOPP – the OpenMS proteomics pipeline. Bioinformatics 23(2):e191–e197
37. Junker J, Bielow C, Bertsch A et al (2012) TOPPAS: a graphical workflow editor for the analysis of high-throughput proteomics data. J Proteome Res 11(7):3914–3920
38. Cox J, Mann M (2008) MaxQuant enables high peptide identification rates, individualized p.p.b.-range mass accuracies and proteome-wide protein quantification. Nat Biotechnol 26:1367–1372
39. Mortensen P, Gouw JW, Olsen JV et al (2010) MSQuant, an open source platform for mass spectrometry-based quantitative proteomics. Proteome Res 9(1):393–403
40. Lu P, Vogel C, Wang R et al (2007) Absolute protein expression profiling estimates the relative contributions of transcriptional and translational regulation. Nat Biotech 25 (1):117–124
41. Searle BC (2010) Scaffold: a bioinformatic tool for validating MS/MS-based proteomic studies. Proteomics 10(6):1265–1269
42. Lundgren DH, Martinez H, Wright ME, Han DK (2009) Protein identification using Sorcerer 2 and SEQUEST. Curr Protoc Bioinform Chapter 13:Unit 13.3
43. Vizcaíno JA, Csordas A, del-Toro N et al (2015) 2016 Update of the PRIDE database and its related tools. Nucleic Acids Res 44 (D1):D447–D456

44. Desiere F, Deutsch EW, King NL (2006) The PeptideAtlas Project. Nucleic Acids Res 34: D655–D658
45. Schwämmle V, Sidoli S, Ruminowicz C et al (2016) Systems level analysis of histone H3 post-translational modifications (PTMs) reveals features of PTM crosstalk in chromatin regulation. Mol Cell Proteomics 15:2715–2729
46. Lam H, Deutsch EW, Eddes JS et al (2007) Development and validation of a spectral library searching method for peptide identification from MS/MS. Proteomics 7(5):655–657
47. Lam H, Deutsch EW, Eddes JS et al (2008) Building consensus spectral libraries for peptide identification in proteomics. Nat Methods 5(10):873–875
48. The Gene Ontology Consortium (2000) Gene ontology: tool for the unification of biology. Nat Genet 25:25–29
49. The Gene Ontology Consortium (2015) Gene ontology consortium: going forward. Nucleic Acids Res 43:D1049–D1056
50. Mi H, Muruganujan A, Casagrande JT, Thomas PD (2013) Large-scale gene function analysis with the PANTHER classification system. Nat Protoc 8(8):1551–1566
51. Zeeberg BR, Feng W, Wang G et al (2003) GoMiner: a resource for biological interpretation of genomic and proteomic data. Genome Biol 4(4):R28
52. Huang DW, Sherman BT, Lempicki RA (2009) Systematic and integrative analysis of large gene lists using DAVID bioinformatics resources. Nat Protoc 4(1):44–57
53. Huang DW, Sherman BT, Lempicki RA (2009) Bioinformatics enrichment tools: paths toward the comprehensive functional analysis of large gene lists. Nucleic Acids Res 37(1):1–13
54. Hosack DA, Dennis G Jr, Sherman BT et al (2003) Identifying biological themes within lists of genes with EASE. Genome Biol 4(10): R70
55. Mootha VK, Lindgren CM, Eriksson K-F et al (2003) PGC-1alpha-responsive genes involved in oxidative phosphorylation are coordinately downregulated in human diabetes. Nat Genet 34(3):267–273
56. Subramanian A, Tamayo P, Mootha VK et al (2005) Gene set enrichment analysis: a knowledge-based approach for interpreting genome-wide expression profiles. Proc Natl Acad Sci U S A 102(43):15545–15550
57. Shannon P, Markiel A, Ozier O (2003) Cytoscape: a software environment for integrated models of biomolecular interaction networks. Genome Res 13:2498–2504
58. Snel B, Lehmann G, Bork P, Huynen MA (2000) STRING: a web-server to retrieve and display the repeatedly occurring neighbourhood of a gene. Nucleic Acids Res 28 (18):3442–3444
59. Szklarczyk D, Franceschini A, Wyder S et al (2015) STRING v10: protein-protein interaction networks, integrated over the tree of life. Nucleic Acids Res 43:D447–D452
60. Carazzolle MF, de Carvalho LM, Slepicka HH et al (2014) IIS-integrated interactome system: a Web-based platform for the annotation, analysis and visualization of protein-metabolite-gene-drug interactions by integrating a variety of data sources and tools. PLoS One 9(6):e100385
61. Antonov AV, Dietmann S, Rodchenkov I, Mewes HW (2009) PPI spider: a tool for the interpretation of proteomics data in the context of protein protein interaction networks. Proteomics 9(10):2740–2749
62. McDowall MD, Scott MS, Barton GJ (2009) PIPs: human protein-protein interactions prediction database. Nucleic Acids Res 37: D651–D656
63. Scott MS, Barton GJ (2007) Probabilistic prediction and ranking of human protein protein interactions. BMC Bioinformatics 8:239–260
64. Stark C, Breitkreutz BJ, Reguly T et al (2006) BioGRID: a general repository for interaction datasets. Nucleic Acids Res 34:D535–D539
65. Breitkreutz BJ, Stark C, Tyers M (2003) The GRID: the general repository for interaction datasets. Genome Biol 4(3):R23
66. Prasad TSK, Goel R, Kandasamy K et al (2009) Human protein reference database – 2009 update. Nucleic Acids Res 37: D767–D772
67. Hermjakob H, Montecchi-Palazzi L, Lewington C et al (2004) IntAct: an open source molecular interaction database. Nucleic Acids Res 32:D452–D455
68. Orchard S, Ammari M, Aranda B et al (2014) The MIntAct project—IntAct as a common curation platform for 11 molecular interaction databases. Nucleic Acid Res 42: D358–D363
69. Licata L, Briganti L, Peluso D et al (2012) Nucleic Acids Res 40:D857–D861
70. Pagel P, Kovac S, Oesterheld M et al (2005) The MIPS mammalian protein-protein interaction database. Bioinformatics 21 (6):832–834
71. Xenarios I, Rice DW, Salwinski L et al (2000) DIP: the database of interacting proteins. Nucleic Acids Res 28(1):289–291

72. Gaulton A, Bellis LJ, Bento AP et al (2012) ChEMBL: a large-scale bioactivity database for drug discovery. Nucleic Acids Res 40: D1100–D1107
73. Knox C, Law V, Jewison T et al (2011) DrugBank 3.0: a comprehensive resource for 'omics' research on drugs. Nucleic Acids Res 39:D1035–D1041
74. Croft D, Mundo AF, Haw R et al (2014) The reactome pathway knowledgebase. Nucleic Acids Res 42:D472–D477
75. Maere S, Heymans K, Kuiper M (2005) BiNGO: a cytoscape plugin to assess overrepresentation of gene ontology categories in biological networks. Bioinformatics 21:3448–3449
76. Mostafavi S, Ray D, Warde-Farley D et al (2008) GeneMANIA: a real-time multiple association network integration algorithm for predicting gene function. Genome Biol 9 (Suppl 1):S4
77. Ogata H, Goto S, Sato K et al (1999) KEGG: Kyoto encyclopedia of genes and genomes. Nucleic Acids Res 27:29–34
78. Pratt D, Chen J, Welker D et al (2015) NDEx, the network data exchange. Cell Systems 1 (4):302–305
79. Kutmon M, Riutta A, Nunes N et al (2016) WikiPathways: capturing the full diversity of pathway knowledge. Nucleic Acids Res 44: D488–D494
80. Kelder T, van Iersel MP, Hanspers K et al (2012) WikiPathways: building research communities on biological pathways. Nucleic Acids Res 40:D1301–D1307
81. Kutmon M, van Iersel MP, Bohler A et al (2015) PathVisio 3: an extendable pathway analysis toolbox. PLoS Comput Biol 11(2): e1004085
82. Cerami EG, Gross BE, Demir E et al (2011) Pathway commons, a web resource for biological pathway data. Nucleic Acids Res 39:D685–D690
83. Horn H, Schoof EM, Kim J et al (2014) KinomeXplorer: an integrated platform for kinome biology studies. Nat Methods 11 (6):603–604
84. Artimo P, Jonnalagedda M, Arnold K et al (2012) ExPASy: SIB bioinformatics resource portal. Nucleic Acids Res 40(W1): W597–W603
85. Finn RD, Attwood TK, Babbitt PC et al (2017) InterPro in 2017 – beyond protein family and domain annotations. Nucleic Acids Res 45:D190–D199
86. Finn RD, Coggill P, Eberhardt RY et al (2016) The Pfam protein families database: towards a more sustainable future. Nucleic Acids Res 44(D1):D279–D285
87. Sigrist CJA, Cerutti L, Hulo N et al (2002) PROSITE: a documented database using patterns and profiles as motif descriptors. Brief Bioinform 3:265–274
88. Schultz J, Milpetz F, Bork P, Ponting CP (1998) SMART, a simple modular architecture research tool: identification of signaling domains. Proc Natl Acad Sci U S A 95:5857–5864
89. Ruepp A, Waegele B, Lechner M et al (2010) CORUM: the comprehensive resource of mammalian protein complexes—2009. Nucleic Acids Res 38:D497–D501
90. Hornbeck PV, Zhang B, Murray B et al (2015) PhosphoSitePlus, 2014: mutations, PTMs and recalibrations. Nucleic Acids Res 43:D512–D520
91. Gnad F, Ren S, Cox J et al (2007) PHOSIDA (phosphorylation site database): management, structural and evolutionary investigation, and prediction of phosphosites. Genome Biol 8:R250
92. Dinkel H, Chica C, Via A et al (2011) Phospho.ELM: a database of phosphorylation sites – update 2011. Nucleic Acids Res 39: D261–D267
93. Huang HD, Lee TY, Tseng SW, Horng JT (2005) KinasePhos: a web tool for identifying protein kinase-specific phosphorylation sites. Nucleic Acids Res 33:W226–W229
94. Amanchy R, Periaswamy B, Mathivanan S et al (2007) A curated compendium of phosphorylation motifs. Nat Biotechnol 25 (3):285–286
95. Cooper CA, Gasteiger E, Packer N (2001) GlycoMod – a software tool for determining glycosylation compositions from mass spectrometric data. Proteomics 1:340–349
96. Cooper CA, Gasteiger E, Packer N (2003) Predicting glycan composition from experimental mass using GlycoMod. In: Conn PM (ed) Handbook of proteomic methods. Humana, Totowa, NJ, pp 225–232
97. Hayes CA, Karlsson NG, Struwe WB et al (2011) UniCarb-DB: a database resource for glycomic discovery. Bioinformatics 27 (9):1343–1344
98. Campbell MP, Nguyen-Khuong T, Hayes CA et al (2014) Validation of the curation pipeline of UniCarb-DB: building a global glycan reference MS/MS repository. Biochim Biophys Acta 1844(1 Pt A):108–116
99. Cooper CA, Harrison MJ, Wilkins MR, Packer NH (2001) GlycoSuiteDB: a new curated relational database of glycoprotein

glycan structures and their biological sources. Nucleic Acids Res 29(1):332–335
100. Zhang H, Loriaux P, Eng J et al (2006) UniPep, a database for human N-linked glycosites: a resource for biomarker discovery. Genome Biol 7:R73
101. Han X, He L, Xin L et al (2011) PEAKS PTM: mass spectrometry based identification of peptides with unspecified modifications. J Proteome Res 10(7):2930–2936
102. Goecks J, Nekrutenko A, Taylor J, Team G (2010) Galaxy: a comprehensive approach for supporting accessible, reproducible, and transparent computational research in the life sciences. Genome Biol 11(8):R86
103. Blankenberg D, Kuster GV, Coraor N et al (2010) Galaxy: a web-based genome analysis tool for experimentalists. Curr Protoc Mol Biol Ch 19:Unit 19.10.1–Unit 19.1021
104. Taylor J, Schenck I, Blankenberg D, Nekrutenko A (2007) Using galaxy to perform large-scale interactive data analyses. In: Baxevanis AD (ed) Current protocols in bioinformatics, pp 1–77, Unit 10-5, Wiley Online Library
105. Wolstencroft K, Haines R, Fellows D et al (2013) The Taverna workflow suite: designing and executing workflows of Web services on the desktop, web or in the cloud. Nucleic Acids Res 41(W1):W557–W561
106. Tiwari A, Sekhar AK (2007) Workflow based framework for life science informatics. Comput Biol Chem 31(5–6):305–319
107. Gillette MA, Carr SA (2013) Quantitative analysis of peptides and proteins in biomedicine by targeted mass spectrometry. Nat Methods 10(1):28–34
108. Tenga MJ, Lazar IM (2014) Proteomic study reveals a functional network of cancer markers in the G1-stage of the breast cancer cell cycle. BMC Cancer 14:710
109. Giansanti P, Tsiatsiani L, Low TY, Heck AJR (2016) Six alternative proteases for mass spectrometry–based proteomics beyond trypsin. Nat Protoc 11(5):993–1006
110. Lazar IM (2009) Recent advances in capillary and microfluidic platforms with MS detection for the analysis of phosphoproteins. Electrophoresis 30(1):262–275
111. Lazar IM, Deng J, Ikenishi F, Lazar AC (2015) Exploring the glycoproteomics landscape with advanced MS technologies. Electrophoresis 36(1):225–237
112. Deracinois B, Flahaut C, Duban-Deweer S, Karamanos Y (2013) Comparative and quantitative global proteomics approaches: an overview. Proteomes 1:180–218
113. Kuzyk MA, Parker CE, Domanski D, Borchers CH (2013) Development of MRM-based assays for the absolute quantitation of plasma proteins. Methods Mol Biol 1023:53–82
114. Zubarev RA, Makarov A (2013) Orbitrap mass spectrometry. Anal Chem 85:5288–5296
115. Wells MJ, McLuckey SA (2005) Collision-induced dissociation (CID) of peptides and proteins. Methods Enzymol 402:148–185
116. Syka JE, Coon JJ, Schroeder MJ et al (2004) Peptide and protein sequence analysis by electron transfer dissociation mass spectrometry. Proc Natl Acad Sci U S A 101 (26):9528–9533
117. Olsen JV, Macek B, Lange O et al (2007) Higher-energy C-trap dissociation for peptide modification analysis. Nat Methods 4:709–712
118. Pinho AJ, Pratas D (2014) MFCompress: a compression tool for FASTA and multi-FASTA data. Bioinformatics 30(1):117–118
119. Elias JE, Gygi SP (2010) Target-decoy search strategy for mass spectrometry-based, proteomics. Methods Mol Biol 604:55–71
120. Cappadona S, Baker PR, Cutillas PR et al (2012) Current challenges in software solutions for mass spectrometry-based quantitative proteomics. Amino Acids 43 (3):1087–1108
121. Chen Y, Wang F, Xu F, Yang T (2016) Mass spectrometry-based protein quantitation (-Chapter 15). In: Mirzaei H, Carrasco M (eds) Modern proteomics-sample preparation analysis and practical application. Springer, New York, NY, pp 255–279
122. Li Z, Adams RM, Chourey K (2012) Systematic comparison of label-free, metabolic labeling, and isobaric chemical labeling for quantitative proteomics on LTQ Orbitrap Velos. J Proteome Res 11(3):1582–1590
123. Schwanhausser B, Busse D, Li N et al (2011) Global quantification of mammalian gene expression control. Nature 473:337–342
124. Rappsilber J, Ryder U, Lamond AI, Mann M (2002) Large-scale proteomic analysis of the human spliceosome. Genome Res 12:1231–1245
125. Ishihama Y, Oda Y, Tabata T et al (2005) Exponentially modified protein abundance index (emPAI) for estimation of absolute protein amount in proteomics by the number of sequenced peptides per protein. Mol Cell Proteomics 4:1265–1272
126. Choi H, Fermin D, Nesvizhskii AI (2008) Significance analysis of spectral count data in

label-free shotgun proteomics. Mol Cell Proteomics 7(12):2373–2385
127. Carvalho PC, Lima DB, Leprevost FV et al (2016) Integrated analysis of shotgun proteomic data with PatternLab for proteomics 4.0. Nat Protoc 11(1):102–117
128. Kolmogorov M, Liu X, Pevzner PA (2016) SpectroGene: a tool for proteogenomic annotations using top-down spectra. J Proteome Res 15:144–151
129. Paik Y-K, Omenn GS, Uhlen M et al (2012) Standard guidelines for the chromosome-centric human proteome project. J Proteome Res 11(4):2005–2013
130. Taylor CF, Paton NW, Lilley KS et al (2007) The minimum information about a proteomics experiment (MIAPE). Nat Biotechnol 25:887–893
131. Lavallée-Adam M, Rauniyar N, McClatchy DB, Yates JR 3rd (2014) PSEA-Quant: a protein set enrichment analysis on label-free and label-based protein quantification data. J Proteome Res 13(12):5496–5509
132. Carnielli CM, Winck FV, Paes Leme AF (2015) Functional annotation and biological interpretation of proteomics data. Biochim Biophys Acta 1854(1):46–54
133. Yang X, Lazar IM (2014) XMAn: a *Homo sapiens* mutated-peptide database for MS analysis of cancerous cell states. J Proteome Res 13(12):5486–5495
134. Gehlenborg N, O'Donoghue SI, Baliga NS et al (2010) Visualization of omics data for systems biology. Nat Methods 7(3s):S56–S68
135. Cho CR, Labow M, Reinhardt M et al (2006) The application of systems biology to drug discovery. Curr Opin Chem Biol 10:294–302
136. Yin N, Ma W, Pei J et al (2014) Synergistic and antagonistic drug combinations depend on network topology. PLoS One 9(4): e93960

Erratum to: Probing Protein Kinase-ATP Interactions Using a Fluorescent ATP Analog

Leslie E.W. LaConte, Sarika Srivastava, and Konark Mukherjee

Erratum to:
Chapter 11 in: Iulia M. Lazar et al. (eds.), *Proteomics for Drug Discovery: Methods and Protocols*, Methods in Molecular Biology, vol. 1647, https://doi.org/10.1007/978-1-4939-7201-2_11

When the volume originally published the author neglected to acknowledge the laboratory that funded the research. The updated chapter now contains that information.

The updated online version of this chapter can be found at
https://doi.org/10.1007/978-1-4939-7201-2_11

Iulia M. Lazar et al. (eds.), *Proteomics for Drug Discovery: Methods and Protocols*, Methods in Molecular Biology, vol. 1647, https://doi.org/10.1007/978-1-4939-7201-2_20,

INDEX

Iulia M. Lazar et al. (eds.), *Proteomics for Drug Discovery: Methods and Protocols*, Methods in Molecular Biology, vol. 1647, DOI 10.1007/978-1-4939-7201-2, © Springer Science+Business Media LLC 2017

Q

R

S

T

MIX
Papier aus verantwortungsvollen Quellen
Paper from responsible sources
FSC® C105338

If you have any concerns about our products,
you can contact us on
ProductSafety@springernature.com

In case Publisher is established outside the EU,
the EU authorized representative is:
Springer Nature Customer Service Center GmbH
Europaplatz 3, 69115 Heidelberg, Germany

Printed by Libri Plureos GmbH
in Hamburg, Germany